THE THEORY OF GRAPHS

Claude Berge

DOVER PUBLICATIONS, INC.
Mineola, New York

Bibliographical Note

This Dover edition, first published in 2001, is an unabridged reprint of the 1966 printing of the work first published in English by Methuen & Co Ltd., London, in 1962. The text was translated by Alison Doig from the French edition published by Dunod, Paris, in 1958.

Library of Congress Cataloging-in-Publication Data

Berge, Claude.
[Théorie des graphes et ses applications. English]
The theory of graphs / Claude Berge ; [translated by Alison Doig].
p. cm.
Originally published: The theory of graphs and its applications. London : Methuen ; New York : Wiley, 1962.
Includes bibliographical references and index.
ISBN 0-486-41975-4 (pbk.)
1. Graph theory. I. Title.

QA166 .B413 2001
511'.5—dc21

2001032367

Manufactured in the United States of America
Dover Publications, Inc., 31 East 2nd Street, Mineola, N.Y. 11501

Contents

Introduction

There are many occasions when we use a group of points joined either by lines or by arrows to depict some situation which interests us; the points may stand for people or places or atoms, and the arrows or lines may represent kinship relations or pipelines or chemical bonds. Diagrams like these are met with everywhere, under different names: they are called variously sociograms (psychology), simplexes (topology), circuit diagrams (physics, engineering), organizational structures (economics), communication networks, family trees, etc. D. König was the first person to suggest that the generic name 'graph' be used for all of them, and to undertake a systematic study of their properties.

It is a matter of common observation that different disciplines often use analogous theorems; the concept of the 'incidence matrix' which was introduced by Kirchhoff to help in the study of electrical circuits, was taken up again in topology by Henri Poincaré as a basis for his 'analysis situs'; the idea of an 'articulation point' which has long been known in sociology has appeared more recently in electronics, and there is no end to examples such as these. The theory of graphs itself needs to be developed in a formal and abstract manner in order that it may be usefully applied in all these different domains.

In fact, even though we may define the fundamental concepts such as 'chain', 'path' or 'centre' in abstract terms, the basic ideas are closely linked with reality, and can easily be identified in any specific case. This is one reason why the theory of graphs should not be confused with the theory of algebraic relations, in which the emphasis and interests are given an entirely different slant. On the other hand, modern combinatorial topology, which ignores the orientations of the edges and which is almost entirely concerned with ideas capable of being generalized to n dimensions, has so far made few contributions to the theory. Some of the results on alternating chains, for instance, show little resemblance to either algebraic or topological theorems.

In this development of the theory of graphs, our aim has been to provide the reader with a mathematical tool which can be used in the behavioural sciences, in the theory of information, cybernetics, games, transport networks, etc., as well as in set theory, matrix theory and any other appropriate abstract discipline.

Some of the theorems which we shall state are very simple; they are the ones which a mathematician uses almost unconsciously, since he often meets them in a form peculiar to his current problem. We have used the most general statement possible for such theorems, hoping thereby to achieve economy of thought and of exposition.

In addition to these 'soft' theorems, there are some 'hard' theorems, the proofs of which are complicated and exacting, and which represent the outcome of much detailed research. Where possible, several brief theorems have been used in place of one long theorem, in order to reduce the disparity between the two groups.

Similar reasons were used in deciding on the order in which to investigate different problems, so that although it may not be immediately apparent to the reader, he will find that most of the earlier theorems are used subsequently in the proofs of those which appear later.

Speaking more generally, 'oriented' concepts are considered in the first half of the book, and 'non-oriented' concepts in the second. Definitions from algebra and from the theory of sets are given at the beginning and are supplemented later as and when it is necessary. Some knowledge of matrices will be required for chapters 14, 15 and 16.

The first French edition of this book was published by Dunod in 1958 as Number 2 in the Collection Universitaire de Mathématiques. Some changes from the original have been made in this translation in agreement with the author.

1. General Definitions†

Sets and Multivalued Functions

Like all modern mathematical theories, the theory of graphs has its own shorthand notation which allows considerable economy of thought and makes it both more effective and easier to manipulate. Although the reader is doubtless familiar with this notation, since it is that used in the classical theory of sets, it is appropriate to recall it.

A *set* is a collection of objects of any nature whatsoever, which are called its points (or its elements). Our arguments will be independent of the nature of the objects which could be people, towns, numbers, functions, etc. For convenience, sets are represented by upper case Roman letters: A, B, X, ...; and their elements by lower case Roman letters: a, b, x, ... The set A which has as its elements a, b, c, d is written $A = \{a, b, c, d\}$.

It is also convenient to regard 'that which contains no elements' as a set; this set is represented by the symbol $\varnothing$ and is called the *null* or *empty set*. Sometimes a set A will be defined not by enumerating its elements, but by a property characteristic of its elements; for example, the set of prime numbers is written: $A = \{x \mid x \text{ is prime}\}$.

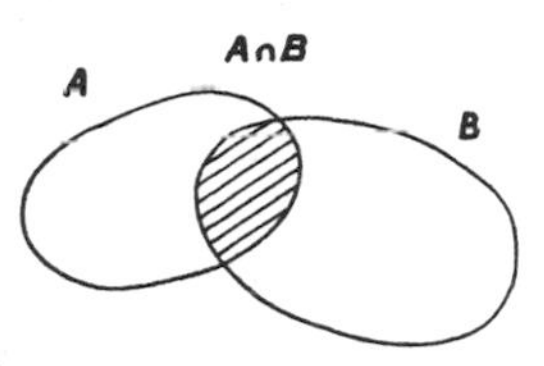

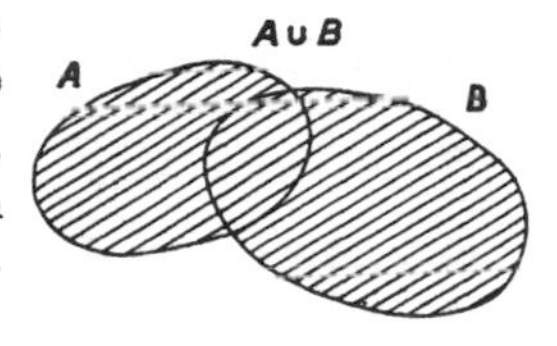

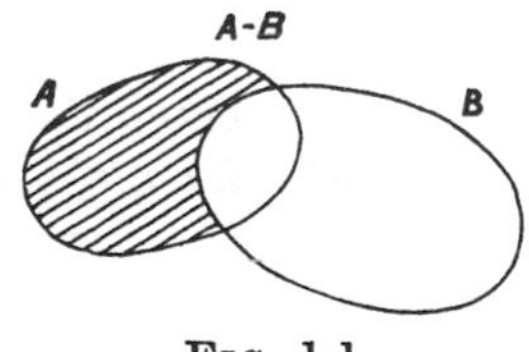

Fig. 1.1

If a property (1) implies a property (2), we write (1) $\Rightarrow$ (2); and if these two properties are equivalent, we write (1) $\Leftrightarrow$ (2).

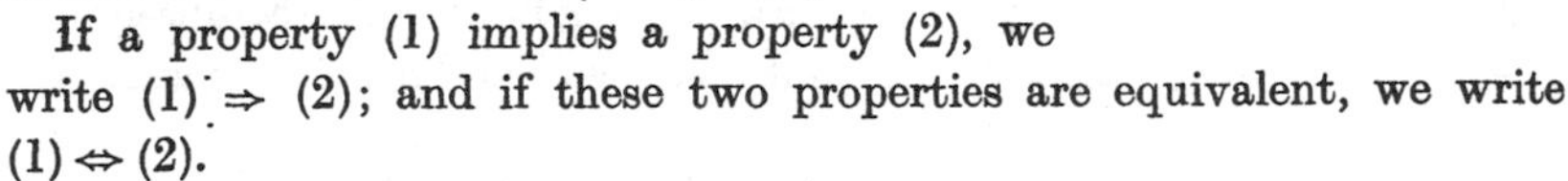

If A and B are two given sets, the following symbols are used:

$a \in A$: a is an element of the set A.

$a \notin A$: a is not an element of A.

† This chapter consists only of a list of the most fundamental terms which we shall use; it states, so to speak, the rules of the game.... The sense of most of the words can be deduced intuitively, but in order to prevent confusion, it is essential to define them axiomatically.

$A \subset B$: A is *contained* in B, or A is a *sub-set* of B (every element of A is an element of B).

$A = B$: A is equal to B. ($A \subset B$ and $B \subset A$).

$A \neq B$: A is not equal to B.

$A \subset\subset B$: A is *strictly contained* in B ($A \subset B$ and $A \neq B$).

$A \cap B$: *intersection* of A and B (the set of elements which belong both to A and to B).

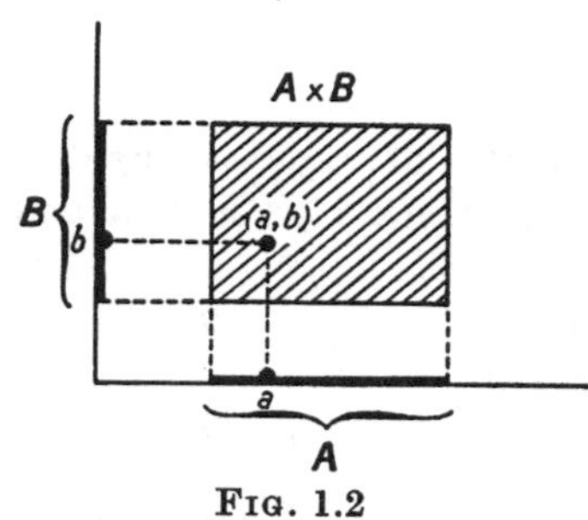

FIG. 1.2

$A \cup B$: *union* of A and B (the set of elements which belong to A or to B or to both A and B).

$A - B$: the set of elements which belong to A but not to B (cf. Fig. 1.1).

$A \times B$: the *Cartesian product* of A and B [the set of all pairs (a, b) such that $a \in A$, $b \in B$] (cf. Fig. 1.2).

The *union, intersection* and *Cartesian product* are operations which can be defined over a family of sets $(A_1, A_2, \ldots)$ instead of simply over two sets A and B.

The *union* of the A_i is the set of elements which belong to at least one of the A_i; it is written:

$$\bigcup_i A_i$$

The *intersection* of the A_i is the set of elements which belong to all the A_i; it is written:

$$\bigcap_i A_i$$

The *Cartesian product* of the A_i is the set of sequences (a_i) where $i = 1, 2, \ldots$, such that $a_i \in A_i$ for every i; it is written:

$$\prod_i A_i$$

A sequence of n elements is called an *n-tuple*; it must not be confused with a 'set' of n elements (in which the order of the elements is irrelevant).

EXAMPLE. *Real 3-dimensional space* is the set of 3-tuples (a_1, a_2, a_3), where a_1, a_2, a_3 are real numbers; it is the Cartesian product of 3 sets, each of which is equal to the set $\boldsymbol{R}$ of real numbers; consequently, real 3-dimensional space can be designated by:

$$\boldsymbol{R}^3 = \boldsymbol{R} \times \boldsymbol{R} \times \boldsymbol{R}$$

Likewise, real n-dimensional space can be written $\boldsymbol{R}^n$.

Another important notion, which we shall often use, is that of *partition*. Given a set X, a *partition of* X is, by definition, a family of sets $A_1, A_2, \ldots, A_k \subset X$, such that:

$$A_i \neq \emptyset \quad (\text{for } i = 1, 2, \ldots, k) \tag{1}$$

$$i \neq j \Rightarrow A_i \cap A_j = \emptyset \tag{2}$$

$$\bigcup_i A_i = X \tag{3}$$

Given a partition $(A_1, A_2, \ldots)$, let us write $x \equiv y$ if x and y are elements of the same set A_i. We have thus defined a relation, written $\equiv$, which has the following properties:

(1) reflexivity: $x \equiv x$
(2) symmetry: $x \equiv y \Rightarrow y \equiv x$
(3) transitivity: $x \equiv y, y \equiv z \Rightarrow x \equiv z$

Any relation which obeys the properties (1), (2), (3) is called an *equivalence*. Inversely, an equivalence $\equiv$ defines a partition of X, the sets of which are of the form:

$$A_i = \{x \mid x \in X, x \equiv x_0\}$$

(where x_0 is a given element of X).

The set A_i is sometimes called the *equivalence class* containing x_0.

EXAMPLE 1. Let p be an integer; let us write

$$x \equiv y \quad (\text{mod. } p)$$

if x and y are two integers which yield the same remainder on division by p. It follows immediately that this relation is an equivalence.

EXAMPLE 2. If two straight lines D and D' are parallel or collinear, let us write $D \parallel D'$; it is immediately clear that $\parallel$ is an equivalence relation.

EXAMPLE 3. If two individuals x and y have eyes of the same colour, let us write $x \equiv y$; it follows immediately that $\equiv$ is an equivalence relation.

Given two sets X and Y, a law σ which associates to each element $x \in X$ a well-defined element $\sigma x \in Y$ is called a *single-valued mapping of* X *into* Y or a *function defined on* X *whose values lie in* Y; for example: a real-valued function of a real variable is a single-valued mapping of $\mathbf{R}$ into $\mathbf{R}$. A *multi-valued function* Γ *mapping* X *into* Y is a law which associates to each element $x \in X$ a well-defined subset $\Gamma x \subset Y$; we can have $\Gamma x = \emptyset$.

Most authors intend a *single-valued function* when they speak simply of a *function*. As the functions which we shall be using here are not single-valued, we shall not make use of this convention and the word 'function' will be used in its wider sense.

Let Γ be a function mapping X into X. If $A \subset X$, the *image* of A is the set:

$$\Gamma A = \bigcup_{x \in A} \Gamma x$$

It can be proved that, if $A_1, A_2, \ldots, A_n$ are subsets of X, we have

$$\Gamma\left(\bigcup_{i=1}^{n} A_i\right) = \bigcup_{i=1}^{n} \Gamma A_i$$

$$\Gamma\left(\bigcap_{i=1}^{n} A_i\right) \subset \bigcap_{i=1}^{n} \Gamma A_i$$

If Δ is another function mapping X into X, the *composite function* $\Gamma\Delta$ is the function defined by:

$$(\Gamma\Delta)x = \Gamma(\Delta x)$$

The functions $\Gamma^2, \Gamma^3, \ldots$ are defined by:

$$\Gamma^2 x = \Gamma(\Gamma x)$$
$$\Gamma^3 x = \Gamma(\Gamma^2 x)$$
$$\cdots\cdots\cdots$$

The *transitive closure* of Γ is a function $\hat{\Gamma}$ mapping X into X defined by:

$$\hat{\Gamma} x = \{x\} \cup \Gamma x \cup \Gamma^2 x \cup \Gamma^3 x \cup \ldots$$

The *inverse* of Γ is a function Γ^{-1} defined by:

$$\Gamma^{-1} y = \{x \mid y \in \Gamma x\}$$

If B is a sub-set of X, we can therefore write:

$$\Gamma^{-1} B = \{x \mid \Gamma x \cap B \neq \emptyset\}$$

EXAMPLE 1. Let X be a set of people, and, if $x \in X$, let Γx be the set of his children. We have:

$\Gamma^2 x$: the set of grandchildren of x,
$\hat{\Gamma} x$: the set comprising x and all his descendants,
$\Gamma^{-1} x$: the parents of x, etc....

It is often convenient to represent individuals by points, and to show that x is y's father by an arrow going from x to y; this gives a *family tree*.

EXAMPLE 2. Consider the game of chess: at any one time, a position in chess is specified by a diagram showing the locations of the pieces on the board and by a statement of the player whose turn it is next to move. Let X be the set of all possible positions, and, if $x \in X$, let Γx be the set of positions which, according to the rules of the game, can follow immediately from the position x; we have $\Gamma x = \emptyset$ whenever x is a position of checkmate or stalemate. Then we have:

$\Gamma^3 x$: the set of positions which can be reached in three moves from the position x;

$\hat{\Gamma} x$: the set of all positions which can be reached from the position x (including x itself);

$\Gamma^{-1} A$ (where $A \subset X$): the set of positions which can, after one move, yield a position in A.

Under this system of notation, an expression can be formulated for the set of winning positions and certain properties of this set can be deduced.†

In the case of chess, the set X and the function Γ completely determine all possible moves; however, owing to the enormous number of possible positions, it would be impossible to represent them by points and the function Γ by arrows connecting certain of these points. Certain properties common to play in chess and to kinship relations amongst a group of people can nevertheless be demonstrated by using an axiomatic method; we shall do this in the following paragraphs.

Paths and Circuits of a Graph

We say that we have a *graph* whenever we have:

1. a set X;
2. a function Γ mapping X into X.

Strictly speaking, a *graph*, which is denoted by $G = (X, \Gamma)$, is the pair consisting of the set X and the function Γ. The parenthood relationships amongst a group of people define a graph, as do the rules of chess, the connections between several pieces of electrical apparatus, the victories of the competitors in a tournament, etc....

Whenever possible, the elements of a set X will be represented by points in the plane, and if x and y are two points such that $y \in \Gamma x$, they will be joined by a continuous line with an arrowhead pointing from x to y. Hence, an element of X is called a *point* or *vertex* of the graph, while the pair (x, y),

† Cf. BERGE [1].

with $y \in \Gamma x$, is called an *arc* of the graph. In the following, the set of the arcs of a graph will be designated by U, the arcs themselves being labelled u, v or w (with an index if necessary).

EXAMPLE. In the graph of Fig. 1.3, the set X consists of the vertices: a, b, c, d, e, x; the set U consists of the arcs: (a,b), (b,a), (b,x), (x,x), (c,x), (x,c), (x,d), (e,e). The function Γ can easily be determined; we have, for example

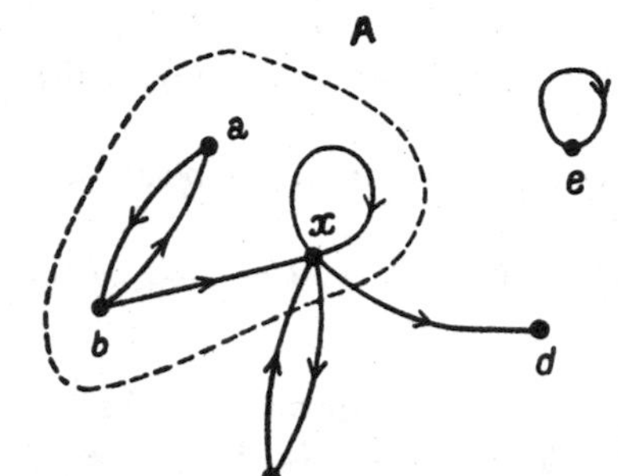

FIG. 1.3

$$\Gamma x = \{x, c, d\}$$

$$\Gamma d = \emptyset, \text{ etc.} \ldots$$

As can be seen, the set of the arcs completely determines the function defining the graph, just as this function completely defines the set U; because of this, we may equally well use the forms (X, Γ) or (X, U) to describe the graph G.

A *subgraph* of a graph (X, Γ) is defined to be a graph of the form (A, Γ_A), where $A \subset X$, and in which the function Γ_A is defined by

$$\Gamma_A x = \Gamma x \cap A$$

A *partial graph* of (X, Γ) is defined to be a graph of the form (X, Δ), where $\Delta x \subset \Gamma x$ for all x.

A *partial subgraph* of (X, Γ) is defined to be a graph of the form (A, Δ_A), where $A \subset X$ and Δ_A is such that $\Delta_A x \subset \Gamma x \cap A$.

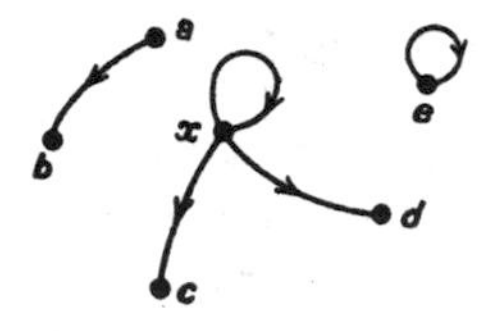

FIG. 1.3*a*

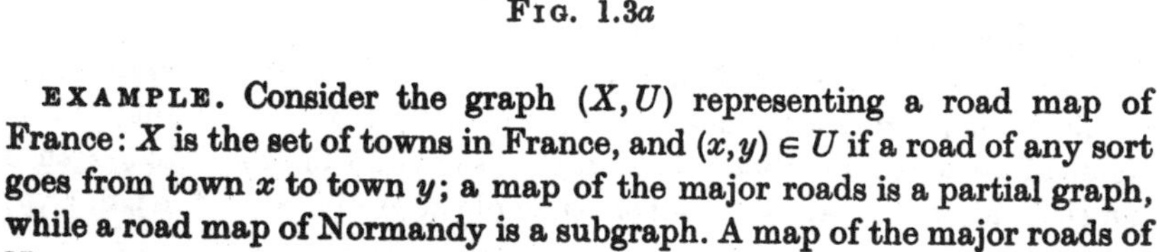

EXAMPLE. Consider the graph (X, U) representing a road map of France: X is the set of towns in France, and $(x,y) \in U$ if a road of any sort goes from town x to town y; a map of the major roads is a partial graph, while a road map of Normandy is a subgraph. A map of the major roads of Normandy is a partial subgraph.

For an arc $u = (a, b)$, the vertex a is called its *initial vertex*, and b its *terminal vertex*. Two arcs u and v are said to be *adjacent* if

1, they are distinct;
2, they have a vertex in common (without specifying if this vertex is initial or terminal for either u or v).

Likewise, two vertices x and y are said to be *adjacent* if

1, they are distinct;
2, there exists an arc $u = (x, y)$ going from x to y or from y to x.

Finally, an arc whose initial vertex is x and whose terminal vertex is not x is said to be *incident out from* x. In a similar way, we define an arc which is *incident into* a vertex x. This concept can easily be extended; if A is a given set of vertices, we say that an arc u is *incident* out from A if

$$u = (a, x), \quad a \in A, \quad x \notin A$$

The set of arcs incident into a set A is written U_A^- and the set of arcs incident out from A is written U_A^+. The set of arcs incident to A is written

$$U_A = U_A^- \cup U_A^+$$

It is easy to prove that in the graph of Fig. 1.3, for a set $A = \{a, b, x\}$, we have

$$U_A^- = \{(c, x)\} \quad U_A^+ = \{(x, c), (x, d)\}$$

The name *path* is given to a sequence $(u_1, u_2, \ldots)$ of arcs of a graph (X, U) such that the terminal vertex of each arc coincides with the initial vertex of the succeeding arc. A path is *simple* if it does not use the same arc twice, and *composite* otherwise.

If a path μ meets in turn the vertices $x_1, x_2, \ldots, x_k, x_{k+1}$, we can also use the notation $\mu = [x_1, x_2, \ldots, x_k, x_{k+1}]$; a path which uses the same arcs as μ to go from x_2 to x_k will be written

$$\mu[x_2, x_k] = [x_2, x_3, \ldots, x_k]$$

A path is *elementary* if it does not meet the same vertex twice; and a path can be finite or infinite.

A *circuit* is a finite path $\mu = [x_1, x_2, \ldots, x_k]$ in which the initial vertex x_1 coincides with the terminal vertex x_k; as above, we shall say that a circuit is *elementary* if, apart from the coincident initial and terminal vertices, every vertex which it meets is distinct.

The *length* of a path $\mu = (u_1, u_2, \ldots, u_k)$ is the number of arcs in the sequence ($l(\mu) = k$, say); if the path is infinite, we put $l(\mu) = \infty$. Finally, we define a *loop* to be a circuit of length 1, which consists of the single arc (x, x).

EXAMPLE. Hierarchy within an organization. Let X be a set of people belonging to some organization, for example the army, and let us denote by Γx the set of people directly under the command of officer x. An officer is thus connected to each of his men by a path of the graph (X,Γ); it is necessary that the graph should contain no circuits in order to prevent the possibility of contradictory orders.

These concepts of *arc, path* and *circuit* allow us to characterize certain important types of graphs. Firstly, a graph (X,U) is said to be *symmetric* if we have

$$(x,y) \in U \Rightarrow (y,x) \in U$$

In a symmetric graph two adjacent vertices x and y are always connected by two oppositely directed arcs; to simplify the representation of such arcs, we shall set up the rule that *two adjacent vertices (of a symmetric graph) are joined by a single continuous line which carries no arrowheads.*

A graph (X,U) is said to be *anti-symmetric* if

$$(x,y) \in U \Rightarrow (y,x) \notin U$$

(every pair of adjacent vertices is connected in exactly one direction.)

A graph (X,U) is said to be *complete* if

$$(x,y) \notin U \Rightarrow (y,x) \in U$$

(every pair of vertices is connected in at least one of the two possible directions.)

Finally, a graph is said to be *strongly connected* if there is a path joining any pair of arbitrary distinct vertices.

EXAMPLE. Communication system. X is a set of people, and we put $(x,y) \in U$ if x can communicate directly with y. In general, such a graph is symmetric: this will be the case if the messages are sent by radio, telephone or tom-tom, etc.; but symmetry is not essential as is shown by considering a system in which beacons or carrier pigeons are used.

In addition, in a well-designed communication network, it must be possible for every member to send a message to any other member, either directly or by means of successive retransmissions: in other words, the graph must be strongly connected.

Chains and Cycles of a Graph

In a graph $G = (X,U)$ we define an *edge* to be a set of two elements x and y such that $(x,y) \in U$ or $(y,x) \in U$; this concept must not be confused with that of an *arc* which implies an orientation. In the graph of Fig. 1.3 for example, there are eight arcs but there are only six edges.

An edge will be denoted by a roman letter in bold type: **u** or **v**, and the set of all edges is written **U**. The edge joining the vertices x and y is written $\mathbf{u} = [x, y]$.

A *chain* $\boldsymbol{\mu}$ is a sequence of edges $(\mathbf{u}_1, \mathbf{u}_2, \ldots \mathbf{u}_9)$, in which each edge $\mathbf{u}_k$ $(1 < k < 9)$ has one vertex in common with the preceding edge $\mathbf{u}_{k-1}$, and the other vertex in common with the succeeding edge $\mathbf{u}_{k+1}$. A chain is *simple* if the edges used are all different, and *composite* otherwise.

A *cycle* is a finite chain which begins at a vertex x and terminates at the same vertex. If the edges used are all different the cycle is *simple*; in all other cases it is *composite*. Finally we say that a cycle is *elementary* if every vertex belonging to the cycle appears only once.

A graph is *connected* if, for every pair of distinct vertices, there is a chain going from one to the other. A strongly connected graph is connected, but the converse is not necessarily true.

Given a vertex a, let us denote by C_a the set of vertices which can be connected by a chain to a, together with the vertex a itself; a *connected component* (or more simply, a *component*) is the subgraph determined by a set of the form C_a.

We give here two very simple theorems:

Theorem 1. *The different components of the graph (X, Γ) constitute a partition of X; that is:*

(1) $$C_a \neq \varnothing$$

(2) $$C_a \neq C_b \textit{ implies } C_a \cap C_b = \varnothing$$

(3) $$\bigcup C_a = X$$

Since $a \in C_a$, (1) holds.

To prove (2), let us suppose that $C_a \cap C_b \neq \varnothing$, and hence show that $C_a = C_b$.

Let $x \in C_a \cap C_b$; this vertex x can be connected by a chain to a and to b; therefore a can be connected to b, and, consequently, $b \in C_a$. Therefore

$$C_b \subset C_a$$

As we also have $C_a \subset C_b$ (from symmetry), it is certainly true that $C_a = C_b$.

Condition (3) holds, for

$$X \supset \bigcup_{a \in X} C_a \supset \bigcup_{a \in X} \{a\} = X$$

and hence

$$\bigcup C_a = X$$

THEOREM 2. *A graph is connected if, and only if, it possesses only one component.*

If the graph has two distinct components, C_a and C_b, it is not connected, since a and b cannot be joined by a chain.

If the graph is not connected, there exist two vertices a and b which cannot be joined by any chain, and C_a and C_b are then two distinct components.

Remark. A graph can be studied as a function of its oriented arcs or as a function of its edges; in the first case, we use the concepts: paths, circuits, strong connectedness, etc.; in the second, we use those of chain, cycle, connectedness, etc. From now on, each time that a concept '...' is defined for a graph by means of its arcs, it will be possible equally to define a new concept by means of its edges, which we shall call: '*semi-*...'. The adverb 'semi' possesses here a special significance.

2. Descendance Relations

Weak Ordering Associated with a Graph

In algebra, a relation $\leqslant$, defined on a set X, is a *weak ordering* if we have:

(1) *reflexivity:* $x \leqslant x$ (for all $x \in X$)
(2) *transitivity:* $x \leqslant y,\ y \leqslant z \Rightarrow x \leqslant z$

Weak orderings are often encountered: 'x is an integer multiple of the integer y', 'x is a real number greater than or equal to y', 'the situation x is either equivalent to the situation y, or else is preferred to it' are examples of weak ordering.

A relation $\leqslant$ will be associated with a graph $G = (X, \Gamma)$ in the following way: for two vertices x and y of the graph, write $x \leqslant y$ if $x = y$ or if a path exists going from x to y (in other words, if $y \in \hat{\Gamma}x$).

This relation $\leqslant$, which certainly satisfies (1) and (2), is a weak ordering which is called *the weak ordering associated with the graph G*.

$x \leqslant y$ states: the vertex x *precedes* the vertex y

If $y \leqslant x$, we can also put $x \geqslant y$; in other words, the vertex x *follows* the vertex y.

If $x \leqslant y$ and $y \leqslant x$, we write $x \equiv y$; and this states: the vertex x is *equivalent* to the vertex y; in this case x and y coincide, or else they are on the same circuit. It can be proved immediately that $\equiv$ is an equivalence relation, that is that we have:

(1) *reflexivity:* $x \equiv x$ $(x \in X)$
(2) *transitivity:* $x \equiv y,\ y \equiv z \Rightarrow x \equiv z$
(3) *symmetry:* $x \equiv y \Rightarrow y \equiv x$

If we have $x \leqslant y$, but not $x \equiv y$, we write $x < y$, and state: the vertex x is an *ancestor* of the vertex y. We may also write $y > x$, and state: the vertex y is a *descendant* of the vertex x.

If $B \subset X$, a vertex z which follows all the vertices of B is called a *majorant* of B; we may then write:

$$z \geqslant b \quad (b \in B)$$

If B contains an element which is a majorant of B, this element is called a *maximum* of B. If b and b' are two maxima of B, we have $b \leqslant b'$, and $b' \leqslant b$, and hence $b \equiv b'$: all maxima are equivalent to one another.

Similarly, a *minorant* of B is a vertex z such that

$$z \leqslant b \quad (b \in B)$$

A minorant of B which is an element of B is called a *minimum* of B.

These well-known algebraic concepts are of considerable use in classifying different types of graphs.

EXAMPLE 1. *Graph without circuits.* If (X, Γ) contains no circuits, its associated relation $\leqslant$ has the following property:

$$x \leqslant y, y \leqslant x \Rightarrow x = y$$

In this case, we say that $\leqslant$ is a *partial order.*

Conversely, if the relation associated with a graph is a partial order, the graph does not contain any circuits.

As an example, let us consider a set X and a family $\mathscr{F}$ of subsets of X; the *Hasse diagram* of $\mathscr{F}$ is a graph G whose vertices are the different elements of $\mathscr{F}$, an arc going from $F \in \mathscr{F}$ to $F' \in \mathscr{F}$ if

1. $F \subset F'$;

2. no $F'' \in \mathscr{F}$ exists such that $F \subset F'' \subset F'$. The weak order associated with the graph G is recognized to be the relation $\subset$, which is a partial order: G therefore contains no circuits. It would be interesting to investigate for $X = \{1, 2, \ldots, n\}$, how many families $\mathscr{F}$ yield a given Hasse diagram; this problem, which is sometimes called Rainey's problem, has been solved only for $n \leqslant 5$.

EXAMPLE 2. *Transitive graph.* A graph (X, U) is said to be *transitive* if

$$x \leqslant y \Rightarrow (x, y) \in U$$

If, for example, the vertices of the graph stand for people, and its arcs represent the ranking of these people in a hierarchy, we get a transitive graph.

EXAMPLE 3. *Total graph.* A weak ordering relation is said to be *total* if for all pairs (x, y), we have $x \leqslant y$ or $y \leqslant x$; a graph is *total* if its associated weak order is total. This gives the following result: *all strongly connected graphs are total.*

EXAMPLE 4. The *lattice property* and *arborescences*. Consider a graph (X,Γ) without circuits and its associated partial order $\leqslant$; if the set of majorants of a set $B \subset X$ possesses a minimum c, c is said to be a *least upper bound* of B; if every set B has a least upper bound, the partial order $\leqslant$ exhibits the (*upper*) *lattice property*; we can also say that the graph (X,Γ) has the (*upper*) *lattice property*.

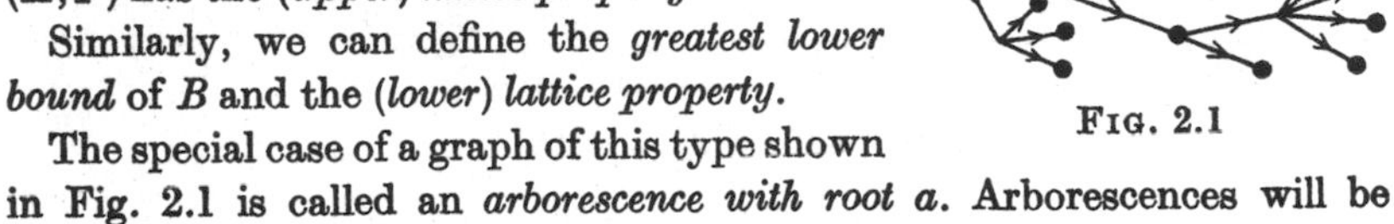

FIG. 2.1

Similarly, we can define the *greatest lower bound* of B and the (*lower*) *lattice property*.

The special case of a graph of this type shown in Fig. 2.1 is called an *arborescence with root a*. Arborescences will be examined later (Chapter 16).

Inductive Graphs and Bases

A set $B \subset X$ is a *basis* of the graph (X,Γ) if the two following conditions hold:

(1) $b_1 \in B$, $b_2 \in B$, $b_1 \neq b_2 \Rightarrow$ neither $b_1 \leqslant b_2$ nor $b_2 \leqslant b_1$;
(2) $x \notin B \Rightarrow b \in B$ exists such that $b \geqslant x$.

EXAMPLE. Consider the graph in Fig. 2.2.

The set $B = \{b_1, b_2, \ldots\}$ is a basis; on the other hand, the subgraph determined by $X - B$ does not possess a basis.

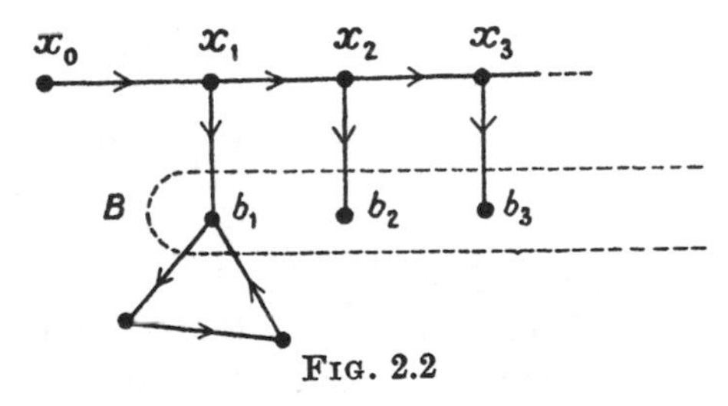

FIG. 2.2

The concept of the basis of a graph arises in numerous practical problems concerning communication networks, and we shall see later (Chapters 5 and 6) that it belongs to the theory of games. Further, it may be used to establish wholly theoretical results of topology or linear analysis. We intend here simply to determine existence conditions for a basis of as general a graph as possible, with a finite or infinite number of vertices.

A graph is called *inductive* if every path $\mu = [x_1, x_2, \ldots]$ possesses a majorant, that is if a vertex z exists such that

$$z \geqslant x_n \quad (n = 1, 2, \ldots)$$

EXAMPLE 1. A graph which has a finite number of vertices is inductive, for if μ is a finite path, its last vertex is a majorant; and if μ is infinite, it contains at least one vertex which is repeated an infinite number of times, and is therefore a majorant of μ.

EXAMPLE 2. The graph of Fig. 2.2, however, is not inductive.

In certain problems, the concept of an inductive graph is not sharp enough; we shall say that a graph is *totally inductive* if, for every infinite path $\mu = [x_0, x_1, \ldots]$, the weak order relation $\leqslant^{\mu}$ of the sub-graph determined by the set $[x_0, x_1, \ldots]$ allows us to write $x_i \equiv^{\mu} x_j$ for i and j sufficiently large.

EXAMPLE. In the graph of Fig. 2.3, we have:

$$x_i \equiv^{\mu} x_j \quad (i \geqslant 3, j \geqslant 3)$$

and it is therefore totally inductive.

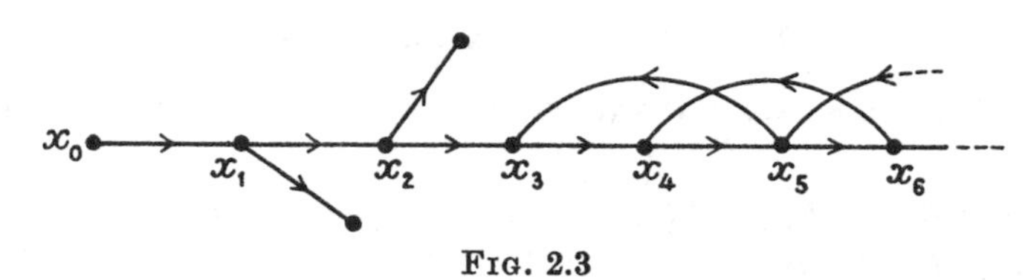

FIG. 2.3

Clearly, *a totally inductive graph is inductive*; further, we have the fundamental property: *every subgraph of a totally inductive graph is totally inductive.* (The analogous proposition for inductive graphs does not exist.)

ZORN'S LEMMA. *For every vertex x of an inductive graph (X, Γ), there exists a vertex z without descendants such that $z \geqslant x$.*

(This is a well-known result of the theory of sets, a proof of which is given in [E.T.F.M.], Chapter 3.)

THEOREM 1. *Every inductive graph possesses a basis*†.

Let Z be the set of vertices without descendants, and within Z let us consider the equivalence $\equiv$ defined by the associated weak order $\leqslant$ ($z_1 \equiv z_2$ if z_1 and z_2 are on the same circuit). A representative b_i is selected from each equivalence class C_i and we shall show that the set $B = \{b_i \mid i \in I\}$ is a basis of the graph.

1. We cannot have $b_i \leqslant b_j$, with $i \neq j$, for this would imply $b_i \equiv b_j$ (the vertex b_i having no descendants) and hence that b_i and b_j belong to the same equivalence class (which is impossible since the equivalence classes are pairwise disjoint).

† It is also possible to show that this enunciation is equivalent to that of Zorn; a direct proof of this theorem would therefore be long and tedious.

2. If $x \notin B$, then from Zorn's Lemma, a point z_0 without descendants exists such that $z_0 \geqslant x$; let b_i be the representative of the equivalence class $C_i = \{z \mid z \in Z,\ z \equiv z_0\}$: then certainly

$$x \leqslant b_i, \qquad b_i \in B$$

THEOREM 2. *If a graph (X, Γ) has a finite basis B, it is an inductive graph.*

Let $\mu = [x_1, x_2, \ldots]$ be an infinite path of the graph (if one exists), and let $b_1, b_2, \ldots$ be vertices of B such that $b_n \geqslant x_n$ for all n; since B is a finite set, there is at least one vertex b_k which appears an infinity of times in the sequence (b_n); it is then certainly true that

$$b_k \geqslant x_n \quad (n = 1, 2, \ldots)$$

Q.E.D.

(similarly: if a graph has a basis, it is a totally inductive graph).

THEOREM 3. *In the set Z of vertices without descendants, consider the equivalence classes $\{z \mid z \in Z,\ z \equiv z_0\}$; a basis B, if it exists, is formed by taking one element from each equivalence class.*

If B is a basis, a vertex $b \in B$ has no descendants; for if $x > b$, $b' \in B$ exists such that $b' \geqslant x$, and hence $b' \geqslant b$, which contradicts the hypothesis that B is a basis.

B possesses at least one vertex in the class $\{z \mid z \in Z,\ z \equiv z_0\}$, for otherwise there is no path from z_0 to a point in B.

B cannot contain two distinct elements b and b' from the same equivalence class, for then we would have $b \equiv b'$, and a path in B would exist going from one point of B to another.

COROLLARY. *If a graph has a basis, all its bases are of the same cardinality; in particular, a finite graph has a basis (since it is inductive), and all its bases contain the same number of elements.*

This follows since the equivalence classes considered in Theorem 3 form a partition of Z.

3. The Ordinal Function and the Grundy Function on an Infinite Graph

General Remarks Concerning Infinite Graphs

We write $|A|$ for the number of elements in a finite set A; if A contains an infinite number of elements, we put $|A| = \infty$.

A graph (X, Γ) is called *finite* if $|X| < \infty$; it is *Γ-finite* if $|\Gamma x| < \infty$ for all x; it is *Γ^{-1}-finite* if $|\Gamma^{-1} x| < \infty$ for all x. A graph which simultaneously possesses these last two properties is called *locally finite*. Clearly, *a finite graph is locally finite.*

Finally, a graph is called *Γ-bounded* if an integer m exists such that $|\Gamma x| \leqslant m$ for all x.

EXAMPLE. The graph shown in Fig. 3.1 is Γ-finite but not Γ-bounded. It is not Γ^{-1}-finite.

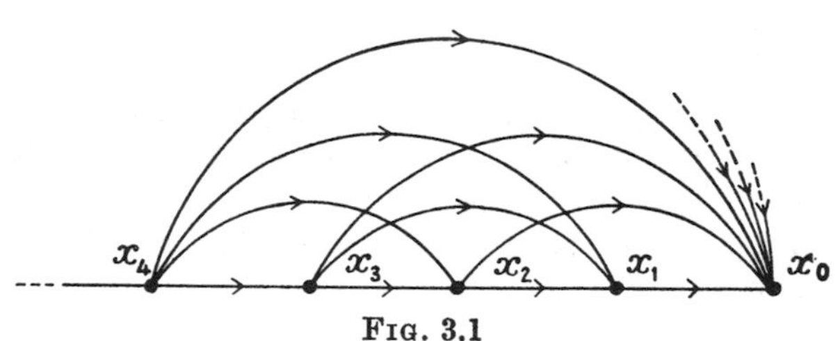

FIG. 3.1

If no paths of infinite length start from a vertex x_0 of a graph, the graph is said to be *progressively finite at* x_0; a graph which is progressively finite at each of its points is called *progressively finite.*

A graph is called *progressively bounded at the vertex* x_0, if a number m exists such that the length of all paths starting from x_0 is less than or equal to m; a graph is *progressively bounded* if it is progressively bounded at each of its points. A progressively bounded graph is also progressively finite but the converse does not necessarily hold.

Finally, a graph (X, Γ) is *regressively finite* (*regressively bounded*) if the graph (X, Γ^{-1}) is progressively finite (progressively bounded).

EXAMPLE. The graph in Fig. 3.1 is progressively finite, but not regressively finite; it is progressively bounded even though there exist paths of any given length.

We shall now establish some very general theorems which will subsequently allow us to extend results for finite graphs to certain infinite graphs. The importance of this connection is clear: although we usually meet finite graphs in psychology or operational research, geometry and the theory of sets deal almost exclusively with infinite graphs.

THEOREM 1. *If a graph is finite, the properties 'progressively finite', 'progressively bounded' and 'without circuits' are equivalent.*

This is obviously true.

THEOREM 2. *If a graph (X, Γ) is progressively finite and Γ-finite, we have*

$$|\hat{\Gamma}x| < \infty \quad (x \in X)$$

Suppose, in fact, that for one vertex x_0, we had $|\hat{\Gamma}x_0| = \infty$; since $|\Gamma x_0| < \infty$, Γx_0 contains at least one vertex x_1 for which $|\hat{\Gamma}x_1| = \infty$; similarly, Γx_1 must contain a vertex x_2 for which $|\hat{\Gamma}x_2| = \infty$; etc.... The path $[x_0, x_1, x_2, \ldots, x_n, \ldots]$ is therefore of infinite length, which contradicts our hypothesis.

COROLLARY 1. *A Γ-finite graph which is progressively finite at a vertex x_0 is also progressively bounded at x_0.*

Since the subgraph determined by the set $\hat{\Gamma}x_0$ is progressively finite, Theorem 2 shows that it is also finite; from Theorem 1, it contains no circuits and is progressively bounded. The graph G is therefore progressively bounded at x_0.

COROLLARY 2 (König). *Let $(A_1, A_2, \ldots)$ be a sequence of finite, non-empty sets which are pairwise disjoint, and let $\prec$ be a relation defined between the elements of two consecutive sets; if for all $x_n \in A_n$, an element x_{n-1} of A_{n-1} exists such that $x_{n-1} \prec x_n$, a sequence $(a_1, a_2, \ldots)$ exists with $a_n \in A_n$ for all n such that:*

$$a_1 \prec a_2 \prec a_3 \prec \cdots \prec a_n \prec \cdots$$

Let X be the union of the A_n, augmented by an arbitrary point x_0, and define a function Γ on X by

$$\Gamma x_0 = A_1$$

$$\Gamma a_n = \{x \mid x \in A_{n+1}, a_n \prec x\} \qquad (a_n \in A_n)$$

The graph (X, Γ) is Γ-finite; it is not progressively bounded at x_0, for given any integer m, a path of length m can easily be constructed by taking an a_m from A_m, an a_{m-1} from A_{m-1} with $a_{m-1} \prec a_m$, etc. From Corollary 1, the graph is therefore not progressively finite at x_0, and an infinite path $[x_0, a_1, a_2, \ldots]$ exists; the points a_1, a_2, ... make up the required sequence.

THEOREM 3. *If a graph is locally finite and connected, the set of its vertices and of its chains can be enumerated.*

Choose an arbitrary vertex x_0; list the chains of length 1 going from x_0 and their terminal vertices, then list the chains of length 2 and their terminal vertices (that is, those which have not already been considered), etc. Since every vertex can be reached by a chain from x_0, we shall thus enumerate X and the set of chains issuing from x_0.

RADO'S THEOREM. *Consider a locally finite graph (X, Γ), a finite set of integers K and a law which defines for all $I \subset K$ a corresponding set $K(I) \subset K$; if an integer-valued function $\phi_A(x)$ is defined on every finite subgraph (A, Γ_A) such that*

$$\phi_A(x) \in K[\phi_A(\Gamma_A x)] \qquad (x \in A)$$

then an integer-valued function $\phi(x)$ will exist on X such that

$$\phi(x) \in K[\phi(\Gamma x)] \qquad (x \in X)$$

For convenience, the set $I = \{\phi(y) \mid y \in \Gamma x\}$ is written $\phi(\Gamma x)$.

Select a finite set $A_0 \subset X$, and consider in turn the sets:

$$A_1 = A_0 \cup \Gamma A_0 \cup \Gamma^{-1} A_0$$

$$A_2 = A_1 \cup \Gamma A_1 \cup \Gamma^{-1} A_1$$

$$\cdots\cdots\cdots\cdots$$

$$A_{n+1} = A_n \cup \Gamma A_n \cup \Gamma^{-1} A_n$$

$$\cdots\cdots\cdots\cdots$$

The sets A_n are finite and non-decreasing; in addition, we can assume that the graph is connected (if not, we can consider each component separately), and hence we have

$$X = \bigcup_{n=0}^{\infty} A_n$$

To simplify the notation, let $\phi_n = \phi_{A_n}$ be the function corresponding to the finite set A_n; its values over the set A_0 determine a function ϕ_n^0, and since these values can only range over the set K, the number of different functions obtained as n varies will be finite. Hence we may say that there is an infinity of values of n for which the function ϕ_n^0 is the same; let these values be:

$$k_0 < k_1 < k_2 < \dots$$

Let the function determined by restricting the function ϕ_{k_n} to the set A_1 be written ϕ_n^1; using the same argument, this function ϕ_n^1 will be the same for an infinity of values of n, say:

$$l_1 < l_2 < l_3 < \dots$$

Similarly, let ϕ_n^2 be the function determined by restricting ϕ_{l_n} to the set A_2, etc....

The functions ϕ_0^0, ϕ_1^1, ϕ_2^2, ... are defined over the sets A_0, A_1, A_2, ..., respectively, and, if $p > q$, we have

$$\phi_p^p(x) = \phi_q^q(x) \quad (x \in A_q)$$

Hence we may conclude that for all $x \in X$, the number $\phi_n^n(x)$ tends to a limit $\phi(x)$ for n sufficiently large, thus defining a function ϕ over X.

For every vertex x_0 an integer n exists such that $A_n \supset \{x_0\} \cup \Gamma x_0$ and we have

$$\phi_n^n(x) = \phi(x) \qquad (x \in A_n)$$

We then have:

$$\phi(x_0) \in K[\phi(\Gamma x_0)]$$

Q.E.D.

Ordinal Function

It is well known that the concept of an *integer* can be generalized to that of an *ordinal number*; we shall recall its definition. To take a concrete example, consider some objects: x_1, x_2, ..., x_{12}. By placing them in this order, we have defined a 12-tuple, in which the rank of x_6 is represented by the symbol: 6.

Now suppose that we have an infinite group of objects x_1, x_2, ..., and that they have been placed in a certain order, say:

$$(x_4, x_5, x_6, \ldots; x_1, x_2, x_3)$$

The ordinary integers may be used to represent the ranks of x_4, x_5, x_6, ..., but they cannot be used for the ranks of x_1, x_2 or x_3; new symbols must therefore be invented: $\omega+1$ for x_1, $\omega+2$ for x_2, $\omega+3$ for x_3. The symbol ω which does not itself represent a rank, is called an *ordinal limit number*, while the symbols 1, 2, ..., $\omega+1$, $\omega+2$, $\omega+3$, are *ordinal numbers*.

Other ordinals can be defined; for example, in the ordering:

$$(x_3, x_5, x_7, x_9, \ldots; x_2, x_4, x_6, x_8, \ldots; x_1)$$

the rank of x_8 will be: $\omega+4$; that of x_1 will be: $2\omega+1$, etc.... If α and β are any two ordinal numbers, we say that $\alpha < \beta$ if the object of rank α is placed before that of rank β.

Naturally, ordinal numbers can be rigorously defined using algebraic concepts†.

In the theory of graphs, the concept of ordinal number is very useful, since statements concerning finite graphs can thereby be extended to certain infinite graphs.

Given a graph (X, Γ), let us consider the sets:

$$X(0) = \{x \mid \Gamma x = \emptyset\}$$

$$X(1) = \{x \mid \Gamma x \subset X(0)\}$$

$$X(2) = \{x \mid \Gamma x \subset X(1)\}$$

$$\cdots\cdots\cdots\cdots$$

$$X(k) = \{x \mid \Gamma x \subset X(k-1)\}$$

$$\cdots\cdots\cdots\cdots$$

$$X(\omega) = \bigcup_{\alpha<\omega} X(\alpha)$$

$$X(\omega+1) = \{x \mid \Gamma x \subset X(\omega)\}$$

$$\cdots\cdots\cdots\cdots$$

† A rigorous definition of the concept of an ordinal number may be found, for example, in [E.T.F.M.], Chapter 3. The pair formed by a set X and a partial order relation $\leqslant$ is a *well-ordered set* if every subset of X possesses a minimum. Two well-ordered sets $(X, \leqslant)$ and $(Y, \leqslant)$ are said to be *equivalent* if a one–one correspondence σ exists between X and Y with

$$x \leqslant x' \qquad \sigma x \leqslant \sigma x'$$

An *ordinal number* is then the equivalence class containing a well-ordered set. We shall keep in mind the result: the ordinal numbers are themselves well-ordered by the relation $<$.

This method of definition can clearly be continued indefinitely: if α is not an ordinal limit, we put

$$X(\alpha) = \{x \mid \Gamma x \subset X(\alpha - 1)\}$$

If α is an ordinal limit, we put

$$X(\alpha) = \bigcup_{\beta<\alpha} X(\beta)$$

Note that if α and β are two ordinals with $\alpha < \beta$, we have $X(\alpha) \subset X(\beta)$.

The *order* of a vertex x is by definition the smallest ordinal α such that

$$x \in X(\alpha)$$

$$x \notin X(\beta) \quad \text{for all } \beta < \alpha$$

We then put $\alpha = o(x)$; naturally there may be some vertices which do not possess an order: for example, those which are on a circuit.

The function $o(x)$, when it may be defined over X, is called the *ordinal function* of the graph.

EXAMPLE. In the graph of Fig. 3.2, every vertex x possesses an order (which is shown); y is the only vertex for which the order is transfinite. We have $X = X(\omega + 1)$.

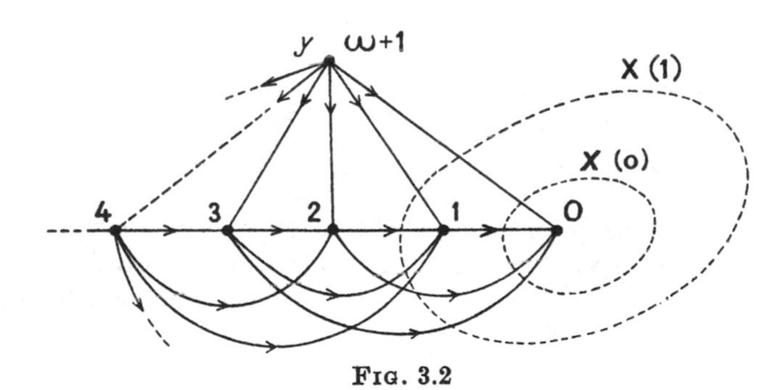

FIG. 3.2

THEOREM 4. *An ordinal function defined on X exists if and only if the graph is progressively finite.*

1. If $o(x)$ exists, let us show that the graph is progressively finite. This we shall do by imagining that an infinite path $a_1, a_2, \ldots$, exists and showing that this leads to a contradiction.

Put $A = \{a_1, a_2, \ldots\}$: then $A \cap X(0) = \emptyset$.

If we have $A \cap X(\beta) = \emptyset$ for every ordinal $\beta < \alpha$, then we also have $A \cap X(\alpha) = \emptyset$. Hence by induction

$$A \cap X = \emptyset$$

This implies $A = \emptyset$, which is absurd.

2. Conversely, let us suppose that a vertex x exists which does not possess an order; let B be the set of such vertices. Then $B \neq \emptyset$.

If $x_1 \in B$, $\Gamma x_1 \neq \emptyset$ [for $x_1 \notin X(0)$], and therefore Γx_1 contains a vertex $x_2 \in B$; likewise Γx_2 contains a vertex $x_3 \in B$, etc....

As the path $[x_1, x_2, x_3, \ldots]$ is of infinite length, the graph cannot be progressively finite.

Grundy Functions

Let us consider a finite graph (X, Γ) and a function g which associates an integer $g(x) \geqslant 0$ with every vertex x. We call $g(x)$ a *Grundy function* on the given graph if, for every vertex x, $g(x)$ is the smallest integer $\geqslant 0$ which is not in the set:

$$g(\Gamma x) = \{g(y) \mid y \in \Gamma x\}$$

A Grundy function may be defined on an infinite graph with the help of transfinite ordinal numbers: $g(x)$ is the smallest ordinal which is not an element of the set:

$$\{g(y) \mid y \in \Gamma x\}$$

It follows from the definition that if $\Gamma x = \emptyset$, $g(x)$ must be zero.

A graph may possess no Grundy function (for example if there is a loop); it may also possess several.

EXAMPLE 1. The graph of Fig. 3.3 has two Grundy functions, whose values are the integers shown on the several vertices; it can be confirmed that, if $\Gamma x = \{y_1, y_2, \ldots\}$, $g(x)$ is the smallest integer different from $g(y_1)$, $g(y_2), \ldots$.

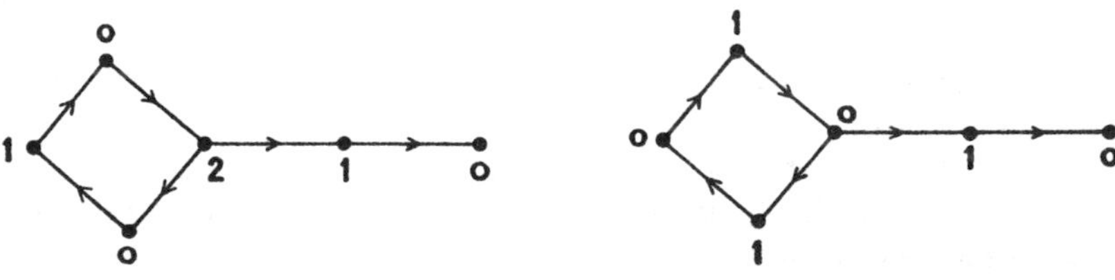

FIG. 3.3

EXAMPLE 2. The graph of Fig. 3.2 has a Grundy function $g(x) = o(x)$; this Grundy function takes a transfinite value at y.

THEOREM 5. *If a graph is progressively finite, it has a unique Grundy function; in addition we have*

$$(1) \qquad g(x) \leqslant o(x)$$

This may be proved immediately by induction over the sets:

$$X(0) = \{x \mid \Gamma x = \emptyset\}$$
$$X(1) = \{x \mid \Gamma x \subset X(0)\}$$
$$X(2) = \{x \mid \Gamma x \subset X(1)\}$$
$$\cdots\cdots\cdots\cdots$$

THEOREM 6. *If* $|\Gamma x| < \infty$, *then* $g(x) \leqslant |\Gamma x|$.

If $g(x) = n$, the function g must take the values $0, 1, 2, \ldots, n-1$, in Γx, and hence $|\Gamma x| \geqslant n = g(x)$.

Theorems 5 and 6 show that $g(x)$ does not easily take large values; in particular, if a graph is Γ-finite or progressively bounded, *all values of* $g(x)$ *are finite.*

Operations over Graphs

Consider n graphs $G_1 = (X_1, \Gamma_1)$, $G_2 = (X_2, \Gamma_2)$, $\ldots$, $G_n = (X_n, \Gamma_n)$, and then consider the graph $G = (X, \Gamma)$ defined by:

$$(1) \qquad X = X_1 \times X_2 \times \ldots \times X_n = \{(x_1, x_2, \ldots, x_n) \mid x_1 \in X_1, \text{etc.}\}$$

(the Cartesian product of X_i).

$$(2) \qquad \Gamma(x_1, x_2, \ldots, x_n) = \Gamma_1 x_1 \times \Gamma_2 x_2 \times \ldots \times \Gamma_n x_n$$

This graph is written $G = G_1 \times G_2 \times \ldots \times G_n$, and is called the *product* of the graphs G_i.

Similarly, the *sum* of the graphs G_i is a graph (X, Γ) defined by:

$$(1) \qquad X = X_1 \times X_2 \times \ldots \times X_n$$

$$(2) \quad \Gamma(x_1, x_2, \ldots, x_n) = [\Gamma_1 x_1 \times \{x_2\} \times \ldots \times \{x_n\}] \cup [\{x_1\} \times \Gamma_2 x_2 \times \ldots \{x_n\}] \ldots \cup [\{x_1\} \times \{x_2\} \ldots \times \Gamma_n x_n]$$

This graph is written $G = G_1 + G_2 + \ldots + G_n$.

The operations $\times$ and $+$ which have just been defined are well known in the theory of games; they are encountered in many other fields where use is made of graphs.

EXAMPLE. We have two machines; the first of these can occupy a certain number of states, the set of which is written X_1, and we put $x_1' \in \Gamma_1 x_1$ if the state x_1' can follow the state x_1. This defines a graph $G_1 = (X_1, \Gamma_1)$; a graph $G_2 = (X_2, \Gamma_2)$ may likewise be defined for the second machine.

If the first machine is in position x_1 and the second is in position x_2, an operator who is free to use either machine is faced with a situation which may be described by the pair (x_1, x_2). If he can operate both machines *simultaneously*, the graph of the possible situations will be $G_1 \times G_2$. If he can operate only one machine at a time, the graph will be $G_1 + G_2$.

Is the property of possessing a Grundy function conserved under these operations on graphs? Before tackling this problem, we must define the notion of the *digital sum* of ordinal numbers; first let us recall that for an integer c its *binary expansion* is a sequence $(c^1, c^2, \ldots, c^k)$ of numbers c^i (which must be either zero or one), such that

$$c = c^1 + 2c^2 + 4c^3 + \ldots + (2)^{k-1} c^k$$

This is more often written

$$c = c^k c^{k-1} \ldots c^2 c^1$$

This *binary representation* is built up in the same way as decimal representation, and the correspondence between these three forms is easily established:

DECIMAL FORM		BINARY FORM		BINARY EXPANSION	DECIMAL FORM		BINARY FORM		BINARY EXPANSION
0	=	0	=	(0)	6	=	110	=	(0, 1, 1)
1	=	1	=	(1)	7	=	111	=	(1, 1, 1)
2	=	10	=	(0, 1)	8	=	1 000	=	(0, 0, 0, 1)
3	=	11	=	(1, 1)	9	=	1 001	=	(1, 0, 0, 1)
4	=	100	=	(0, 0, 1)	10	=	1 010	=	(0, 1, 0, 1)
5	=	101	=	(1, 0, 1)	11	=	1 011	=	(1, 1, 0, 1)

In arithmetic, the remainder (which is equal to zero or one) on dividing n by 2, is written $[n]_{(2)}$, or 'n modulo 2'.

The *digital sum* of the integers c_1, c_2, ..., c_n is by definition an integer $c = c_1 \dot{+} c_2 \dot{+} c_3 \dot{+} \ldots \dot{+} c_n$ which is obtained by writing each number c_k as a binary expansion $(c_k^1, c_k^2, \ldots)$, and putting

$$c = \left(\left[\sum_{k=1}^{n} c_k^1 \right]_{(2)}, \left[\sum_{k=1}^{n} c_k^2 \right]_{(2)}, \left[\sum_{k=1}^{n} c_k^3 \right]_{(2)}, \ldots \right)$$

We have thus:

$$3 \dot{+} 7 = (1,1) \dot{+} (1,1,1) = (0,0,1) = 4$$

$$1 \dot{+} 3 \dot{+} 11 = (1) \dot{+} (1,1) \dot{+} (1,1,0,1) = (1,0,0,1) = 9, \text{ etc.}$$

These ideas may be extended without difficulty to transfinite ordinals; the binary expansion of $\alpha = 4\omega^2 + 3\omega + 7$, for example, is a well-ordered set $(c^1, c^2, \ldots; c^{\omega+1}, c^{\omega+2}, \ldots; c^{2\omega+1}, c^{2\omega+2}, \ldots)$ of numbers equal to zero or one, which is constructed in the following manner: the first sequence (from c^1 up to, but not including, $c^{\omega+1}$) is the binary expansion of 7 followed by an infinity of zeros; the second sequence (from $c^{\omega+1}$ to $c^{2\omega+1}$) is the binary expansion of 3 followed by an infinity of zeros; the third sequence is the binary expansion of 4 (the succeeding numbers are all zero and hence need not be written down); therefore

$$4\omega^2 + 3\omega + 7 = (1,1,1,0,0,\ldots;1,1,0,0,\ldots;0,0,1)$$

The digital sum of ordinals is formed in the same way as that for integers by adding, modulo 2, terms of corresponding rank; by using a digital addition table for finite numbers, it is easy to form the digital sum of transfinite ordinals; for example:

$$(3\omega+1) \dotplus (7\omega+3) \dotplus 11 = (3 \dotplus 7)\omega + (1 \dotplus 3 \dotplus 11) = 4\omega + 9$$

THEOREM 7. *The digital sum* $\dotplus$ *has the following properties:*

(1) *it is associative:* $(\alpha \dotplus \beta) \dotplus \gamma = \alpha \dotplus (\beta \dotplus \gamma)$;

(2) *an identity element* 0 *exists, satisfying:* $\alpha \dotplus 0 = \alpha$;

(3) *for all* α, *an element* $(\dot{-}\alpha)$ *exists, satisfying:* $\alpha \dotplus (\dot{-}\alpha) = 0$;

(4) *it is commutative:* $\alpha \dotplus \beta = \beta \dotplus \alpha$.

(in algebraic terms, the first three properties express that the ordin. numbers form a '*group*', and the fourth that this group is 'Abelian').

The proof is immediate.

COROLLARY. *If* α *and* β *are two given ordinals, there exists one and only one ordinal* x, *such that* $\alpha \dotplus x = \beta$.

This is a well-known property of groups; it is sufficient to take

$$x = (\dot{-}\alpha) \dotplus \beta = \beta \dot{-} \alpha$$

THEOREM 8. *If the graphs* $G_1, G_2, \ldots, G_n$ *possess Grundy functions* $g_1, g_2, \ldots, g_n$, *the graph sum* $G = G_1 + G_2 + \ldots + G_n$ *also possesses a Grundy function, the value of which at* $x = (x_1, x_2, \ldots, x_n)$ *is given by the digital sum:*

$$g(x_1, x_2, \ldots, x_n) = g_1(x_1) \dotplus g_2(x_2) \dotplus \ldots \dotplus g_n(x_n)$$

We shall show that the function $g(x)$ thus defined on

$$X = X_1 \times X_2 \times \ldots \times X_n$$

is a Grundy function for the sum $G = (X, \Gamma)$.

1. *For every ordinal $\beta < g(x)$, Γx contains a vertex y such that $g(y) = \beta$.*
We shall use the binary expansions:

$$g_k(x_k) = (c_k^1, c_k^2, \dots)$$

$$g(x) = (c^1, c^2, \dots) \quad \text{where } c^r = \left[\sum_{k=1}^{n} c_k^r\right]_{(2)}$$

$$\beta = (\beta^1, \beta^2, \dots) \quad \text{where } \beta < g(x)$$

Let r be the largest ordinal such that $\beta^r \neq c^r$ [r exists, since $\beta \neq g(x)$]; as $\beta < g(x)$, we have

$$\beta^r = 0; \qquad c^r = \left[\sum_{k=1}^{n} c_k^r\right]_{(2)} = 1$$

Therefore among the numbers c_1^r, c_2^r, ..., c_n^r, there is at least one, c_1^r say, which is equal to one. Let us put:

$$d^s = \begin{cases} c_1^s & \text{if } \beta^s = c^s \\ [c_1^s + 1]_{(2)} & \text{if } \beta^s \neq c^s \end{cases}$$

Since $d^r = 0$, there is a vertex $y_1 \in \Gamma_1 x_1$ such that $g_1(y_1) = (d^1, d^2, \dots)$, and it is also true that

$$g(y_1, x_2, x_3, \dots, x_n) = g_1(y_1) + g_2(x_2) + \dots + g_n(x_n) = \beta$$

2. *The set Γx does not contain an element y such that $g(y) = g(x)$.*
Consider a vertex $y = (y_1, x_2, x_3, \dots, x_n)$, where $y_1 \in \Gamma_1 x_1$, and let us put

$$g_1(y_1) = (d^1, d^2, \dots)$$

Since $g_1(y_1) \neq g_1(x_1)$, $d^r \neq c_1^r$ for some ordinal r; therefore

$$[d^r + c_2^r + c_3^r + \dots + c_n^r]_{(2)} \neq [c_1^r + c_2^r + \dots + c_n^r]_{(2)}$$

Therefore $g(y) \neq g(x)$.

This result, which gives a rule for calculating the Grundy function of a sum of graphs, is going to be fundamental in the following chapters.

4. The Fundamental Numbers of the Theory of Graphs

Cyclomatic Number

The concept which we shall now introduce is *unoriented*: we shall speak of *edges* and not of *arcs*. We are concerned only with finite graphs but, for greater generality, we shall extend the definition to include s-graphs. An s-graph (X, U) is defined to be the pair formed by a set X of vertices, and by a set U of edges connecting certain vertices; but, contrary to graphs, *there may be as many as s distinct edges connecting the same initial and terminal vertices.*

There are many problems where it is more convenient to use s-graphs than graphs.

EXAMPLE (*Chemistry*). A molecule can be represented by an s-graph, the vertices of which may be identified by a symbol from Mendeleef's periodic classification. (The theory of graphs has been applied in this way to organic chemistry by POLYA [6], in the enumeration of the isomers of a

H$_2$C=CH$_2$ H—C≡C—H

Ethylene   *Acetylene*

FIG. 4.1

chemical compound.) Hence we say that ethylene is a 2-graph, acetylene is a 3-graph, etc....

Let us consider an s-graph G, with n vertices, m edges and p connected components. Put

$$\rho(G) = n - p$$
$$\nu(G) = m - \rho(G) = m - n + p$$

$\nu(G)$ is defined to be the *cyclomatic number* of the s-graph G. It has important properties.

THEOREM 1. *Let G be an s-graph and let G' be an s-graph formed from G by adding to it a new edge connecting two vertices a and b of G; if a and b are the same or are connected by a chain of G, we have*

$$\rho(G') = \rho(G) \qquad \nu(G') = \nu(G)+1$$

in all other cases, we have

$$\rho(G') = \rho(G)+1 \qquad \nu(G') = \nu(G)$$

These results follow immediately from the definitions.

COROLLARY. $\rho(G) \geqslant 0$, $\nu(G) \geqslant 0$.

Starting from a graph containing the vertices of G, each as an isolated point, let us build up G edge by edge; to begin with, we have $\rho = 0$, $\nu = 0$; each time that we add an edge, either ρ is increased, ν remaining unchanged, or else ν is increased in which case ρ remains unaltered. Both ρ and ν are therefore non-decreasing quantities.

It is convenient now to associate a vector with every cycle according to the following procedure: assign an arbitrary orientation to every edge of the s-graph G; if the edge $\boldsymbol{u}_k$ is included r_k times in the sense of this orientation and s_k times in the opposite sense in a cycle μ, we write $c^k = r_k - s_k$. (If $\boldsymbol{u}_k$ is a loop, then by convention we always put $s_k = 0$.) The vector

$$(c^1, c^2, \ldots, c^k, \ldots, c^m)$$

in the m-dimensional space $\boldsymbol{R}^m$ is the *vector-cycle corresponding to* μ, and will be denoted by $\boldsymbol{\mu}$ (or by μ, if there is no possibility of confusion).

The cycles $\mu, \mu', \mu'', \ldots$ are said to be *independent* if the corresponding vectors are linearly independent†. Note that this property is invariant with respect to the orientation selected for the edges.

† We shall restate some classical definitions of linear algebra. If $\boldsymbol{c} = (c^1, c^2, \ldots, c^m)$ and $\boldsymbol{d} = (d^1, d^2, \ldots, d^m)$ are two vectors of $\boldsymbol{R}^m$, and if $\alpha \in \boldsymbol{R}$, we put

$$\begin{aligned} \alpha\boldsymbol{c} &= (\alpha c^1, \alpha c^2, \ldots, \alpha c^m) \\ -\boldsymbol{c} &= (-c^1, -c^2, \ldots, -c^m) \\ \boldsymbol{c}+\boldsymbol{d} &= (c^1+d^1, c^2+d^2, \ldots, c^m+d^m) \\ \boldsymbol{0} &= (0, 0, \ldots, 0) \end{aligned}$$

A set $E \subset \boldsymbol{R}^m$ is a *vector sub-space* if

$$\begin{aligned} \alpha \in \boldsymbol{R}, \boldsymbol{c} \in E &\Rightarrow \alpha\boldsymbol{c} \in E \\ \boldsymbol{c}, \boldsymbol{d} \in E &\Rightarrow \boldsymbol{c}+\boldsymbol{d} \in E \end{aligned}$$

The vectors $\boldsymbol{c}_1, \boldsymbol{c}_2, \ldots, \boldsymbol{c}_k$ of $\boldsymbol{R}^m$ are said to be *linearly independent*, if

$$\alpha_1 \boldsymbol{c}_1 + \alpha_2 \boldsymbol{c}_2 + \ldots + \alpha_k \boldsymbol{c}_k = 0 \Rightarrow \alpha_1 = \alpha_2 = \ldots = \alpha_k = 0$$

In the contrary case, we have $\alpha_1 \boldsymbol{c}_1 + \alpha_2 \boldsymbol{c}_2 + \ldots + \alpha_k \boldsymbol{c}_k = 0$ for numbers α_i not all of which are zero, and the vectors are said to be *linearly dependent*. If $\alpha_1 \neq 0$, we can also write

$$-\boldsymbol{c}_1 = \frac{\alpha_2}{\alpha_1}\boldsymbol{c}_2 + \ldots + \frac{\alpha_k}{\alpha_1}\boldsymbol{c}_k$$

Thus $\boldsymbol{c}_1$ has been *expressed linearly* in terms of $\boldsymbol{c}_2, \boldsymbol{c}_3, \ldots, \boldsymbol{c}_k$. A *basis* of a vector

We have:

THEOREM 2. *The cyclomatic number $\nu(G)$ of an s-graph G is equal to the maximum number of independent cycles.*

As before, starting with a graph consisting of the vertices of G, isolated one from the other, let us build up G edge by edge. From Theorem 1 the cyclomatic number will increase by one if the addition of an edge closes new cycles and will not alter in all other cases. Suppose that before adding the edge u_k we had obtained a fundamental basis containing the cycles: $\mu_1, \mu_2, \ldots$; and that after the edge u_k has been added we have formed the new cycles: $\nu_1, \nu_2, \ldots$. Clearly ν_1 cannot be expressed linearly in terms of the μ_i (since $\nu_1^k \neq 0$ whereas $\mu_1^k = \mu_2^k = \ldots = 0$); on the other hand, the vectors $\nu_2, \nu_3, \ldots$ can be expressed linearly in terms of the μ_i and ν_1. To sum up, each time the cyclomatic number is increased by one, the maximum number of linearly independent cycles increases by one. This gives the required result.

COROLLARY 1. *The graph G contains no cycles if and only if $\nu(G) = 0$.*

COROLLARY 2. *The graph G possesses a unique cycle if and only if $\nu(G) = 1$.*

THEOREM 3. *In a strongly connected graph G, the cyclomatic number is equal to the maximum number of linearly independent circuits.*

Let us consider the unoriented 2-graph consisting of the arcs of G, and an elementary cycle μ; among the vertices encountered by the cycle μ, we can distinguish a set S of points for which one arc of μ is incident into S and another is incident out from S; a set S' of points for which both arcs of μ are incident out from S'; a set S'' of points for which both arcs of μ are incident into S''. The number of terminal vertices being equal to the number of initial vertices, we have

$$|S'| = |S''|$$

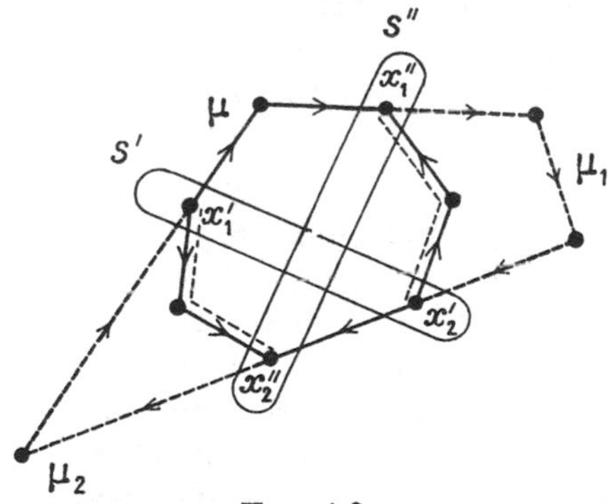

FIG. 4.2

Let $x_1', x_2', \ldots, x_k'$ be the elements of S', and let $x_1'', x_2'', \ldots, x_k''$ be those of S''.

sub-space E is a set of vectors $\{e_1, e_2, \ldots, e_k\}$ of E such that every vector of E can be expressed linearly in terms of the e_i; the *dimension* of E is the minimum number k.

In $E = R^m$, one possible basis is the set of vectors

$$e_1 = (1, 0, 0, \ldots, 0),\ e_2 = (0, 1, 0, 0, \ldots, 0), \ldots, e_m = (0, 0, 0, \ldots, 1)$$

Over the cycle μ, the elements of S' alternate with those of S'', and we can suppose that following x'_i, the first vertex encountered which does not belong to S will be x''_i; finally if μ_0 is a path which meets the vertex x before the vertex y, we denote by $\mu_0[x,y]$ the partial path going from x to y. As the graph is strongly connected, a circuit μ_i exists passing through x'_{i+1} and x''_i, and which uses the arcs of μ in order to go from x'_{i+1} to x''_i. The cycle μ is a linear combination of circuits for we can write:

$$\begin{aligned}\mu &= \mu[x'_1, x''_1] - \mu_1[x'_2, x''_1] + \mu[x'_2, x''_2] - \dots \\ &= \mu[x'_1, x''_1] + \mu_1[x''_1, x'_2] + \mu[x'_2, x''_2] + \dots \\ &\qquad - (\mu_1 + \mu_2 + \dots)\end{aligned}$$

Therefore every elementary cycle can be expressed as a linear combination of circuits, and the same applies to any cycle whatever (since any cycle can be built up as a linear combination of elementary cycles).

In $\boldsymbol{R}^m$, the circuits make up a basis of the vector sub-space determined by the cycles, and from Theorem 2, this basis is of dimension $\nu(G)$: therefore the maximum number of linearly independent circuits is $\nu(G)$.

Chromatic Number

Given a positive integer p, a graph G is said to be *p-chromatic* if the vertices can be painted with p distinct colours in such a way that no two adjacent vertices are of the same colour. The smallest number p for which the graph is p-chromatic is called the *chromatic number* of the graph G and is written $\gamma(G)$.

EXAMPLE. *Map-making.* A map of a set X of countries is drawn on a plane and we put $(x,y) \in U$ if the countries x and y have a common frontier. The graph (X, U) is symmetric and it possesses a further remarkable property: it can be drawn in such a manner that no two edges cut one another (except at their terminals). Such graphs are called *planar* and it is known that the chromatic number of a planar graph is less than or equal to 5 (cf. Chapter 21): 5 colours are therefore sufficient to colour the map so that no two adjacent regions are of the same colour.

The *chromatic index* of a graph is an integer q with the following properties:

1. By using q distinct colours, the edges of the graph can be coloured in such a way that no two adjacent edges are of the same colour.

2. This would not be possible with $q-1$ colours.

FIG. 4.3. *4-chromatic map (the choice of the colours 0, 1, 2, 3 corresponds to the values of a Grundy function of the symmetric planar graph represented by the heavy lines; in addition, the vertices of colour '0' are the largest set which can all be given the same colour)*

The chromatic index of a graph (X, U) is the chromatic number of a graph $(U, \bar{\Gamma})$ defined in the following manner: its vertices are the edges of the first graph; $u' \in \bar{\Gamma}u$ if the edges u and u' are adjacent in the first graph (cf. Fig. 4.4).

The problem of finding the chromatic index is therefore associated with the study of the chromatic number.

We shall now give some basic results concerning chromatic numbers.

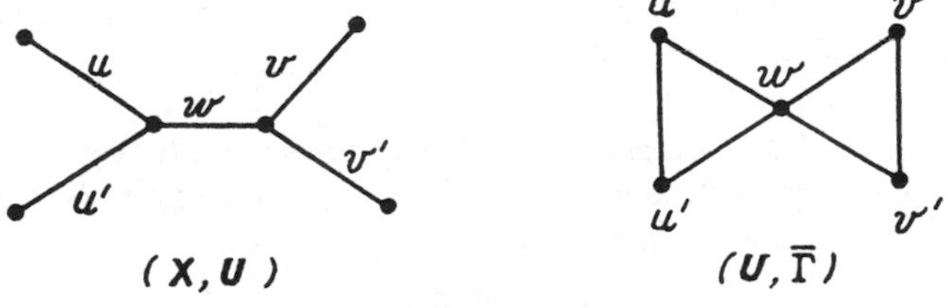

FIG. 4.4

KÖNIG'S THEOREM. *A graph is bi-chromatic if and only if it contains no cycles of uneven length.*

(1) We show first that a graph (X, U) with no uneven cycles is bi-chromatic. We can suppose that the graph is connected (if not, we treat each connected component separately). We shall colour the vertices according to the following rule:

1. An arbitrary vertex a is coloured blue;
2. If a vertex x is coloured blue, the vertices adjacent to x are coloured red; if a vertex y is coloured red, the vertices adjacent to it are coloured blue.

Since the graph is connected, every vertex will eventually be coloured; a vertex x cannot be coloured both blue and red, for this would imply that x and a are on a cycle of uneven length. The graph is therefore bi-chromatic.

(2) If a graph is bi-chromatic, it clearly cannot contain any cycles of uneven length for it would not be possible to colour the vertices of such a cycle with two colours according to the given rule.

Remark. The property:

(1) *The graph G possesses no cycles of uneven length,*

is equivalent to:

(2) *The graph G contains no elementary cycles of uneven length.*

Clearly (1) $\Rightarrow$ (2); to show (2) $\Rightarrow$ (1), suppose a cycle

$$\mu = [x_0, x_1, \ldots, x_p = x_0]$$

of uneven length p exists. Whenever two vertices x_j and x_k are found for which $j < k < p$ and $x_j = x_k$, μ is decomposed into two partial cycles $\mu[x_j, x_k]$ and $\mu[x_0, x_j] + \mu[x_k, x_0]$; further one of these cycles must be of uneven length (for otherwise μ would be even in length).

If the cycle μ is decomposed in this fashion for as long as possible, there will always remain one cycle of uneven length; but since at the end there will only be elementary cycles, statement (2) has been contradicted.

In virtue of these results, bi-chromatic graphs can easily be recognized; but no graphical method is known for finding the chromatic number of other graphs. Nevertheless it should be noted that in many cases the concept of a Grundy function proves to be a useful tool by reason of the following theorem:

THEOREM 4. *Let G be a graph, which may be assumed to be symmetric; the necessary and sufficient condition for G to be p-chromatic is that G should possess a Grundy function g(x) with*

$$\max_{x \in X} g(x) \leqslant p - 1$$

1. If such a function $g(x)$ exists, the graph is p-chromatic: it is sufficient to allot a colour to each integer from 0 to $p-1$, and then to give each vertex x the colour corresponding to the value $g(x)$.

2. If the graph is p-chromatic, we shall show that a Grundy function with maximum value less than or equal to $p-1$ exists on G. Let $S_0, S_1, \ldots, S_{p-1}$, be the different sets of vertices of the same colour. Starting from the set S_0, we build up a set $\bar{S}_0$ by adding to S_0 any vertex $x \notin S_0$ which is not adjacent to S_0 or to any vertex already incorporated in $\bar{S}_0$. Similarly, we form the set $\bar{S}_1$ by starting with the set $S_1 - (S_1 \cap \bar{S}_0)$ and adding to this set any vertex $x \notin \bar{S}_0 \cup S_1$ which is not adjacent to $S_1 - (S_1 \cap \bar{S}_0)$ or to any vertex subsequently added to it. In the same way, we next define $\bar{S}_2$, then $\bar{S}_3$, etc....

The function $g(x)$ which is equal to k when $x \in \bar{S}_k$, is a Grundy function on G which obeys the condition of the theorem.

Methods for Finding the Chromatic Number

Consider a graph G with n vertices and m edges; to find its chromatic number we can use an empirical procedure which is very simple and capable of direct application, but which is not always effective; or we can use an analytic method which gives a solution systematically but generally requires the services of an electronic computer.

The *empirical procedure* consists of starting with an arbitrary colouring using the colours 1, 2, ..., p, and attempting step by step to eliminate one of these colours (which we shall call the 'critical' colour). Consider a vertex x of the critical colour, and the connected components C_1^{jk}, C_2^{jk}, ... of the subgraph determined by two non-critical colours j and k. We succeed in changing the colour of the vertex x immediately if the sets $C_1^{jk} \cap \Gamma x$, $C_2^{jk} \cap \Gamma x$, etc., are not bicoloured: Consider separately each component C^{jk} for which $C^{jk} \cap \Gamma x$ is of colour j, and in this component permute the colours j and k (leaving the colours of all other vertices unaltered), and then change the colour of the vertex x (which is no longer adjacent to any vertices of colour j) to j.

The *analytic method* consists of testing analytically whether the graph G can be coloured with p colours; with any scheme using p colours, we can associate numbers ξ_q^i (where $i = 1, 2, \ldots, n$; $q = 1, 2, \ldots, p$) by writing:

$$\xi_q^i = \begin{cases} 1 \text{ if the vertex } x_i \text{ is of colour } q \\ 0 \text{ otherwise} \end{cases}$$

Let us write:

$$r_j^i = \begin{cases} 1 \text{ if the edge } u_j \text{ has } x_i \text{ as a vertex} \\ 0 \text{ otherwise} \end{cases}$$

The problem then reduces to finding integers ξ_q^i such that:

1. $$\xi_q^i \geqslant 0 \qquad (i = 1,2,\ldots,n; q = 1,2,\ldots,p)$$

2. $$\sum_{q=1}^{p} \xi_q^i = 1 \qquad (i = 1,2,\ldots,n)$$

3. $$\sum_{k=1}^{n} r_j^k \xi_q^k \leqslant 1 \qquad (j = 1,2,\ldots,m; q = 1,2,\ldots,p)$$

Thus we have a system of *linear inequalities* whose compatibility can easily be investigated by using the usual methods of 'linear programming'; however the solution is meaningful only if it is in integers, and systematic techniques for finding such solutions have only recently been developed†.

Consider now two distinct graphs G and H; it is reasonable to ask if the chromatic number of their sum $G+H$ or of their product $G \times H$ (cf. Chapter 3) can be determined. We have the following results:

THEOREM 5. *Let G be a $(p+1)$-chromatic graph and H a $(q+1)$-chromatic graph, and let r be the largest of the digital sums $p' \dot{+} q'$, with $p' \leqslant p$, $q' \leqslant q$; then the graph $G+H$ is $(r+1)$-chromatic.*

We can always suppose the graphs G and H to be symmetric (since this does not affect the edges of $G+H$); using the preceding theorem, construct a Grundy function $g(x)$ with maximum value p on G, and a Grundy function $h(x)$ with maximum value q on H. From Theorem 8 (Chapter 3), the graph $G+H$ possesses a Grundy function with maximum value r, which proves the theorem.

Thus, the reader can prove that if G is 6-chromatic and H is 7-chromatic, the graph $G+H$ is 8-chromatic, since

$$r = 6 \dot{+} 1 = (1,1,1) = 7$$

THEOREM 6. *If G and H are two distinct graphs with chromatic numbers p and q, and if we put $r = min\{p,q\}$, the graph $G \times H$ is r-chromatic.*

To be definite, suppose $p \leqslant q$, and let us colour the graph G with p colours; in the graph $G \times H$, we shall give the vertex $\xi = (x,y)$, the same colour as the vertex x in G. Two adjacent vertices of $G \times H$ are therefore of different

† The first published method for integer programming was derived by R. E. GOMORY and is based on the simplex method of G. B. DANTZIG. It consists essentially of finding a solution to the ordinary linear programme and, if one of the variables of this solution is not integral, of deriving new linear inequalities which are satisfied by the desired integral solution but not by the solution already found. An alternative algorithm which starts from the optimum solution of the problem without discrete variable constraints and then successively constrains the variables to take integral values has been developed by LAND and DOIG.

colours (for otherwise G would contain two adjacent vertices of the same colour).

Q.E.D.

This theorem can also be expressed by

$$\gamma(G \times H) \leqslant \min\{\gamma(G), \gamma(H)\}$$

Coefficient of Internal Stability

Consider a graph $G = (X, \Gamma)$; a set $S(\subset X)$ is said to be *internally stable* if no two vertices of S are adjacent; in other words, if

$$\Gamma S \cap S = \emptyset$$

Let us denote by $\mathscr{S}$ the family of internally stable sets of a graph; we have

$$\emptyset \in \mathscr{S}$$
$$S \in \mathscr{S}, A \subset S \Rightarrow A \in \mathscr{S}$$

The *coefficient of internal stability* of the graph G is defined to be

$$\alpha(G) = \max_{S \in \mathscr{S}} |S|$$

EXAMPLE 1 (Gauss). *Problem of the eight queens.* In the game of chess, is it possible to place eight queens on the chessboard so that no one queen can be taken by any other? This celebrated problem reduces to that of finding the maximum internally stable set of a symmetric graph with 64 vertices (the squares of the board) where $y \in \Gamma x$ if the squares x and y are on the same rank or file or on the same diagonal. This is more difficult than would appear at first sight, since Gauss initially believed that there were 76 solutions, and the *Schachzeitung*, the chess journal of Berlin, published in 1854 only 40 positions which had been discovered by different players.

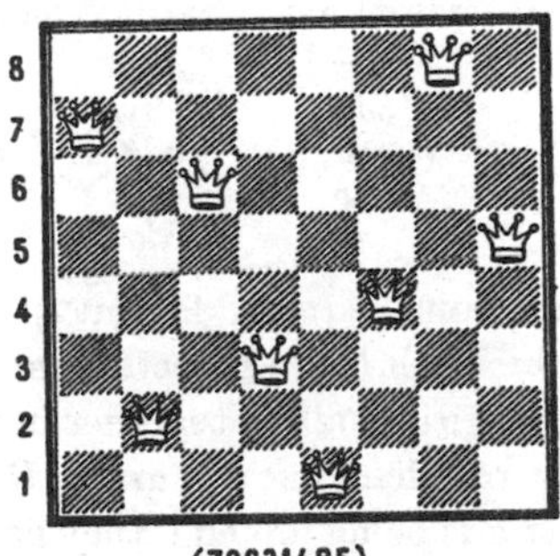

(72631485)

(61528374)

FIG. 4.5

There are in fact, 92 solutions which are indicated by the following 12 diagrams:

(72631485)	(61528374)	(58417263)
(35841726)	(46152837)	(57263148)
(16837425)	(57263184)	(48157263)
(51468273)	(42751863)	(35281746)

A diagram like that of Fig. 4.5 corresponds to each permutation, and from each diagram we can deduce eight solutions, four by rotations through 90°, and the remaining four by reflecting these with respect to the principal diagonal; the last permutation (35281746) yields the same diagram after rotation through a half-circle, and hence gives only four solutions.

EXAMPLE 2. A *clique* of a graph is a set $C \subset X$ such that

$$x \in C, y \in C \Rightarrow y \in \Gamma x \text{ or } x \in \Gamma y$$

If X is a set of people, and if $y \in \Gamma x$ signifies that there is an agreement between the individuals x and y, we are often led to look for the maximum clique, that is, the alliance containing the largest number of people. Consider a function Γ' defined by:

$$y \in \Gamma' x \Leftrightarrow y \notin \Gamma x$$

The problem reduces to finding the maximum internally stable set of the graph (X, Γ').

EXAMPLE 3. The *problem of the schoolgirls* which has been the subject of many mathematical studies, may be described as follows: fifteen schoolgirls whom we represent by the letters $a, b, c, d, e, f, g, h, i, j, k, l, m, n$ and o are at a boarding-school; if they go for a walk each day in groups of three, is it possible to choose the groups so that on seven consecutive days no two girls walk together more than once?

Using considerations of symmetry, CAYLEY found the following solution:

SUNDAY	MONDAY	TUESDAY	WEDNESDAY	THURSDAY	FRIDAY	SATURDAY
afk	*abe*	*alm*	*ado*	*agn*	*ahj*	*aci*
bgl	*cno*	*bcf*	*bik*	*bdj*	*bmn*	*bho*
chm	*dfl*	*deh*	*cjl*	*cek*	*cdg*	*dkm*
din	*ghk*	*gio*	*egm*	*fmo*	*efi*	*eln*
ejo	*ijm*	*jkn*	*fhn*	*hil*	*klo*	*fgj*

The schoolgirls problem is akin to a second, equally famous problem, which we shall call the *auxiliary problem*: with these fifteen girls, can 35 distinct triads be formed such that no two girls shall be together in a triad more than once? To solve the auxiliary problem, form a graph G whose vertices are the 455 possible triads, two triads being joined if they have two girls in common: then find a maximum internally stable set. We have

$\alpha(G) \leqslant 35$, for the same girl cannot appear in more than seven distinct triads, which implies a total of at most $15 \times 7 \times \frac{1}{3} = 35$ triads; any internally stable set of 35 triads is therefore maximal.

To see if a solution of the auxiliary problem also satisfies the schoolgirls problem, construct the graph G' whose vertices are the chosen triads, two triads being joined if they have one girl in common; if the chromatic number $\gamma(G')$ is equal to 7, the solution satisfies both problems; if $\gamma(G') > 7$, another set of triads must be selected. It is easy to show that solutions of the auxiliary problem exist which do not satisfy the schoolgirls problem.

Remark 1. The chromatic number $\gamma(G)$ and the coefficient of internal stability $\alpha(G)$ are connected by the inequality:

$$\alpha(G)\,\gamma(G) \geqslant n$$

We can, in fact, partition X into $\gamma(G)$ stable sets, consisting of vertices all of the same colour, and containing $m_1, m_2, \ldots, m_\gamma$ vertices respectively. Therefore we have

$$n = m_1 + m_2 + \ldots + m_\gamma \leqslant \alpha(G) + \alpha(G) + \ldots + \alpha(G) = \gamma(G)\,\alpha(G)$$

Remark 2. We can ask if there is not a closer connection between the two concepts, and if it would not be possible to find the chromatic number by colouring first with one colour (1) the maximum internally stable set S_1, then with a second colour the maximum internally stable set S_2 of the subgraph determined by $X - S_1$, then with a third colour the maximum internally stable set of the remaining subgraph, etc.... This cannot be done, as will be seen from the graph of Fig. 4.6 (whose chromatic number is clearly 4): the only maximum internally stable set consists of the vertices marked in white, and if these are all given the same colour, we are left with only three colours for the vertices a, b, c and d, which is obviously insufficient†.

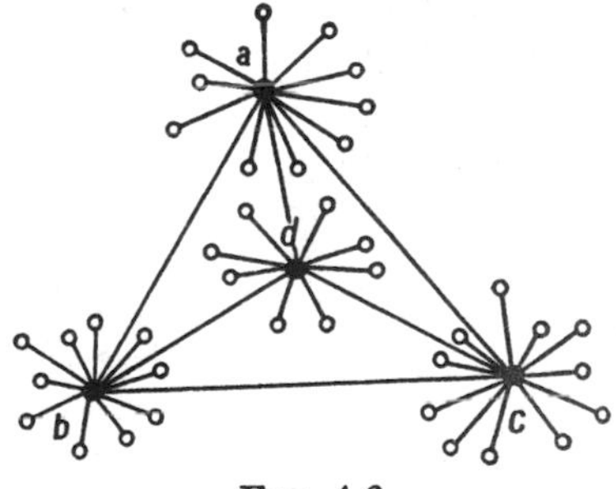

FIG. 4.6

† The following interesting question was recently raised: What is the smallest possible coefficient of internal stability $f(m, n)$ for a graph with n vertices and m edges? The solution may be derived from a theorem of P. TURAN, *Mat. Fiz. Lapok*, **48**, 1941, p. 436; the minimum number of edges for a graph G with n vertices and $\alpha(G) = p$ ($p \geqslant 2$, $p \leqslant n$) is exactly

$$\frac{(q+1)qr}{2} + \frac{q(q-1)\,(p-r)}{2}$$

where q and r are defined by $n = pq + r$.

If we have two graphs G and H, we sometimes wish to find the coefficient of internal stability of the product graph $G \times H$.

EXAMPLE (C. E. SHANNON [7]). *Problem of the capacity of a set of signals.* Consider the very simple case of a transmitter which can send five signals a, b, c, d, e; at the receiving end, each signal can give rise to two different interpretations: signal a can give p or q, signal b can give q or r, etc..., as shown in the diagram of Fig. 4.7. What is the maximum number of signals which can be used so that there is no possibility of confusion on reception? The problem reduces to finding a maximum internally stable set S of a graph G (Fig. 4.8) where two vertices are adjacent if they represent two signals which can be confused; we take $S = \{a,c\}$ and $\alpha(G) = 2$.

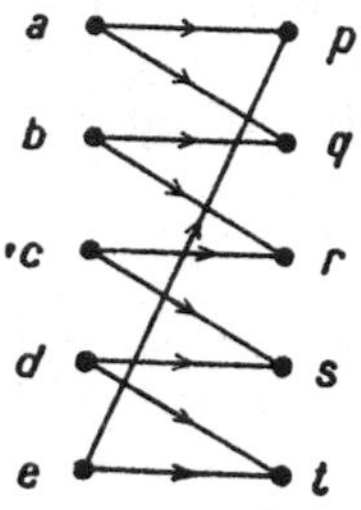

FIG. 4.7

In place of single letter signals, we could use 'words' of two letters, on condition that these do not lead to confusion on reception. Using the letters a and c which cannot be confused, we form the code, *aa*, *ac*, *ca*, *cc*. But a richer code is: *aa*, *bc*, *ce*, *db*, *ed*. (It may be proved immediately that no two of these words can be confused at the receiving end.)

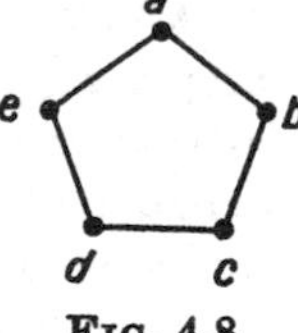

FIG. 4.8

We shall here call the *product* of two graphs $G = (X, U)$ and $H = (Y, V)$, a graph $G \times H$ whose vertices are the pairs xy with $x \in X$, $y \in Y$, two vertices xy and $x'y'$ being joined if one of the following conditions holds:

(1) $$x = x';\ (y,y') \in V$$

(2) $$(x,x') \in U;\ y = y'$$

(3) $$(x,x') \in U;\ (y,y') \in V$$

(This definition differs slightly from that of Chapter 3, but does not require different notation, since it represents an ordinary product of graphs after the addition of loops.) With the graph G of Fig. 4.8, two words xy and $x'y'$ can be confused if they represent two adjacent vertices of the product graph $G \times G = G^2$, and the richness of the code using two-letter words is $\alpha(G^2) = 5$; in general, the number of words which cannot be confused in a code using n-letter words, is the coefficient of internal stability of the product graph

$$G^n = G \times G \times \ldots \times G$$

LEMMA 1. *Given two graphs G and H, we have*

$$\alpha(G\times H) \geqslant \alpha(G)\,\alpha(H)$$

If S and T are maximum internally stable sets of G and H respectively, the Cartesian product $S\times T$ is internally stable for the graph $G\times H$, and hence

$$\alpha(G\times H) \geqslant |S\times T| = |S|\times|T| = \alpha(G)\,\alpha(H)$$

This lemma suggests the following definition: we call the *capacity* of a graph G the number

$$\theta(G) = \sup_n \sqrt[n]{[\alpha(G^n)]}$$

We have $\theta(G) \geqslant \alpha(G)$, and we shall show that we almost always have

$$\theta(G) = \alpha(G)$$

Shannon has shown elsewhere that the graph G of Fig. 4.8 is the only one with fewer than six vertices for which we have $\theta(G) \neq \alpha(G)$; in fact, its capacity has not yet been determined, and we only know that

$$\sqrt{5} \leqslant \theta(G) \leqslant \tfrac{5}{2}$$

Consider a single-valued function σ which maps X into itself and transforms each vertex x into a vertex $\sigma(x)$; this function is called *preserving* if

$$y \neq x,\, y \notin \Gamma x \Rightarrow \sigma(y) \neq \sigma(x),\, \sigma(y) \notin \Gamma\sigma(x)$$

This function preserves the property that a pair of vertices are 'distinct and not adjacent'.

LEMMA 2. *An internally stable set S is transformed by a preserving function σ into an internally stable set $\sigma(S)$ such that $|S| = |\sigma(S)|$.*

The function σ being single-valued, we have $|\sigma(S)| \leqslant |S|$; since it is also preserving, we have $|\sigma(S)| = |S|$.

LEMMA 3. *If the set $\sigma(X)$ is internally stable, the coefficient of internal stability of the graph G is $\alpha(G) = |\sigma(X)|$.*

If $\sigma(X)$ is internally stable, we have

$$|\sigma(X)| \leqslant \max_{S\in\mathscr{S}} |S| = \alpha(G)$$

On the other hand, if S_0 is a maximum internally stable set, we have, from Lemma 2,

$$|\sigma(X)| \geqslant |\sigma(S_0)| = |S_0| = \alpha(G)$$

Therefore

$$\alpha(G) = |\sigma(X)|$$

THEOREM 7 (SHANNON). *If one of the graphs G or H can be transformed by a preserving function σ into an internally stable set, we have*

$$\alpha(G \times H) = \alpha(G)\,\alpha(H)$$

It is only necessary to show that $\alpha(G \times H) \leqslant \alpha(G)\,\alpha(H)$; let σ be a preserving function defined on G such that $\sigma(X)$ is internally stable, and let σ_0 be the function mapping $G \times H$ into itself defined by

$$\sigma_0(x, y) = [\sigma(x), y]$$

The function σ_0 which transforms two non-adjacent, distinct vertices $\xi = (x,y)$ and $\xi' = (x',y')$ into two non-adjacent, distinct vertices $(\sigma x, y)$ and $(\sigma x', y')$, is preserving.

If S_0 is a maximum internally stable set of $G \times H$, then from Lemma 2, $\alpha(G \times H) = |S_0| = |\sigma_0(S_0)|$; let us distribute the elements of $\sigma_0(S_0)$ among different classes according to the value of σx for each element. From Lemma 3, this gives $|\sigma(X)| = \alpha(G)$ distinct classes. Further, since no two elements of $\sigma_0(S_0)$ are adjacent, each class has at most $\alpha(H)$ elements, and therefore

$$\alpha(G \times H) = |\sigma_0(S_0)| \leqslant \alpha(G)\,\alpha(H)$$

COROLLARY. *If a graph G can be transformed by a preserving function σ into an internally stable set, its capacity is equal to its coefficient of internal stability.*

We have

$$\alpha(G \times G) = [\alpha(G)]^2$$
$$\alpha(G \times G \times G) = [\alpha(G)]^2\,\alpha(G) = [\alpha(G)]^3$$
$$\text{etc.}$$

and hence:

$$\theta(G) = \sup_n \sqrt[n]{[\alpha(G^n)]} = \alpha(G)$$

Coefficient of External Stability

Given a graph $G = (X, U)$, a set $T (\subset X)$ is *externally stable* if for every vertex $x \notin T$, we have $\Gamma x \cap T \neq \emptyset$; in other words, we have $\Gamma^{-1} T \supset X - T$.

If $\mathscr{T}$ denotes the family of externally stable sets of a graph, we have

$$X \in \mathscr{T}$$
$$T \in \mathscr{T}, A \supset T \Rightarrow A \in \mathscr{T}$$

By definition, the *coefficient of external stability* of the graph G is

$$\beta(G) = \min_{T \in \mathscr{T}} |T|$$

We now consider the problem of constructing an externally stable set which contains the minimum possible number of elements.

EXAMPLE 1. *Problem of the warders.* In the prison of ..., each cell is provided with a post for a warder, but the warder at cell x_0, for example, can also overlook cells x_1, x_2, x_3, and x_4, which are connected with cell x_0 by a straight corridor, as shown in Fig. 4.9. What is the minimum number of warders needed in order to survey all the cells ? This is the same as finding the coefficient of external stability for the symmetric graph of Fig. 4.9, which for this very simple plan is $\beta = 2$.

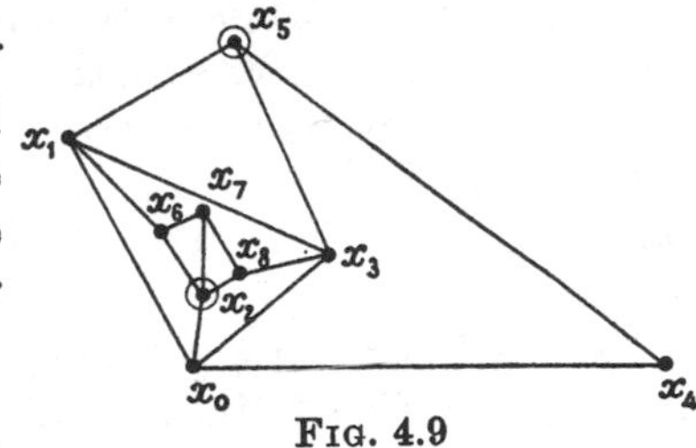

FIG. 4.9

EXAMPLE 2. *Problem of the five queens.* In the game of chess, what is the smallest number of queens which can be placed on the board so that every square is dominated by at least one of the queens ? This problem is the same as finding a minimum externally stable set for a graph with 64 vertices (the squares on the board), with $(x,y) \in U$ if and only if the squares x and y are on the same rank or file or diagonal.

$\beta(G_1) = 5$

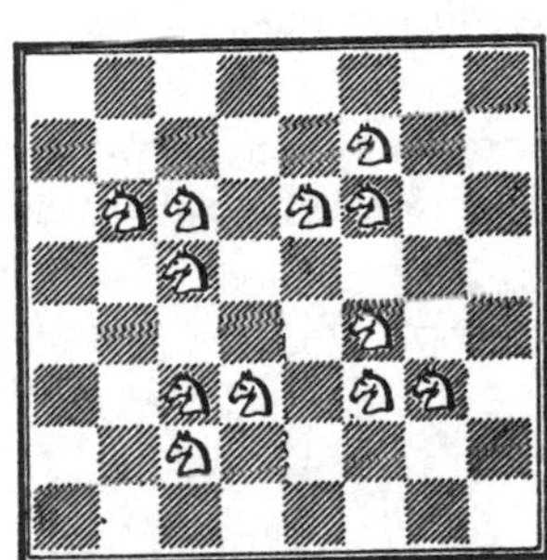

$\beta(G_2) = 12$

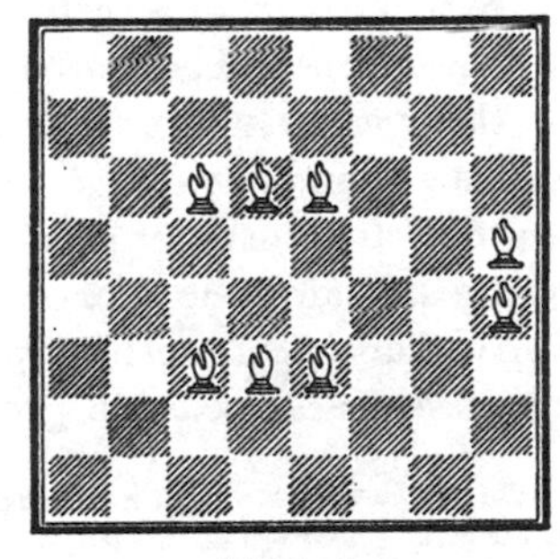

$\beta(G_3) = 8$

FIG. 4.10

The coefficient of external stability is $\beta = 5$ for the queens; $\beta = 8$ for the rooks; $\beta = 12$ for the knights; $\beta = 8$ for the bishops.

EXAMPLE 3. *Problem of the six red discs.* The following game is well-known to lovers of funfairs: a large white disc (of radius one) is placed on a table, and it is required to cover it completely with six small red discs (of radius $\rho < 1$) which are successively laid in position and which, once placed, cannot again be moved. What is the least radius ρ for which the problem is possible?

The problem reduces to finding a minimum externally stable set T for an infinite graph (X, U), where X is the set of points of the white disc, and where $(x, y) \in U$ if the distance from x to y, $d(x, y) \leqslant \rho$.

To determine a maximum internally stable set or a minimum externally stable set, one could consider successively each set $T \subset X$ and test if it satisfied the appropriate conditions. Naturally, this method of elimination is in general impracticable; we therefore have recourse to an *algorithm*, that is, to a routine which makes such calculations possible (the use of high-speed electronic computers is another possibility).

ALGORITHM FOR FINDING A MINIMUM EXTERNALLY STABLE SET. The algebraic form of the algorithm will be developed by the use of the concepts of 'logical sum' and 'logical multiplication'†.

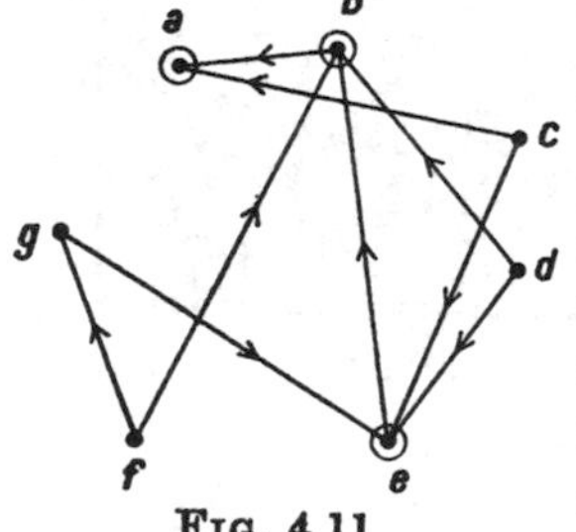

FIG. 4.11

Consider the graph of Fig. 4.11 and let us denote by $a, b, c, \ldots$ the operations consisting of using the vertex $a, b, c, \ldots$ in forming a minimal externally stable set. The *logical sum* $x + y$ denotes the operation of using 'x or y', and *logical multiplication* xy denotes the operation of using 'x and y'. If we regard $a, b, c, \ldots$ as sets each containing exactly one element, these two operations correspond to the operations $\cup$ and $\cap$ of the theory of sets and hence enjoy the same associative and commutative properties, etc. Let us now form a logical expression ϕ for the externally stable sets of the graph in which, for each vertex x, we include either x or one of the elements of Γx; by means of the logical operations, this can be expressed by an algebraic expression; we shall next use the associative property and the law of

† These operations are familiar in the theory of 'switching circuits' (cf. for example, S. M. CALDWELL, Switching circuits and logical design, J. Wiley & Sons, New York, 1958); the treatment we have given here is due to K. MAGHOUT, *C. R. Acad. Sci. Paris*, **248**, p. 3522.

absorption ($ax = a$ whenever $x \supset a$) to simplify the expression and remove any redundancies. For the graph of Fig. 4.11 this gives:

$$\begin{aligned}\phi &\supset a(b+a)(c+a+e)(d+b+e)(e+b)(f+b+g)(g+e)\\ &= a(b+e)(b+f+g)(g+e)\\ &= ab(g+e)+ae(b+f+g)\\ &= abg+abe+aef+aeg\end{aligned}$$

Therefore, there are in all four minimal externally stable sets, namely *abe*, *abg*, *aef* and *aeg*.

APPLICATION TO COLOURING A GRAPH. If the number of vertices is not too large, the same algebraic method can be used to find all the maximal internally stable sets. If, for every vertex x, we take either x or all vertices adjacent to it, we determine the complement of such a set; for the graph of Fig. 4.11, this gives:

$$\psi = (a + bc)\,(b + adef)\,(c + ae)\,(d + be)\,(e + bcdg)\,(f + bg)\,(g + fe)$$

Using the law $(a+bx)(b+ay) = (a+bx)(b+y)$, this becomes:

$$\begin{aligned}\psi &= (a+bc)(b+def)(c+e)(d+e)(e+dg)(f+g)(g+f)\\ &= (a+bc)(b+def)(c+e)(e+dg)(f+g)\\ &= (a+bc)\,b(c+e)(ef+eg+dg)+(a+bc)\,def\\ &= (ab+bc)(ef+eg+cdg)+adef+bcdef\\ &= abef+abeg+abcdg+bcef+bceg+bcdg+adef+bcdef\\ &= abef+abeg+bcef+bceg+bcdg+adef\end{aligned}$$

By taking the complements of the several sets shown in this sum, we get the maximal internally stable sets:

$$A = cdg, \quad B = cdf, \quad C = adg, \quad D = adf, \quad E = aef, \quad F = bcg$$

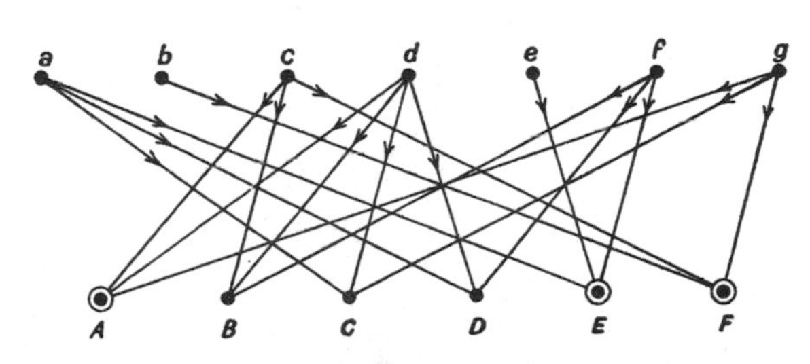

FIG. 4.12

To find the chromatic number of the graph, we require the minimum number of these sets which contain all the vertices; alternatively, we want a minimum externally stable set of the auxiliary graph illustrated in Fig. 4.12. As before, write down the logical function:

$$\begin{aligned}\phi &= (C+D+E)(F)(A+B+F)(A+B+C+D)E(B+D+E)(A+C+F)\\ &= EF(A+B+C+D)\\ &= EFA+EFB+EFC+EFD\end{aligned}$$

Therefore three colours suffice to colour the vertices of the graph: for example, with colours (A), (E) and (F), we could colour each vertex by considering to which of the sets A, E and F it belonged.

5. Kernels of a Graph

Existence and Uniqueness Theorems

We consider here a graph (X, Γ) which may or may not be finite. A set $S \subset X$ is called a *kernel* of the graph if S is internally and externally stable; we have therefore

$$(1) \qquad x \in S \Rightarrow \Gamma x \cap S = \emptyset$$

$$(2) \qquad x \notin S \Rightarrow \Gamma x \cap S \neq \emptyset$$

From condition (1), we deduce that the kernel S is free from loops; and from condition (2), we deduce that S contains every vertex x for which $\Gamma x = \emptyset$; further $\emptyset$ is not a kernel.

EXAMPLE 1 (VON NEUMANN–MORGENSTERN). The notion of a kernel was first introduced into the theory of games under the name of 'solution'.

Suppose that n players, whom we shall designate (1), (2), ..., (n), can by their choice of action select a situation x from a set X; if player (i) prefers a situation a to a situation b, we shall write $a \geqslant^i b$, and it has been shown (Example 3, Chapter 2) that this is a total weak order relation. Nevertheless the preferences of different individuals may not be compatible: we need therefore to introduce the notion of *effective preference* $\succ$. The situation a is said to be effectively preferred to b, written $a \succ b$, if there exists a set of players who judge a to be better than b, and who can, if they so wish, make their point of view prevail; if in addition $b \succ c$, a set of players exists who can force b to prevail over c, but since these two sets do not necessarily contain the same players, we may not have $a \succ c$: the relation $\succ$ is not transitive!†

Let us now consider the graph (X, Γ), where Γx represents the set of situations which are effectively preferred to x. Let S be a kernel of the graph (if one exists); Von Neumann and Morgenstern propose that the game be limited to the elements of S. Since S is internally stable, no one situation of

† The term *effective preference* is due to G. T. GUILBAUD, who develops it in these terms in [2]; VON NEUMANN and MORGENSTERN use that of *dominance*. For different mathematical formulations of effective preference, see, e.g. BERGE [1], Chapter V.

S can be effectively preferred to another, which makes for a certain consistency; since S is externally stable, to any situation x which is not in S, a situation in S can be found which is effectively preferred to x, so that x can immediately be discarded.

EXAMPLE 2. Let us consider a graph (X, Γ) which possesses a Grundy function $g(x)$; the set $S = \{x \mid g(x) = 0\}$ is a kernel of the graph, since

(1) $$x \in S \Rightarrow \Gamma x \cap S = \emptyset$$

(2) $$x \notin S \Rightarrow g(x) \neq 0, \text{ and hence } \Gamma x \cap S \neq \emptyset$$

EXAMPLE 3. Not every graph has a kernel, as the reader can show for himself on the graph of Fig. 5.1; if a graph possesses a kernel S_0, its coefficients of stability satisfy

$$\alpha(G) = \max_{S \in \mathscr{S}} |S| \geqslant |S_0| \geqslant \min_{T \in \mathscr{T}} |T| = \beta(G)$$

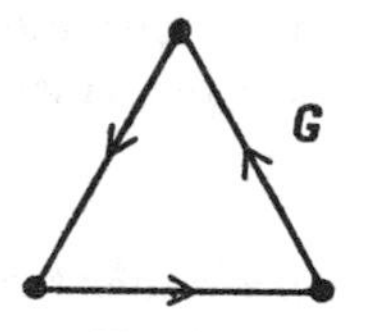

Fig. 5.1

In the graph of Fig. 5.1, there is no kernel, since:

$$\alpha(G) = 1 < 2 = \beta(G)$$

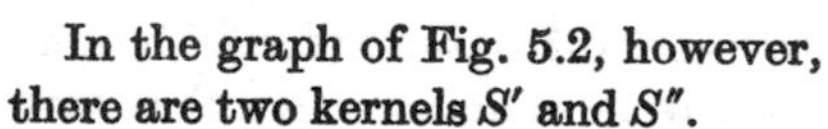

In the graph of Fig. 5.2, however, there are two kernels S' and S''.

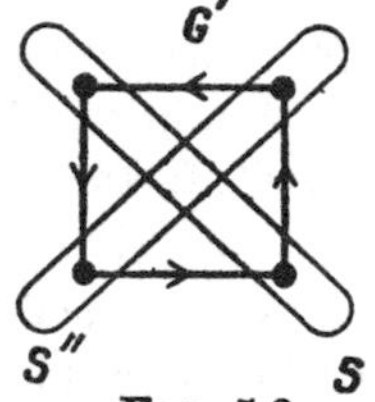

Fig. 5.2

We now intend to look for criteria whereby we can recognize if a graph possesses a kernel, and if this kernel is unique.

THEOREM 1. *If a set S is a kernel of a graph (X, Γ), S is a maximal set of the family $\mathscr{S}$ of internally stable sets; that is, we have:*

$$A \in \mathscr{S}, A \supset S \Rightarrow A = S$$

Let A be an internally stable set containing the kernel S; we shall suppose that S is *strictly* contained in A, and show that this is absurd. In fact, if this is the case, a vertex a exists, with $a \in A$, $a \notin S$, for which $\Gamma a \cap S \neq \emptyset$, which in turn implies $\Gamma a \cap A \neq \emptyset$, which contradicts $A \in \mathscr{S}$.

THEOREM 2. *In a symmetric graph without loops, any maximal set of the family $\mathscr{S}$ of internally stable sets is a kernel.*

Let S be a maximal set of $\mathscr{S}$; it is required to show that any vertex $x \notin S$ satisfies $\Gamma x \cap S \neq \emptyset$; but if this were not so, the set $A = S \cup \{x\}$ would be internally stable (since $x \notin \Gamma x$); since we also have $A \supset S$, $A \neq S$, the hypothesis that S is maximal within $\mathscr{S}$ would be contradicted.

COROLLARY. *A symmetric graph without loops possesses a kernel.*

Let us form an auxiliary graph $(\mathscr{S}, \bar{\Gamma})$ whose vertices are the internally stable sets of the given symmetric graph, and where $S \in \bar{\Gamma}S'$ if and only if $S \supset S'$. The auxiliary graph is inductive and therefore, from Zorn's Lemma (Chapter 2), it contains a vertex $S \in \mathscr{S}$ which has no descendants. This S is a maximal internally stable set and hence, from Theorem 2, it is a kernel.

In the case where the given symmetric graph is finite, this corollary is obvious, and the kernel may be found by using the following procedure:

Choose an arbitrary vertex x_0, and put $S_0 = \{x_0\}$; next select a vertex $x_1 \notin \Gamma S_0$, and put $S_1 = \{x_0, x_1\}$; then a vertex $x_2 \notin \Gamma S_1$, etc.... If the graph is finite, we shall obtain, sooner or later, $\Gamma S_n = X$, and S_n, being a maximal set of $\mathscr{S}$, is a kernel.

THEOREM 3. *A necessary and sufficient condition for S to be a kernel, is that its characteristic function $\phi_s(x)$ satisfies*

$$\phi_s(x) = 1 - \max_{y \in \Gamma x} \phi_s(y)$$

The characteristic function $\phi_s(x)$ of a set S is defined by:

$$\phi_s(x) \begin{cases} = 1 & \text{if } x \in S \\ = 0 & \text{if } x \notin S \end{cases}$$

If $\Gamma x = \emptyset$, it is conventional to put:

$$\max_{y \in \Gamma x} \phi_s(y) = 0$$

1. Let S be a kernel. By reason of its internal stability, we have

$$\phi_s(x) = 1 \Rightarrow x \in S \Rightarrow \max_{y \in \Gamma x} \phi_s(y) = 0$$

By reason of the external stability, we have

$$\phi_s(x) = 0 \Rightarrow x \notin S \Rightarrow \max_{y \in \Gamma x} \phi_s(y) = 1$$

Combining these, we get the required formula.

2. Let $\phi_s(x)$ be the characteristic function of a set S; if $\phi_s(x)$ satisfies the condition of the theorem, we have

$$x \in S \Rightarrow \phi_s(x) = 1 \Rightarrow \max_{y \in \Gamma x} \phi_s(y) = 0 \Rightarrow \Gamma x \cap S = \emptyset$$

$$x \notin S \Rightarrow \phi_s(x) = 0 \Rightarrow \max_{y \in \Gamma x} \phi_s(y) = 1 \Rightarrow \Gamma x \cap S \neq \emptyset$$

Thus S is certainly a kernel.

THEOREM 4. *A progressively finite graph possesses a kernel, and this kernel is unique.*

This may be proved immediately by observing that a characteristic function $\phi_s(x)$, which satisfies the formula of the preceding theorem, can be defined by induction on the sets:

$$X(0) = \{x \mid \Gamma x = \emptyset\}$$
$$X(1) = \{x \mid \Gamma x \subset X(0)\}$$
$$X(2) = \{x \mid \Gamma x \subset X(1)\}$$
$$\cdots\cdots\cdots\cdots$$

Theorem 4 of Chapter 3 shows that by these means ϕ_s is defined for the whole of X.

RICHARDSON'S THEOREM. *If a finite graph has no circuits of uneven length, it possesses a kernel.*

Let (X, Γ) be a graph with no uneven circuits; we now define a sequence $Y_0, Y_1, Y_2, \ldots$ of sets in X in the following manner:

1. We take $Y_0 = \emptyset$; and we denote by B_0 the basis (cf. Chapter 2) of the subgraph determined by $X - Y_0$; from Theorem 1 of Chapter 2, this basis is known to exist. We put $Y_1 = B_0 \cup \Gamma^{-1} B_0$.

2. If Y_n has been defined, we denote by B_n a basis of the subgraph determined by $X - Y_n$, such that $B_n \subset \Gamma^{-1}(\Gamma^{-1} B_{n-1} - Y_{n-1})$.

It can easily be seen that such a basis always exists†; next we put

$$Y_{n+1} = Y_n \cup B_n \cup \Gamma^{-1} B_n$$

† To show that the graph $(X - Y_n,\ \Gamma_{X-Y_n})$ contains a basis

$$B_n \subset \Gamma^{-1}(\Gamma^{-1} B_{n-1} - Y_{n-1})$$

let us consider an arbitrary basis B'_n, and a vertex a_0 which has no descendants which has been selected as an element of B'_n according to Theorem 3 (Chapter 2); we need to show that $X - Y_n$ contains a vertex b which can be substituted for a_0, and which satisfies:

(1) $$b \overset{n}{\equiv} a_0$$

(where $\overset{n}{\geqslant}$ is the weak order relation of the subgraph determined by $(X - Y_n)$.)

(2) $$b \in \Gamma^{-1}(\Gamma^{-1} B_{n-1} - Y_{n-1})$$

In the graph $(X - Y_{n-1}, \Gamma_{X-Y_{n-1}})$, a path $\mu = [a_0, a_1, \ldots, b']$ exists leading from a_0 into B_{n-1}; let a_k be the first vertex of μ belonging to Y_n. Since $a_k \in X - Y_{n-1}$, $a_k \in Y_n$, $a_k \notin B_{n-1}$, we have $a_k \in \Gamma^{-1} B_{n-1} - Y_{n-1}$; the vertex $b = a_{k-1}$ must satisfy (1) and (2).

We have then:

$$\emptyset = Y_0 \subset\subset Y_1 \subset\subset Y_2 \subset\subset \dots$$

Since the graph is finite, an integer m exists such that $Y_m = X$, and we then put

$$S = \bigcup_{n=0}^{m-1} B_n$$

We shall show that S is a kernel of the graph (X, Γ).

1. *S is externally stable* for if $x \notin S$, we have $x \in \Gamma^{-1} B_k - Y_k$ for some integer k; therefore $\Gamma x \cap B_k \neq \emptyset$, and hence $\Gamma x \cap S \neq \emptyset$.

2. *S is internally stable.* Clearly two elements of B_n cannot be adjacent (B_n being a basis of a subgraph); let us see if it is possible to have two

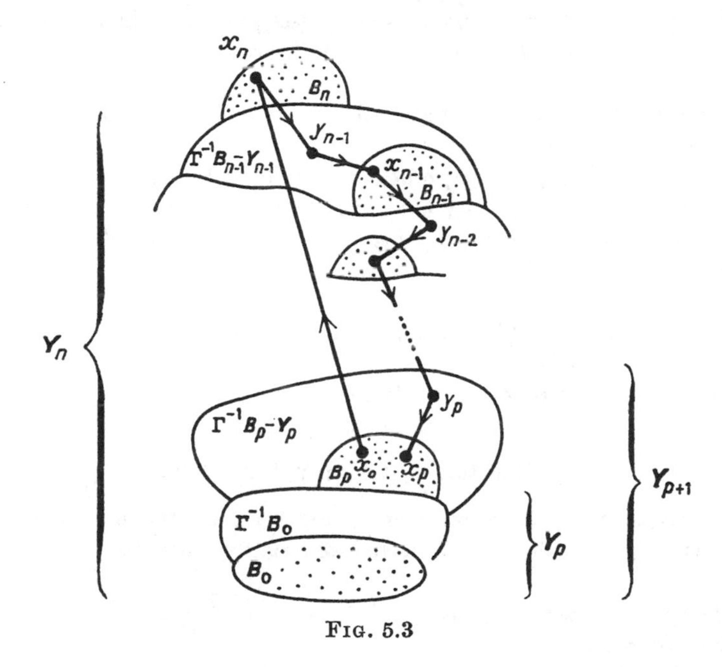

FIG. 5.3

adjacent vertices, one of which is in B_n and the other in B_p, where $p < n$. We have

$$B_n \cap \Gamma^{-1} B_p = \emptyset$$

since

$$\Gamma^{-1} B_p \subset Y_{p+1}$$

and

$$B_n \subset X - Y_n \subset X - Y_{p+1}$$

Likewise we have $B_p \cap \Gamma^{-1} B_n = \emptyset$, for if not we would be able to find a path $\mu = [x_0, x_n, y_{n-1}, x_{n-1}, y_{n-2}, x_{n-2}, \ldots, y_p, x_p]$ with

$$
\begin{aligned}
&x_0 \in B_p, \qquad x_0 \in \Gamma^{-1} B_n \\
&x_n \in B_n \subset \Gamma^{-1}(\Gamma^{-1} B_{n-1} - Y_{n-1}) \\
&y_{n-1} \in \Gamma^{-1} B_{n-1} - Y_{n-1} \\
&x_{n-1} \in B_{n-1} \\
&\ldots\ldots\ldots\ldots\ldots \\
&y_p \in \Gamma^{-1} B_p - Y_p \\
&x_p \in B_p
\end{aligned}
$$

Note that all vertices of the path μ are in $X - Y_p$ and that μ goes from $x_0 \in B_p$ to $x_p \in B_p$; since B_p is a basis of the subgraph determined by $X - Y_p$, we must therefore have $x_0 = x_p$; the path μ is thus a circuit of uneven length, which contradicts the hypothesis.

GENERALIZATION 1. *If a totally inductive graph contains no circuits of uneven length, it possesses a kernel.*

In fact, since every subgraph of a totally inductive graph possesses a basis, the proof will be exactly the same as the preceding; the sets Y_α are defined by induction (where in this case α can be a transfinite ordinal number).

GENERALIZATION 2. *If a locally finite graph has no circuits of uneven length, it possesses a kernel.*

This follows immediately from Rado's theorem (Chapter 3).

Application to Grundy Functions

The following result permits us to extend the existence conditions which have been established for kernels to Grundy functions:

THEOREM 5. *If every subgraph of (X, Γ) possesses a kernel, (X, Γ) possesses a Grundy function.*

Let S_0 be a kernel of (X, Γ), S_1 a kernel of the subgraph determined by $X_1 = X - S_0$, S_2 a kernel of the subgraph determined by $X_2 = X_1 - S_1$, etc. . . . ; and, in general, let us write:

$$X_\alpha = X_{\alpha-1} - S_{\alpha-1} \text{ if } \alpha \text{ is not an ordinal limit number}$$

$$X_\alpha = \bigcap_{\beta<\alpha} X_\beta \text{ if } \alpha \text{ is an ordinal limit number}$$

(notice that $\beta > \alpha$ implies $X_\alpha \supset \supset X_\beta$, and hence from the induction principle, an ordinal γ exists such that $X_\gamma = \emptyset$).

The sets S_α form a partition of X, and to every vertex x we assign an ordinal number $g(x)$ as follows:

$$g(x) = \alpha \Leftrightarrow x \in S_\alpha$$

We shall show that $g(x)$ is a Grundy function.

(1) *Given $g(x) = \alpha$; to show that for every $\beta < \alpha$ a vertex $y \in \Gamma x$ exists such that $g(y) = \beta$.*

Since $x \in S_\alpha$, $\alpha > \beta$, we have $x \in X_\alpha \subset X_\beta$; $x \in S_\beta$.

Since S_β is an externally stable set of the subgraph determined by X_β, S_β contains a vertex y such that $y \in \Gamma x$; in other words, Γx contains a vertex y such that $g(y) = \beta$.

(2) *Given $g(x) = \alpha$; Γx does not contain a vertex y such that $g(y) = \alpha$*, for if this were so, the set S_α would not be internally stable.

[Those two conditions show that $g(x)$ is a Grundy function.]

Corollary 1. *A symmetric graph possesses a Grundy function if and only if it has no loops.*

(It is obvious that a graph with a loop cannot possess a Grundy function.)

Corollary 2. *A graph with no uneven circuits which is either finite, locally finite or totally inductive possesses a Grundy function.*

Remark. Consider the graphs $G_1, G_2, \ldots, G_n$, each of which possesses a kernel; does their sum (or product) also possess a kernel? The perfect tool for determining a kernel of a sum of graphs is the concept of a Grundy function (which is the reason for the stress laid in this work on the theory of Grundy functions). By combining Theorem 8 (Chapter 3) with Theorem 5 above, we see in effect that *if every subgraph of G_i possesses a kernel (for all i), the sum:*

$$G_1 + G_2 + \ldots + G_n$$

possesses a Grundy function g, and hence a kernel $S = \{x \mid g(x) = 0\}$.

In the following chapter we shall observe some applications of this important remark.

6. Games on a Graph

Nim Type Games

We can use a graph (X, Γ) to define a game between two people, whom we shall call (A) and (B), in such a way that a vertex of the graph represents a position in the game; a starting point x_0 is selected at random, and the two players have alternate moves; player (A) first chooses a vertex x_1 from the set Γx_0, then (B) chooses a vertex x_2 from the set Γx_1, then (A) chooses a vertex x_3 from Γx_2, etc. If a player selects a vertex x_k such that $\Gamma x_k = \emptyset$, the game is finished, the winner being the player who chose the last vertex. If the graph is not progressively finite, it is clear that a single play of the game could go on indefinitely.

In honour of the familiar pastime of which this is a generalization, we shall call the game described above a *game of Nim*, and we shall denote both it and the graph which defines it by (X, Γ); the problem now arises of how to characterize the winning positions, that is to say, the vertices which must be chosen in order to win the game irrespective of the responses of our opponent. The main result is the following:

THEOREM 1. *If the graph possesses a kernel S, and if a player chooses a vertex in S, this choice assures him of a win or a draw.*

In fact, if player (A) chooses $x_1 \in S$, either $\Gamma x_1 = \emptyset$ in which case he has certainly won; or else his opponent will be forced to choose a vertex x_2 in $X - S$, and then, at his next move, player (A) can again choose a vertex x_3 in S, and so on. Should the game be finished at some stage by one of the players choosing a vertex x_k such that $\Gamma x_k = \emptyset$, we have $x_k \in S$, and hence the winning player is necessarily (A).

Q.E.D.

Therefore one basic method of good play would be to calculate a Grundy function $g(x)$, if it exists; from this we can deduce a kernel

$$S = \{x \mid g(x) = 0\}$$

for the graph in question. Should the initial vertex x_0 be such that $g(x_0) = 0$, player (A) is in a critical position, for now his opponent can assure himself of either a win or a draw; if on the other hand $g(x_0) \neq 0$, player (A) will be able either to win or to draw by selecting a vertex x_1 such that $g(x_1) = 0$.

COROLLARY. *If the graph is progressively finite, it possesses a unique Grundy function $g(x)$; any choice y such that $g(y) = 0$ is a winning one, and any choice z such that $g(z) \neq 0$ is a losing one.*

The proof is immediate.

EXAMPLE 1 (R. ISAACS). X is the set of points (p,q) in the Cartesian plane for which p and q are non-negative integers; the two players choose turn by turn a point of X; if x is the last point picked, the next player must make his choice either from points on one of the perpendiculars xa or xb from x to the co-ordinate axes, or on the line xc, bisecting the angle bxa; for example, following the vertex $x = (4,1)$, the choice must be made from the vertices $(4,0)$, $(3,1)$, $(2,1)$, $(1,1)$, $(0,1)$ or $(3,0)$. The winner is the first player who manages to reach the vertex $(0,0)$.

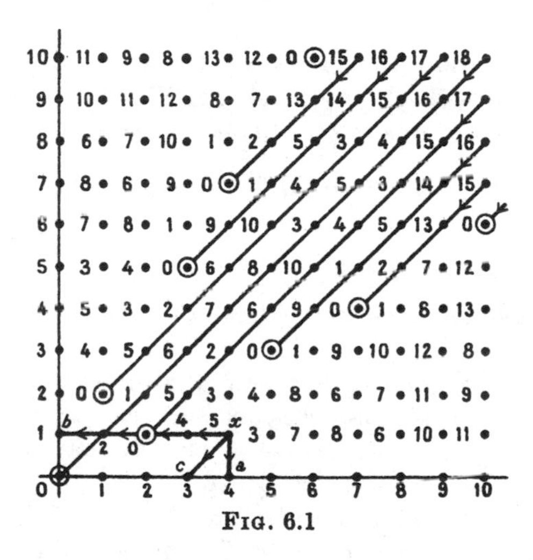

FIG. 6.1

The graph (X, Γ) is progressively finite and consequently possesses a Grundy function $g(x)$, the value of which at each of its vertices is shown in Fig. 6.1; the winning positions have been ringed. Although they form a set which is symmetric with respect to the principal diagonal, their irregular distribution will be noted. It can be proved that the abscissa of the n-th winning position under the main diagonal is the greatest integer m where

$$m \leqslant \frac{3+\sqrt{5}}{2} n$$

EXAMPLE 2. The graph of Fig. 6.2 represents a game of Nim for which neither a Grundy function nor a kernel exists; therefore the preceding method cannot be applied. Nevertheless each player must limit himself to choosing one of the vertices a, b or c if he does not wish to lose the game.

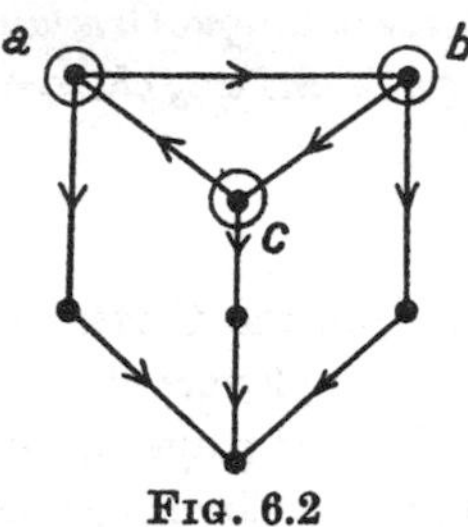

FIG. 6.2

Let us now consider several different games of Nim (X_1,Γ_1), (X_2,Γ_2), ..., (X_n,Γ_n), and suppose that two players wish to play turn and turn about according to the following rule: each one, when it is his turn to play, chooses one of the games of Nim, makes a move in the game which he has chosen, and no move in any of the others; the loser is the first player who cannot play at all. This situation is also a game of Nim on a graph which is simply a sum of the graphs (X_i,Γ_i) (cf. Chapter 3); the following result furnishes a method whereby one may win:

THEOREM 2. *Consider the Nim-games* (X_1,Γ_1), (X_2,Γ_2), ..., (X_n,Γ_n) *which possess Grundy functions* $g_1, g_2, \ldots, g_n$*; if a game is played using their sum, a win or a draw can be assured by choosing positions*

$$x = (x_1, x_2, \ldots, x_n) \in X_1 \times X_2 \times \ldots \times X_n$$

for which:

$$g(x) = g_1(x_1) \dotplus g_2(x_2) \dotplus \ldots \dotplus g_n(x_n) = 0$$

This can be deduced immediately from Theorem 8 (Chapter 3).

EXAMPLE. *Simple game of Nim or Fan Tan.* At the beginning of the game, there are n heaps of matches and two players; they each choose a (non-empty) heap in turn and remove one or more matches from it. The one who removes the last match is the winner.

This situation represents a sum of games (X_1,Γ_1), (X_2,Γ_2), ..., (X_n,Γ_n); in the game (X_k,Γ_k), an element x_k of X_k represents the state of the k-th heap, and $g_k(x_k)$ is equal to the corresponding number of matches.

From Theorem 2, a move is a winning one if and only if the digital sum of the numbers of matches left in the several heaps is zero.

We shall consider some generalizations of Theorem 2; first, if (X_1,Γ_1), (X_2,Γ_2), ..., (X_n,Γ_n) are n given games of Nim, their *sum of order p* is defined to be a game of Nim in which each player, faced with all n games, chooses k of them ($k \neq 0, k \leqslant p$), makes a move in each of the k selected games and does not make a move in any of the others; the sum of order n corresponds to the *product* of the graphs and the sum of order 1 to their *sum*. To select a winning move, the following result may be used:

THEOREM 3. *Let* (X_1, Γ_1), (X_2, Γ_2), ..., (X_n, Γ_n) *be n Nim-games without circuits and such that* $\Gamma_i x = \hat{\Gamma}_i x - \{x\}$, $x \in X$, $i = 1, 2, \ldots, n$, *and let* g_1, g_2, ..., g_n *be their Grundy functions (these always exist); the sum of order p of the n games possesses a Grundy function g, the value of which at* $x = (x_1, x_2, \ldots, x_n)$ *may be found in the following way: form the binary expansions*

$$g_k(x_k) = (c_k^1, c_k^2, \ldots)$$

Take:

$$g(x) = \left[\sum_{k=1}^{n} c_k^1\right]_{(p+1)} + (p+1)\left[\sum_{k=1}^{n} c_k^2\right]_{(p+1)} + (p+1)^2\left[\sum_{k=1}^{n} c_k^3\right]_{(p+1)} + \ldots$$

where $[m]_{(p+1)}$ *denotes the integer remainder after dividing m by* $p+1$ *['m modulo* $(p+1)$*'].*

x is a winning move if and only if $g(x) = 0$.

The proof is rigorously analogous to that of Theorem 8 (Chapter 3).

EXAMPLE. *Games of Nim of order p* (E. H. MOORE). We have two players and n heaps of matches; let p be a given number $< n$; the player whose turn it is to move chooses p heaps and removes one or more matches from each of them. The player who takes the last match is the winner.

To know if $x = (x_1, x_2, \ldots, x_n)$ is a winning choice, it is necessary to express the number of matches in the k-th heap in the binary system giving $g_k(x_k) = (c_k^1, c_k^2, \ldots)$, and to ensure that for every integer r, we have:

$$\left[\sum_{k=1}^{n} c_k^r\right]_{(p+1)} = 0$$

Let us now consider a progressively finite game of Nim, with Grundy function $g(x)$; suppose that an operation $+$ is defined on X, that is, a law which associates a vertex $z = x + y$ to every pair:

$$(x, y) \in X \times X$$

if $S \subset X$, $T \subset X$, we write:

$$S + T = \{s + t \mid s \in S, t \in T\}$$

The following result can sometimes be used to select a winning move:

THEOREM 4 (P. M. GRUNDY). *If a graph (X, Γ) is progressively finite and an operation $\dotplus$ is defined on it, satisfying*

$$\Gamma(x \dotplus y) = (\Gamma x \dotplus y) \cup (x \dotplus \Gamma y)$$

then for $z = x \dotplus y$, its Grundy function satisfies the formula:

$$(1) \qquad g(z) = g(x \dotplus y) = g(x) \dotplus g(y)$$

We shall prove this formula by induction, taking points $z = x \dotplus y$ successively from the sets:

$$X(0) = \{x \mid \Gamma x = \emptyset\}$$
$$X(1) = \{x \mid \Gamma x \subset X(0)\}$$
$$X(2) = \{x \mid \Gamma x \subset X(1)\}$$
etc.

If $z \in X(0)$, the formula (1) is satisfied, because

$$z \in X(0) \Rightarrow (\Gamma x \dotplus y) \cup (x \dotplus \Gamma y) = \Gamma z = \emptyset \Rightarrow x, y \in X(0)$$

and hence:

$$g(z) = 0 = 0 \dotplus 0 = g(x) \dotplus g(y)$$

We suppose that formula (1) is satisfied for all z in $X(\alpha - 1)$, and show now that it is true for a vertex $z = x \dotplus y$ in $X(\alpha)$; if this is not the case, one of the two inequalities:

$$g(z) > g(x) \dotplus g(y)$$
$$g(z) < g(x) \dotplus g(y)$$

must be true.

1. If $g(z) > g(x) \dotplus g(y)$, a vertex $z_1 \in \Gamma z$ exists such that:

$$g(z_1) = g(x) \dotplus g(y)$$

We can write $z_1 = x_1 \dotplus y$, $x_1 \in \Gamma x$ (permuting x and y, if necessary); since $z_1 \in \Gamma z \subset X(\alpha - 1)$, we have:

$$g(z_1) = g(x_1 \dotplus y) = g(x_1) \dotplus g(y)$$

And hence:

$$g(x) \dotplus g(y) = g(x_1) \dotplus g(y)$$

From Theorem 7 (Chapter 3), this implies $g(x) = g(x_1)$, which contradicts $x_1 \in \Gamma x$.

2. If $g(z) < g(x) \dotplus g(y)$, we can determine (as in Theorem 8 [Chapter 3] for example) a number γ such that:

$$g(z) = \gamma \dotplus g(y); \quad \gamma < g(x)$$

Since $x_1 \in \Gamma x$ exists such that $g(x_1) = \gamma$, this gives

$$g(z) = g(x_1) + g(y)$$

Since $x_1 + y \in \Gamma z \subset X(\alpha - 1)$, we have

$$g(x_1 + y) = g(x_1) + g(y) = g(z)$$

But this is absurd, for $x_1 + y \in \Gamma z$.

EXAMPLE. Two players play alternately by dividing a heap of matches selected from n heaps into two unequal heaps; the last one who can do this is the winner.

A position x is represented by the numbers $\bar{x}_1, \bar{x}_2, \ldots, \bar{x}_k$, giving the number of matches in each heap; we put:

$$x + y = (\bar{x}_1, \bar{x}_2, \ldots, \bar{x}_k) + (\bar{y}_1, \bar{y}_2, \ldots, \bar{y}_l) = (\bar{x}_1, \bar{x}_2, \ldots, \bar{x}_k, \bar{y}_1, \ldots, \bar{y}_l)$$

The operation $+$ satisfies:

(1) $\Gamma(x+y) = (\Gamma x + y) \cup (x + \Gamma y)$ (fundamental law);
(2) $x + y = y + x$ (commutative law);
(3) $x + (y + z) = (x + y) + z$ (associative law);
(4) $x + z = y + z \Rightarrow x = y$;
(5) A position e (no matches) exists such that $x + e = x$, for all x.

A vertex z for which we cannot write $z = x + y$, $x \neq e$, $y \neq e$, is said to be *irreducible*; here the irreducible vertices are the configurations $z = (\bar{x}_1)$ of a single heap. We have:

$$\begin{aligned} g(x) &= g(\bar{x}_1, \bar{x}_2, \ldots, \bar{x}_k) = g((\bar{x}_1) + (\bar{x}_2) + \ldots + (\bar{x}_k)) \\ &= g(\bar{x}_1) + g(\bar{x}_2) + \ldots + g(\bar{x}_k) \end{aligned}$$

Therefore if the values of g for the irreducible positions are known, the other values can be found immediately by means of a simple digital sum. We have straightaway:

$\bar{x}_1$	=	1	2	3	4	5	6	7	8	9	10	11	12	13	14	15	16	17	18	19	20
$g(\bar{x}_1)$	=	0	0	1	0	2	1	0	2	1	0	2	1	3	2	1	3	2	4	3	0

More generally, let us consider a progressively finite game possessing only one vertex e such that $\Gamma e = \emptyset$ and an operation $+$ satisfying (1), (2), (3), (4) and (5); E. W. ADAMS and D. C. BENSON [1] have proved that *every* *position x can be decomposed into a finite sum of irreducible positions and that this decomposition is unique.*

The General Definition of a Game (with Perfect Information)

Considerable care must be taken in formulating the general definition of a game if the restriction that a single play of the game be finite is removed. A *player* is an individual who is described as being either *active* or *passive*; for two players, whom for simplicity we shall call (A) and (B), a *game* is defined by:

1. A set X, whose elements are called the *positions of the game*;
2. A multi-valued function Γ mapping X into X, called the *rules of the game*;
3. A single-valued function θ mapping X into $\{0,1,2\}$, called the *turn of play*; we suppose that $\theta(x)=0$ if and only if $\Gamma x = \emptyset$; if $\theta(x)=1$, we say that at position x it is (A)'s *turn to play*, while if $\theta(x)=2$, it is (B)'s *turn to play*;
4. Two real-valued functions $f(x)$ and $g(x)$, defined and bounded on X, called the *preference functions* of the players (A) and (B).

A single play of the game takes place as follows: an initial position x_0 is chosen at random from X; if $\theta(x_0)=1$, for example, it is (A)'s turn to play and he chooses a position x_1 in the set Γx_0; the player whose turn it is now to move [that is, (A) if $\theta(x_1)=1$, or (B) if $\theta(x_1)=2$] chooses a position x_2 from the set Γx_1, and so on. If a player selects a position x_k such that $\Gamma x_k = \emptyset$, the game stops (naturally, it is possible for one game to go on indefinitely). If $S \subset X$ is the set of positions met with in the course of the game, the *gain* of player (A) is defined to be:

$$f(S) = \sup_{y \in S} f(y) \quad \text{if } (A) \text{ is active}$$

$$f(S) = \inf_{y \in S} f(y) \quad \text{if } (A) \text{ is passive}$$

Using the function g, the gain of player (B) may be similarly defined.

Player (A) aims to make as large a gain as possible; in general, $|f(S)|$ is interpreted as a sum of money which (A) will receive if $f(S) \geqslant 0$, and which he must pay out if $f(S) \leqslant 0$.

We shall put $X_A = \{x \mid \theta(x)=1\}$, $X_B = \{x \mid \theta(x)=2\}$; it will be noted that a game is simply a graph (X,Γ) in which each vertex x has associated with it a vector $(\theta(x), f(x), g(x))$.

A subgraph of (X,Γ) defines a *subgame*, a partial graph defines a *partial game*; observe that in terms of the preceding definitions, every subgame and every partial game are themselves games (this would not be true had other, more usual definitions been used); in exact terms, a game is a 'structure' which is given to an abstract set X.

EXAMPLE 1. *The game of Nim.* Consider a game of Nim on a graph $(\bar{X}, \bar{\Gamma})$; a *position of the game* is then a pair $(\bar{x}, i)$, where $\bar{x} \in \bar{X}$, and where i is equal to 1 or 2, according to whether the last player to move was (B) or (A); we have:

$$\theta(\bar{x}, i) = i \quad \text{if } \bar{\Gamma}\bar{x} \neq \emptyset$$
$$\theta(\bar{x}, i) = 0 \quad \text{if } \bar{\Gamma}\bar{x} = \emptyset$$
$$\Gamma(\bar{x}, 1) = \bar{\Gamma}\bar{x} \times \{2\}$$
$$\Gamma(\bar{x}, 2) = \bar{\Gamma}\bar{x} \times \{1\}$$

If $\bar{\Gamma}\bar{x} = \emptyset$, we have:

$$f(\bar{x}, 2) = g(\bar{x}, 1) = 1$$

In every other case, $f(x) = g(x) = 0$; both players are active.

EXAMPLE 2. *Chess.* Players (A) and (B) are 'white' and 'black', respectively; a *position* $x = (\bar{x}_1, \bar{x}_2, \bar{x}_3, \ldots, i)$ consists of the locations $\bar{x}_1, \bar{x}_2, \ldots$ of the different pieces on the chessboard, and of an index i equal to 1 or 2 identifying the player whose turn it is to move at position x; $\bar{x} = (\bar{x}_1, \bar{x}_2, \bar{x}_3, \ldots)$ is also called the *diagram* of the position x.

Player (A) is active, and $f(x) = 1$ if, at position x, the black king is checkmated; in all other cases, $f(x) = 0$.

EXAMPLE 3. *Pursuit game* [2]. A ship is pursuing another on a surface $\bar{S}$ ('the sea'); a possible approximation to this situation is a game in which two players (A) and (B) move alternately, once a second, say, each moving the ship which is in his charge. If the pursuing ship is at the point $\bar{x}_1$ of $\bar{S}$, and the ship being pursued is at the point $\bar{x}_2$ of $\bar{S}$, and if i indicates the turn of play, the position of the game is $x = (\bar{x}_1, \bar{x}_2, i)$. Let $B_i(\bar{x}_i) \subset \bar{S}$ be the circle with centre $\bar{x}_i$ whose radius is the maximum distance which the ship $\bar{x}_i$ can travel in one second; we have:

$$\Gamma(\bar{x}_1, \bar{x}_2, 1) = \bar{B}_1(\bar{x}_1) \times \{\bar{x}_2\} \times \{2\}$$
$$\Gamma(\bar{x}_1, \bar{x}_2, 2) = \{\bar{x}_1\} \times \bar{B}_2(\bar{x}_2) \times \{1\}$$

Let $d(\bar{x}, \bar{y})$ denote the distance between the points $\bar{x}$ and $\bar{y}$; if the aim of the pursuer is to get as close as possible to his quarry, player (A) is active, and

$$f(\bar{x}_1, \bar{x}_2, i) = -d(\bar{x}_1, \bar{x}_2)$$

If the pursuer aims only at catching his quarry, (A) is active, and

$$f(\bar{x}_1, \bar{x}_2, i) = 1 \quad \text{if } \bar{x}_1 = \bar{x}_2$$
$$f(\bar{x}_1, \bar{x}_2, i) = 0 \quad \text{if } \bar{x}_1 \neq \bar{x}_2$$

If the aim of the ship being pursued is not to be caught, player (B) is passive, and

$$g(\bar{x}_1, \bar{x}_2, i) = 1 \quad \text{if } \bar{x}_1 \neq \bar{x}_2$$
$$= 0 \quad \text{if } \bar{x}_1 = \bar{x}_2$$

In this example, (A)'s goal is essentially *active*, just as that of (B) is essentially *passive*. We would get a *subgame* if certain regions of $\bar{S}$ were forbidden to one of the ships, and a *partial game* if the top speed of one of the ships were to be diminished.

Strategies

For a game $(X, \Gamma, \theta, f, g)$, a *strategy* of player (A) is defined to be a single-valued function σ mapping X_A into X such that:

$$\sigma x \in \Gamma x \qquad (x \in X_A)$$

Player (A) *adopts* the strategy σ if he decides *a priori*: 'Whenever during the course of the game it is my turn to move and the game is in position x, I shall choose the position σx.' The set of strategies σ of player (A) is written Σ_A.

If the initial position x_0 is fixed, and if the two players (A) and (B) adopt strategies σ and τ respectively, the game is entirely determined, and the set of positions encountered is written $\langle x_0; \sigma, \tau \rangle$; (A)'s gain is then:

$$f(x_0; \sigma, \tau) = \sup \{f(x) \mid x \in \langle x_0; \sigma, \tau \rangle\} \quad \text{if } (A) \text{ is active}$$
$$f(x_0; \sigma, \tau) = \inf \{f(x) \mid x \in \langle x_0; \sigma, \tau \rangle\} \quad \text{if } (A) \text{ is passive}$$

A pair (σ_0, τ_0) is defined to be an *equilibrium* (relative to the position x) if:

$$f(x; \sigma, \tau_0) \leqslant f(x; \sigma_0, \tau_0) \quad (\sigma \in \Sigma_A)$$
$$g(x; \sigma_0, \tau) \leqslant g(x; \sigma_0, \tau_0) \quad (\tau \in \Sigma_B)$$

This can be described verbally by saying that a decision taken by the two players before beginning the game to adopt strategies σ_0 and τ_0, possesses a certain stability in that if either player then wishes to alter the course of the game, he is 'punished' by a decrease in his gain†.

† J. Nash uses the term 'equilibrium point' in [8] for the situation here described as an 'equilibrium'.

A pair (σ_0, τ_0) which is an equilibrium for every position x, is called an absolute *equilibrium* of the game; if the theorem concerning the existence of a kernel of a progressively finite graph is generalized, we obtain the following fundamental result:

THEOREM OF ZERMELO AND VON NEUMANN†. *If the graph (X, Γ) of a game is progressively finite, and if the sets $f(X) = \{f(x) \mid x \in X\}$ and $g(X)$ are finite, the game possesses an absolute equilibrium (σ_0, τ_0).*

As in the preceding chapter, we shall use sets $X(\alpha)$ which are defined sequentially as follows:

$$X(0) = \{x \mid \Gamma x = \emptyset\}$$
$$X(\alpha) = \{x \mid \Gamma x \subset X(\alpha-1)\} \quad \text{if } \alpha \text{ is not an ordinal limit}$$
$$X(\alpha) = \bigcup_{\beta<\alpha} X(\beta) \quad \text{if } \alpha \text{ is an ordinal limit}$$

The graph is progressively finite, and hence from Theorem 3 (Chapter 3) an ordinal γ exists such that $X = X(\gamma)$; we shall make a stepwise definition of an equilibrium (σ_0, τ_0) for each set $X(\alpha)$.

In $X(0)$, we define arbitrarily:

$$\sigma_0 x = x \qquad (x \in X(0))$$

If α is not an ordinal limit, and if (σ_0, τ_0) has been defined in $X(\alpha-1)$, we shall define an equilibrium $(\bar{\sigma}_0, \bar{\tau}_0)$ in $X(\alpha)$ in the following manner: if $x \in X_A$, we put

1. if $x \in X(\alpha-1)$: $\bar{\sigma}_0 x = \sigma_0 x$;
2. if $x \in X(\alpha) - X(\alpha-1)$: $\bar{\sigma}_0 x = y$, with $y \in \Gamma x$, and

$$f(y; \sigma_0, \tau_0) = \max_{z \in \Gamma x} f(z; \sigma_0, \tau_0)$$

Clearly (σ_0, τ_0) is an equilibrium for the subgame determined by $X(0)$; we assume now that (σ_0, τ_0) is an equilibrium for the subgame determined by $X(\alpha-1)$ and we shall show that $(\bar{\sigma}_0, \bar{\tau}_0)$ is then an equilibrium for the subgame determined by $X(\alpha)$, that is:

$$f(x; \bar{\sigma}, \bar{\tau}_0) \leqslant f(x; \bar{\sigma}_0, \bar{\tau}_0) \quad (\bar{\sigma} \in \Sigma_A, x \in X(\alpha)) \tag{1}$$
$$g(x; \bar{\sigma}_0, \bar{\tau}) \leqslant g(x; \bar{\sigma}_0, \bar{\tau}_0) \quad (\bar{\tau} \in \Sigma_B, x \in X(\alpha)) \tag{1'}$$

(1) and (1′) being satisfied for $x \in X(\alpha-1)$, let us suppose $x \in X(\alpha) - X(\alpha-1)$.

† This theorem was discovered by E. ZERMELO for the game of chess, and, almost simultaneously, by J. KALMAR; J. VON NEUMANN extended it to zero-sum games (where the gain of one player equals his opponent's loss). It was next extended by H. W. KUHN to games in which the interests of the two players were not necessarily opposed; and, in [2], we have generalized it to games of unlimited length, and to the case in which there are several ways of going from one position to the next. It is stated here in its most general form.

If $x \in X_A$, put $\bar{\sigma}x = z$; since $z \in X(\alpha-1)$, we have:

$$f(z;\sigma,\tau_0) \leqslant f(z;\sigma_0,\tau_0) \leqslant f(y;\sigma_0,\tau_0)$$

From this we shall deduce (1) for the different possible cases.

1. If (A) is active, and $f(x) \leqslant f(z;\sigma,\tau_0)$, we have

$$f(x;\bar{\sigma},\bar{\tau}_0) = f(z;\sigma,\tau_0) \leqslant f(y;\sigma_0,\tau_0) \leqslant f(x;\bar{\sigma}_0,\bar{\tau}_0)$$

2. If (A) is active, and $f(x) \geqslant f(z;\sigma,\tau_0)$, we have

$$f(x;\bar{\sigma},\bar{\tau}_0) = f(x) \leqslant f(x;\bar{\sigma}_0,\bar{\tau}_0)$$

3. If (A) is passive, and $f(x) \leqslant f(y;\sigma_0,\tau_0)$, we have

$$f(x;\bar{\sigma},\bar{\tau}_0) \leqslant f(x) = f(x;\bar{\sigma}_0,\bar{\tau}_0)$$

4. If (A) is passive, and $f(x) \geqslant f(y;\sigma_0,\tau_0)$, we have

$$f(x;\bar{\sigma},\bar{\tau}_0) \leqslant f(z;\sigma,\tau_0) \leqslant f(y;\sigma_0,\tau_0) = f(x;\bar{\sigma}_0,\bar{\tau}_0)$$

Inequality (1′) may be proved in the same way by replacing f by g.

COROLLARY 1. *If the graph (X,Γ) of a game is progressively finite, then for any $\epsilon > 0$, a pair (σ_0,τ_0) exists such that:*

$$f(x;\sigma,\tau_0) \leqslant f(x;\sigma_0,\tau_0)+\epsilon \quad (x \in X, \sigma \in \Sigma_A)$$
$$g(x;\sigma_0,\tau) \leqslant g(x;\sigma_0,\tau_0)+\epsilon \quad (x \in X, \tau \in \Sigma_B)$$

Replace the payoff functions f and g by functions f' and g' such that the sets:

$$f'(X) = \{f'(x) \mid x \in X\}$$
$$g'(X) = \{g'(x) \mid x \in X\}$$

are finite, and the functions themselves satisfy:

$$|f'(x)-f(x)| \leqslant \epsilon \quad (x \in X)$$
$$|g'(x)-g(x)| \leqslant \epsilon \quad (x \in X)$$

This is always possible, since f and g are assumed to be bounded on X. Since an equilibrium always exists for f' and g', an ϵ-equilibrium always exists for f and g.

COROLLARY 2. *If the graph (X,Γ) of a game is progressively finite, then for any $\epsilon > 0$ strategies σ_0 and τ_0 exist such that:*

$$f(x_0;\sigma,\tau_0)-\epsilon \leqslant f(x_0;\sigma_0,\tau_0) \leqslant f(x_0;\sigma_0,\tau)+\epsilon$$

(for any $\sigma \in \Sigma_A$, $\tau \in \Sigma_B$).

In other words, in a game with initial position x_0, player (A) can guarantee $f(x_0;\sigma_0,\tau_0)$ (to within ϵ) by means of the strategy σ_0; further, the left-hand inequality shows that there is no other strategy by which he can do better (again to within ϵ). The strategy σ_0 is therefore a *good strategy* for (A).

To prove this corollary, it is sufficient to put $f(x) = -g(x)$, and consider player (A) to be active and player (B) passive; we then have:

$$f(x_0;\sigma,\tau) = -g(x_0;\sigma,\tau)$$

from which the result follows immediately.

The game is said to be *zero-sum* if the gains of the two players are equal in absolute value and of opposite sign; the *value* of the game is:

$$f(\sigma,\tau) = -g(\sigma,\tau)$$

(A) is trying for as large a value as possible, and (B) for as small a value as possible. The greatest value which (A) can guarantee through his choice of strategy is:

$$\alpha_0 = \sup_\sigma \inf_\tau f(\sigma,\tau)$$

The smallest value which (B) can guarantee through his choice of strategy is:

$$\beta_0 = \inf_\tau \sup_\sigma f(\sigma,\tau)$$

Note that $\alpha_0 \leqslant \beta_0$, for if (A) and (B) each adopt a good strategy, the value cannot be less than α_0, nor exceed β_0. But a much stronger result than this is:

THEOREM 5. *If the graph (X,Γ) of a zero-sum two-person game is progressively finite, $\alpha_0 = \beta_0$*†.

From Corollary 2, we have:

$$\sup_\sigma f(\sigma,\tau_0) \leqslant f(\sigma_0,\tau_0)+\epsilon$$

$$\inf_\tau f(\sigma_0,\tau) \geqslant f(\sigma_0,\tau_0)-\epsilon$$

From which we have:

$$\beta_0 \leqslant f(\sigma_0,\tau_0)+\epsilon \leqslant \alpha_0+2\epsilon$$

Since ϵ is an arbitrary number, we have $\beta_0 \leqslant \alpha_0$; and since the reverse inequality is always true, we must have $\alpha_0 = \beta_0$.

† We have shown elsewhere that this statement is still true even if we do not assume that the graph is progressively finite (cf. [2], Chapter 1, pp. 15–18).

EXAMPLE 1. There are many consequences of this theorem. For a progressively finite game of Nim, the value of which must be $+1$ or -1, we discover a result of which we were already aware: for a given initial position x_0, there is either a strategy for (A) by which he is certain to win ($\alpha_0 = +1$), or else there is a strategy for (B) by which he is certain to win ($\alpha_0 = -1$).

EXAMPLE 2. Consider the following game: (A) and (B) play alternately by moving a counter along an arc of the graph of Fig. 6.3. If, when the game finishes, the counter has passed an odd number of times through a circled vertex (a or b), (A) gives a penny to (B); in the opposite case, (B) gives a penny to (A).

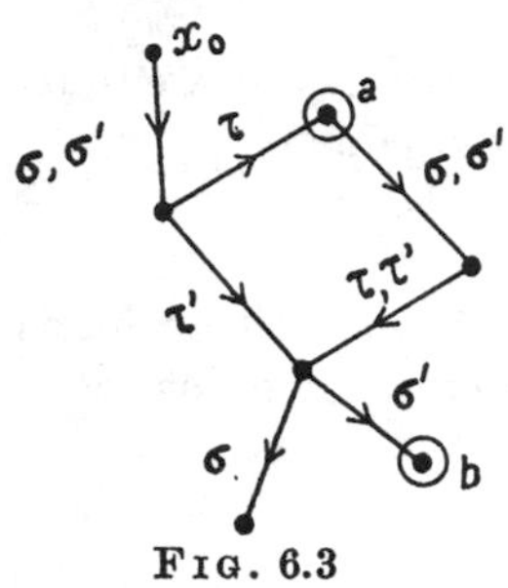

FIG. 6.3

This game can be set up as a paradox in which it is claimed that each player has two strategies (shown in Fig. 6.3). If (A) adopts σ, the value can be $f(\sigma, \tau) = -1$, and if (A) adopts σ', the value can be $f(\sigma', \tau') = -1$; it seems therefore that the best value which (A) can guarantee is $\alpha_0 = -1$. The same reasoning gives $\beta_0 = +1$.

Thus it appears that we have $\alpha_0 < \beta_0$, which contradicts Theorem 5.

We have, in fact, an apparent paradox here, which is often encountered in the literature.

In the present case, the error arises because a *position* of the game is not a vertex of the graph, but a sequence μ of vertices which have been encountered. This then implies that the gain should be of the form $\max\{f(x) \mid x \in \langle\sigma, \tau\rangle\}$. The problem is now correctly stated, and it is easy to prove that $\alpha_0 = \beta_0 = +1$.

7. The Problem of the Shortest Route

Stepwise Procedures

Given a graph (X, Γ) and two vertices a and b, we shall investigate the following problems:

PROBLEM 1. *Find a path of the graph going from a to b.*

PROBLEM 2. *Find a shortest path going from a to b.*
Clearly, Problem 2 includes Problem 1.

EXAMPLE 1. All games for one person (the ring puzzle, solitaire, the 'fifteen' puzzle, etc...) may be reduced to stepwise procedures: given a state a, we endeavour to reach state b by successive steps. A famous example

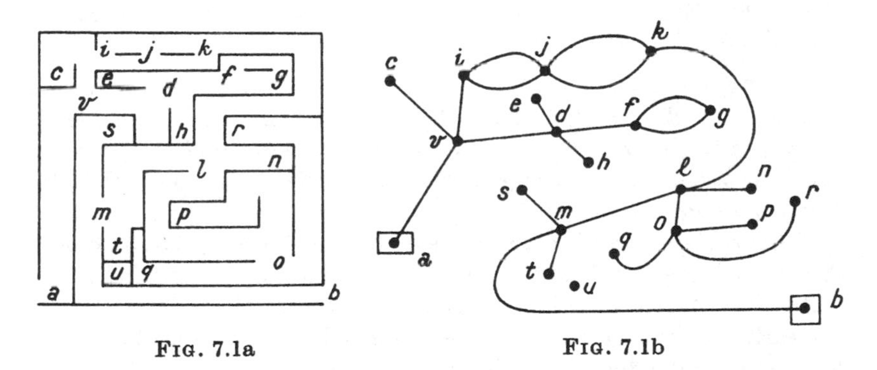

FIG. 7.1a FIG. 7.1b

is the *maze*; an individual at a tries to get out of the maze by following the passageways (Fig. 7.1a); this is the same as solving Problem 1 for the symmetric graph of Fig. 7.1b, where X is the set of passages, and $y \in \Gamma x$ if it is possible to go directly from passage x to passage y†.

† The maze problem can be solved by making a string model of the maze in which each edge is proportional to the length of the corresponding passageway. The required path is found by picking up the model at points a and b and pulling it taut (H. W. KUHN).

EXAMPLE 2. *The problem of the wolf, the cabbage and the goat.* A wolf, a goat and a cabbage are on one bank of a river; a ferryman wants to take them across, but, since his boat is small, he can take only one of them at a time. For obvious reasons, neither the wolf and the goat, nor the goat and the cabbage, can be left unguarded. How is the ferryman going to get them across the river?

Because of the small number of states to be considered, this well-known problem can easily be solved in one's head; nevertheless, we have here a typical example of Problem 1: we form the graph of Fig. 7.2, and look for a path leading from state a (where the cabbage C, the ferryman F, the goat G and the wolf W are all together on the right bank) to state b (where the right bank is empty); the arrows of Fig. 7.2 indicate such a path.

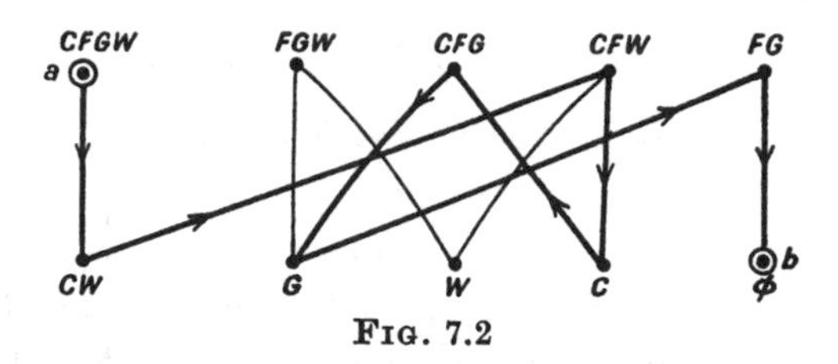

FIG. 7.2

In more complicated cases, a systematic algorithm is necessary, and several methods have been suggested.

ALGORITHM FOR PROBLEM 1 IN THE CASE OF A PLANAR GRAPH. In the case of a maze which, like that of Fig. 7.1a, contains no staircases and exists entirely in the horizontal plane, the graph has a remarkable property: its edges never cut one another (except at the vertices); in other words, the graph is *planar*.

If HG is planar and if a and b are on the infinite face, the solution of Problem 1 is greatly simplified, and the following rule is sufficient:

At a junction, always take the passageway on the extreme right.

Clearly, one can, in this way, reach the end of the maze without ever having to go more than twice along any one edge.

GENERAL ALGORITHM FOR PROBLEM 1. Let us first consider a graph which is an arborescence with root a (cf. Fig. 2.1), and look for a path going from a to a vertex b; one simple method consists of starting from a, following any branch as far as possible, then returning to the nearest junction at which it is possible to make a new start along some previously unexplored direction, etc. In this way, we can be sure of reaching b without having to go more than twice along any edge.

If we have a graph G which is not an arborescence, we can still make use of the preceding argument by 'cutting' certain edges u at one of their vertices x: a cut is made, if, on following an edge u for the first time, we arrive at a junction x which has been encountered before (this procedure will certainly produce an arborescence).

Note that this algorithm can be used in the problem of the maze, not only by those who know its plan, but also by any casual visitor who has strayed into it; all he has to do is make some distinctive mark on the junctions and passageways which he has gone through.

ANOTHER ALGORITHM FOR PROBLEM 1 (G. TARRY [5]). For a traveller lost in a maze, the following rule is sufficient:

Never go twice in the same direction along any one edge; at the point x, only take the edge by which one first arrived at the junction x if no other choice is available.

1. If the traveller is stopped because he cannot obey the rule, we shall show that he must be at a, and that every edge incident to a has been followed in both directions. In fact, if he is at x (with $x \neq a$), let k be the number of times he has previously arrived at this junction; he has been $(k+1)$ times towards x along the edges incident to x, and k times along these edges going away from x, and therefore there is still an edge which has not been used by which he can depart from x.

2. Let $(a_0 = a, a_1, a_2, \ldots)$ be the sequence of junctions encountered, indexed according to the order in which they were met for the first time; we shall show that every edge incident to a_k has been followed twice. Clearly, this is true for $k = 0$ (from (1)); let us assume the statement to be true for $k \leqslant p-1$, and show that it is then also true for $k = p$. Let (a_i, a_p) be the edge by which the traveller reached a_p for the first time; since $i < p$, this edge has been followed in both directions and hence, from Tarry's rule, all other edges incident to a_p must also have been followed in both directions.

Q.E.D.

3. If a path $(a, x_1, x_2, \ldots, x_m, b)$ going from a to b exists, then clearly the junction b must be encountered at some stage; for, from (2), any vertex adjacent to a vertex which has already been encountered, is itself also sure to be encountered.

ALGORITHM FOR PROBLEM 2. Using an iterative procedure, we attach to each vertex x a cost equal to the length of the shortest path going from a to x:

1. We label the vertex a with the index 0.

2. If all vertices labelled with the index m make up a known set $A(m)$, we give the label $(m+1)$ to the vertices of the set:

$$A(m+1) = \{x \mid x \in \Gamma A(m), x \notin A(k) \quad \text{for all } k \leqslant m\}$$

3. We stop as soon as the vertex b has been labelled; if $b \in A(m)$, consider vertices $b_1, b_2, \ldots$, such that:

$$\begin{array}{ll} b_1 \in A(m-1), & b_1 \in \Gamma^{-1} b \\ b_2 \in A(m-2), & b_2 \in \Gamma^{-1} b_1 \\ \cdots\cdots & \cdots\cdots \\ b_m \in A(0), & b_m \in \Gamma^{-1} b_{m-1} \end{array}$$

The path $\mu = [a = b_m, b_{m-1}, \ldots, b_1, b]$ is the one required.

Various Generalizations

We often meet the following problems, which contain Problems 1 and 2:

PROBLEM 3. *Given a graph (X, U), we assign a number $l(u) \geqslant 0$ to each arc u which is called the 'length' of u; find a path μ, going from a vertex a to a vertex b, and such that its total length:*

$$\sum_{u \in \mu} l(u)$$

should be as small as possible.

PROBLEM 4. *With every path $\mu = [x_1, x_2, \ldots, x_m]$, there is associated a real-valued function $f(\mu)$; find the path μ for which $f(\mu)$ is as small as possible.*

PROBLEM 5. *Given a graph G, a number $h(G_1)$ is assigned to every partial subgraph G_1; find a partial subgraph G_1 such that $h(G_1)$ is as small as possible.*

Each of these problems includes the preceding one.

EXAMPLE 1. Let X be a set of localities and U a set of roads connecting these places. We shall suppose that all road intersections have been included in X. For a given road u, the number $l(u)$ may signify its length in miles, or the cost of travelling over u, or, better still, the time required for this journey (with due allowance being made for traffic conditions).

Then we can look for the shortest route, the cheapest route or the quickest route to go from one town to another. In each of these cases, we should have to solve Problem 3.

EXAMPLE 2. *Dynamic programming* (R. BELLMAN). Two gold mines a and b contain α and β tons of ore, respectively; we have only one machine to work them. In one year, it is capable of extracting $r\%$ of the gold content of a (and the probability that it is still in working order at the end of the year is p), or else it is employed at b, where it extracts $s\%$ of the gold content of b (and the probability of still being in working order at the end of the year is q). Let us consider a graph with two vertices a and b, for which $\Gamma a = \Gamma b = \{a, b\}$, and look for a path $[x_1, x_2, \ldots]$ (where $x_i = a$ if it is planned to work mine a in the i-th year) such that the mathematical expectation of the gain is a maximum. The expected total gold yield $F(\alpha, \beta)$ from an optimal strategy is defined by the functional equation:

$$F(\alpha,\beta) = \max\{r\alpha + pF[(1-r)\alpha, \beta], \quad s\beta + qF[\alpha, (1-s)\beta]\}$$

This so-called 'dynamic programming' problem is a special case of Problem 4, where Γx is a constant set, and we have a highly specialized function $f(\mu)$.

We shall concern ourselves solely with problems of type 3: we consider a graph $G = (X, U)$, a 'length' $l(u) \geqslant 0$ associated with each arc u, and we want a path μ between x_0 and x_n, which minimizes the total 'length':

$$l(\mu) = \sum_{u \in \mu} l(u)$$

ALGORITHM FOR PROBLEM 3 (L. FORD). The following method can easily be mechanized for networks with a large number of vertices:

1. Each vertex x_i is given an index λ_i; to begin with we put $\lambda_0 = 0$ and $\lambda_i = \infty$ if $i \neq 0$. In the following, we use the convention: $\infty - \infty = 0$.

2. Look for an arc (x_i, x_j) such that $\lambda_j - \lambda_i > l(x_i, x_j)$; then replace λ_j by $\lambda'_j = \lambda_i + l(x_i, x_j) < \lambda_j$; note that $\lambda'_j > 0$ if $j \neq 0$. Continue thus until no arc remains whereby the λ_j can be further diminished.

3. A vertex x_{p_1} exists such that $\lambda_n - \lambda_{p_1} = l(x_{p_1}, x_n)$; for during this procedure, λ_n has been decreased monotonically, and x_{p_1} is the vertex last used for this purpose. Likewise, let x_{p_2} be the vertex such that:

$$\lambda_{p_1} - \lambda_{p_2} = l(x_{p_2}, x_{p_1}), \text{ etc.}$$

Since the sequence $\lambda_n, \lambda_{p_1}, \lambda_{p_2}, \ldots$ is strictly decreasing, at some stage, we shall have $x_{p_{k+1}} = x_0$. We state that *λ_n is the length of the shortest path from x_0 to x_n, and that $\mu_0 = [x_0, x_{p_k}, x_{p_{k-1}}, \ldots, x_{p_1}, x_n]$ is the shortest path between x_0 and x_n.*

Proof. Let $\mu = [x_0, x_{k_1}, x_{k_2}, \ldots, x_{k_{s+1}} = x_n]$ be any path whatever between x_0 and x_n, and let its length (in the generalized sense) be $l(\mu)$; we have:

$$\begin{aligned} \lambda_{k_1} - 0 &\leqslant l(x_0, x_{k_1}) \\ \lambda_{k_2} - \lambda_{k_1} &\leqslant l(x_{k_1}, x_{k_2}) \\ &\cdots\cdots\cdots \\ \lambda_n - \lambda_{k_s} &\leqslant l(x_{k_s}, x_n) \end{aligned}$$

Summing over all terms, we see therefore that for any path μ, we have:

$$\lambda_n - 0 \leqslant l(\mu)$$

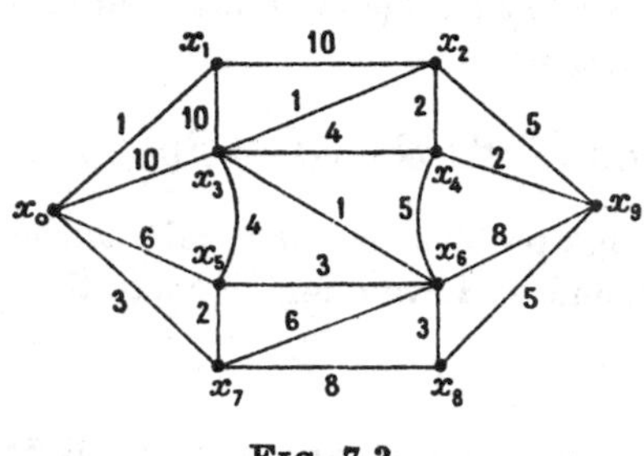

FIG. 7.3

Since we have $\lambda_n = l(\mu_0)$ for the path μ_0, it must certainly be the shortest.

EXAMPLE. We leave it to the reader to prove that in the symmetric graph of Fig. 7.3, the length of the shortest path between x_0 and x_9 is 14.

Remark. For simplicity, put $l_j^i = l(x_i, x_j)$ if $(x_i, x_j) \in U$, $l_j^i = +\infty$ if $(x_i, x_j) \notin U$; for any path μ, we can assign numbers ξ_j^i, where $\xi_j^i = 1$ if μ uses the arc (x_i, x_j), and $\xi_j^i = 0$ otherwise. Problem 3 then requires that a set of integers ξ_j^i, $i = 0, 1, \ldots, n$; $j = 0, 1, \ldots, n$, be found, which minimize the linear form $\sum_{i,j} l_j^i \xi_j^i$, and satisfy the following system of linear inequalities:

$$\begin{cases} 0 \leqslant \xi_j^i \leqslant 1 \\ \sum_j (\xi_j^i - \xi_i^j) = 0 \quad \text{if } i \neq 0, i \neq n \\ \sum_j (\xi_j^0 - \xi_0^j) = +1 \\ \sum_j (\xi_j^n - \xi_n^j) = -1 \end{cases}$$

In other words, we are dealing with a *linear programme*, which can be solved by the usual electronic machines; but in this case, isn't this taking a sledgehammer to crack a nut?

8. Transport Networks

The Problem of Maximum Flow

The name *transport network* is given to a finite connected graph without loops in which associated with each arc u is a number $c(u) \geqslant 0$ (which we shall call the *capacity* of the arc u), and in which:

1. There is one and only one vertex x_0 such that $\Gamma^{-1}x_0 = \emptyset$; this vertex x_0 is called the *source* of the network;
2. There is one and only one vertex z such that $\Gamma z = \emptyset$; this vertex z is called the *sink* of the network.

Let U_x^- be the set of arcs incident to x and oriented towards x, and let U_x^+ be the set of arcs incident to x and oriented away from x. An integer-valued function $\phi(u)$ defined on U, is said to be a *flow* for this transport network if we have:

$$\phi(u) \geqslant 0 \qquad (u \in U) \tag{1}$$

$$\sum_{u \in U_x^-} \phi(u) - \sum_{u \in U_x^+} \phi(u) = 0 \qquad (x \neq x_0;\, x \neq z) \tag{2}$$

$$\phi(u) \leqslant c(u) \qquad (u \in U) \tag{3}$$

$\phi(u)$ may be likened to a volume of material flowing along the arc $u = (x, y)$ from x to y, and not exceeding the capacity of that arc; further, if x is neither the source x_0 nor the sink z, the amount flowing into x must equal the amount flowing out.

From (2), we deduce immediately

$$\sum_{u \in U_z^-} \phi(u) = \sum_{u \in U_{x_0}^+} \phi(u) = \phi_z \tag{4}$$

This number ϕ_z, which represents the amount of material arriving at z, is called the *inflow at the point* z, or the *value* of the flow ϕ; we are concerned here with the problem of finding the maximum value of the flow for a given transport network.

EXAMPLE 1 (merchant shipping). A certain product is shipped from ports $x_1, x_2, \ldots$ to ports $y_1, y_2, \ldots$; the stock available at x_i is s_i, and the demand at y_j is d_j.

Let c_{ij} be the total amount of the product which could be shipped over the route from x_i to y_j by the one or more ships using this route. Can all the demands be satisfied? How should the shipping be organized? To reduce this problem to one of maximum flow, connect x_i to y_j by an arc of capacity c_{ij}, then connect a source x_0 to each vertex x_i by an arc of capacity s_i, and finally connect each y_j to a sink z by an arc of capacity d_j; if ϕ is a maximum flow, $\phi(x_i, y_j)$ shows the amount of the product which should be sent from x_i to y_j to give the best response to all demands.

EXAMPLE 2 (dynamic networks [3]). From a group of n towns, $\bar{x}_1, \bar{x}_2, \ldots, \bar{x}_n$, buses run to a single destination $\bar{y}$; if $(\bar{x}_i, \bar{x}_j)$ is a road connecting the two towns $\bar{x}_i$ and $\bar{x}_j$, let t_{ij} be the time required to go from $\bar{x}_i$ to $\bar{x}_j$ by this road, and let c_{ij} be the number of buses which can use it per unit of time (if there is no such road, $c_{ij} = 0$); let c_{ii} be the number of buses which can be stationed at $\bar{x}_i$, and let a_i be the number of buses originally at $\bar{x}_i$.

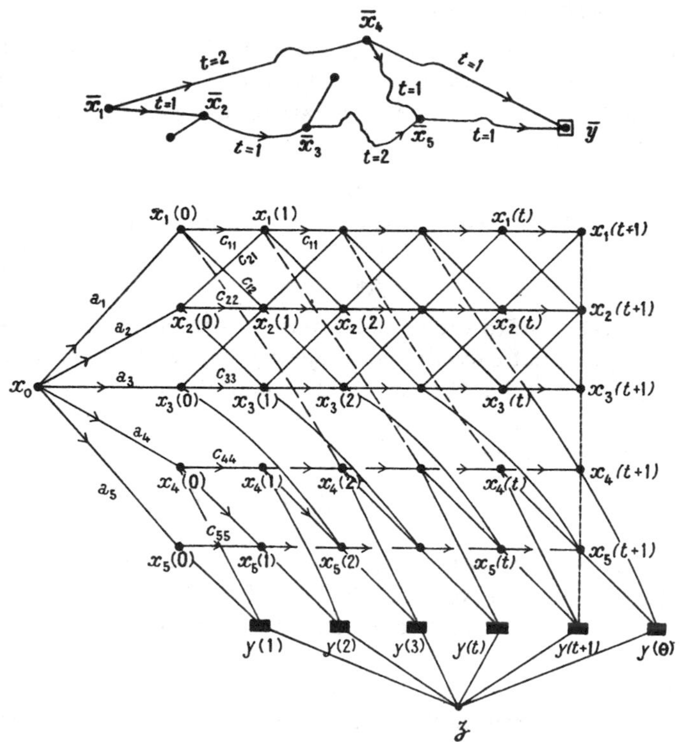

FIG. 8.1. *Road map and corresponding transport network in space-time*

How should the buses be scheduled in order to have as many as possible arrive at $\bar{y}$ in a given time interval θ?

Define a transport network for which the set of vertices is the Cartesian product:

$$X = \{\bar{x}_1, \bar{x}_2, \ldots, \bar{y}\} \times \{0, 1, 2, \ldots, \theta\}$$

At the vertex $x_i(t) = (\bar{x}_i, t)$, define arcs $(x_i(t), x_j(t+t_{ij}))$ of capacity c_{ij}, and arcs $(x_i(t), x_i(t+1))$ of capacity c_{ii}; we denote by:

$$\phi(x_i(t), x_j(t+t_{ij}))$$

the number of buses leaving the town $\bar{x}_i$ at time t to go to the town $\bar{x}_j$, and by $\phi(x_i(t), x_i(t+1))$ the number of buses remaining in $\bar{x}_i$ at time t. Finally, connect a source x_0 to every vertex $x_i(0)$ by an arc of capacity a_i, $i = 1, 2, \ldots, n$, and connect the vertices $y(1), y(2), \ldots, y(\theta)$ to a sink z by arcs with infinite capacity; the optimum schedule is given by the maximum flow (cf. Fig. 8.1).

Remark. For simplicity, put $z = x_{n+1}$, and define numbers c_j^i and ξ_j^i, $i = 0, 1, 2, \ldots, n$, $j = 1, 2, \ldots, n+1$, by:

$$c_j^i = \begin{cases} 0 & \text{if} \quad (x_i, x_j) \notin U \\ c(x_i, x_j) & \text{if} \quad (x_i, x_j) \in U \end{cases}$$

$$\xi_j^i = \phi(x_i, x_j)$$

The problem of finding the maximum flow is then the same as finding numbers ξ_j^i such that:

$$(1) \qquad 0 \leqslant \xi_j^i \leqslant c_j^i \qquad (i = 0, 1, 2, \ldots, n;\, j = 1, 2, \ldots, n+1)$$

$$(2) \qquad \sum_{k=0}^{n} \xi_i^k - \sum_{j=1}^{n+1} \xi_j^i = 0 \qquad (i = 1, 2, \ldots, n)$$

$$(3) \qquad \sum_{j=0}^{n} \xi_{n+1}^j \text{ is maximized}$$

We have therefore a classical *linear programming problem*, despite the fact that the ξ_j^i required must be integers (in fact, because of the form of (1), (2) and (3), an integer solution always exists). It would be possible to use the simplex method, but the algorithms we shall give here are much simpler and allow us to solve very large networks 'by hand' (a graph with 500 vertices and 4000 arcs has been solved in this way at RAND).

The principle behind the algorithms for the problem of the maximum flow is always the same. Let $A \subset X$ be a set such that:

$$x_0 \notin A, \qquad z \in A$$

The set U_A^- of arcs incident into A is sometimes called a *cut* of the network; likewise, the expression:

$$c(U_A^-) = \sum_{u \in U_A^-} c(u)$$

is called the *capacity* of the cut U_A^-. Every unit of material going from x_0 to z must make use of an arc of U_A^- at least once; therefore, *for any flow* ϕ *and any cut* U_A^-, we must have:

$$\phi_z \leqslant c(U_A^-)$$

Thus, if for a flow ϕ and a cut V, we have $\phi_z = c(V)$, we may be sure that the flow is at its maximum value (and that V is a cut of minimum capacity).

ALGORITHM FOR DETERMINING THE MAXIMUM FLOW IN A PLANAR GRAPH (FORD and FULKERSON [2]). In the case of a planar graph, the search for a maximum flow reduces to making a cut in the following manner:

1. A straight horizontal arc is drawn connecting x_0 and z; we assume that the graph is still planar and draw it above this arc (x_0, z) in such a fashion that the other arcs do not cut one another (except at their vertices); since the graph is planar, we can determine the *highest* path going from x_0 to z by always taking the arc at the extreme left (except for retracing one's steps should an impasse be reached);

2. On the highest path μ_1 going from x_0 to z, find the arc $u_1 \in \mu_1$ which has the smallest capacity. Delete u_1, and subtract $c(u_1)$ from the capacity of every other arc of the path μ_1. Begin again, with the new highest path μ_2 and its arc of minimum capacity u_2;

3. Continue in this way until only the arc (x_0, z) connects the vertices x_0 and z; the required flow is found by sending along each path μ_k an amount equal to the capacity of the corresponding arc u_k (this capacity being measured at the stage at which μ_k is the highest path).

We shall show that the value of this flow ϕ is a maximum; in fact, the set A_0 of vertices which cannot be reached by a path issuing from x_0 determines a cut $U_{A_0}^-$. Let U_A^- be a *minimal* cut included in $U_{A_0}^-$. For this set A we have:

$$c(U_A^-) = \sum_{u \in U_A^-} c(u) = \phi_z$$

Therefore from the statement preceding this algorithm, the flow ϕ is a maximum.

GENERAL ALGORITHM FOR DETERMINING THE MAXIMUM FLOW (FORD and FULKERSON [2]). We consider a transport network with a flow $\phi(x,y)$ in which we shall try to increase the value ϕ_z.

1. An arc u is said to be *saturated* if $\phi(u) = c(u)$; a flow is called *complete* if every path going from x_0 to z contains at least one saturated arc.

If the flow is not complete, we can find a path μ which consists entirely of unsaturated arcs (to do this, use the algorithm for Problem 1 (Chapter 7): find a path going from x_0 to z in the partial graph which is made up of the unsaturated arcs). Put:

$$\begin{aligned} \phi'(u) &= \phi(u)+1 \quad \text{if} \quad u \in \mu \\ \phi'(u) &= \phi(u) \qquad\quad \text{if} \quad u \notin \mu \end{aligned}$$

ϕ' is clearly a flow, and its value is:

$$\phi'_z = \phi_z + 1 > \phi_z$$

It is therefore easy to augment the value of an incomplete flow up to the stage at which it becomes complete.

2. Let $\phi(x, y)$ be a complete flow; we shall use an iterative procedure to label successively all the vertices of the graph which may be used to transmit a supplementary flow of one unit:

Label x_0 with the coefficient: 0.

If x_i is a labelled vertex, any unlabelled vertex y such that $(x_i, y) \in U$ and $\phi(x_i, y) < c(x_i, y)$, is given the label $+i$.

If x_i has been labelled, any vertex y which is not yet labelled, for which $(y, x_i) \in U$ and $\phi(y, x_i) > 0$, is given the label $-i$.

If, by this procedure, the vertex z acquires a label, a chain μ exists between x_0 and z, all of whose vertices are distinct and labelled with the index of the preceding vertex (apart from sign). Put:

$\phi'(u) = \phi(u)$ if $u \notin \mu$

$\phi'(u) = \phi(u)+1$ if $u \in \mu$, and u is oriented in the direction of the chain μ going from x_0 to z,

$\phi'(u) = \phi(u)-1$ if $u \in \mu$, and u is oriented in the opposite direction of the chain μ going from x_0 to z (cf. Fig. 8.2).

Clearly, $\phi'(u)$ is still a flow, and since $\phi'_z = \phi_z + 1$, the value of the flow has been increased.

3. If, for a flow ϕ^0, the value cannot be improved by using the preceding method, that is to say if the vertex z cannot be labelled, the flow ϕ^0 has attained its maximum value.

Let A be the set of unlabelled vertices; since $x_0 \notin A$, $z \in A$, the set A defines a cut U_A^-; we have:

$$\phi_z^0 = \sum_{u \in U_A^-} \phi^0(u) - \sum_{u \in U_A^+} \phi^0(u) = \sum_{u \in U_A^-} c(u) + 0 = c(U_A^-)$$

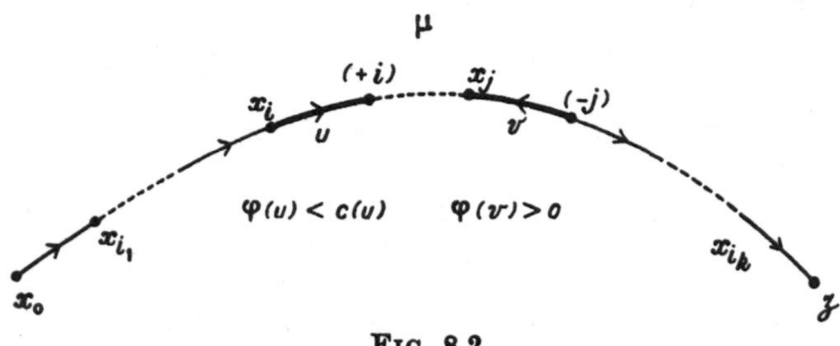

FIG. 8.2

Therefore the flow ϕ^0 must be at its maximum value.

The preceding argument enables us to deduce the fundamental result:

THEOREM OF FORD AND FULKERSON. *In a given transport network, the maximum value of a flow is equal to the minimum capacity of a cut:*

$$\max_{\phi} \phi_z = \min_{\substack{A \not\ni x_0 \\ A \ni z}} c(U_A^-)$$

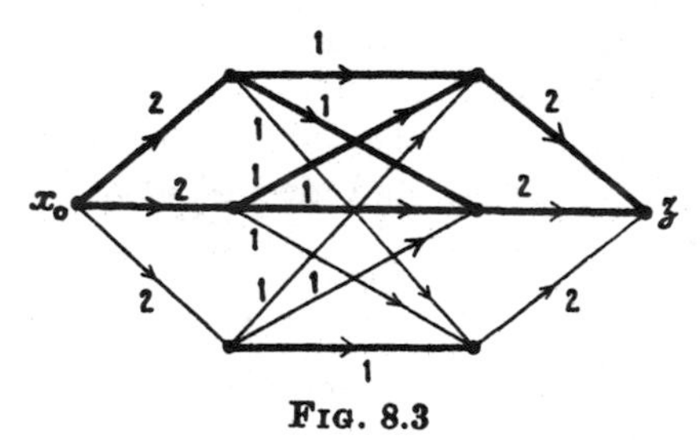

FIG. 8.3

EXAMPLE. In the transport network of Fig. 8.3, the numbers indicate the capacities of the several arcs. A complete flow is shown, in which the saturated arcs are marked with heavy lines; its value is $\phi_z = 5$. From this, the reader can immediately deduce a maximum flow of value 6.

Note that for a network of this type, the following rule determines the maximum flow directly: the quantity of material transported is increased by unit amount at a time, each new unit being sent to the vertex x for which the current value of $\sum_{u \in U_x^+} (c(u) - \phi(u))$ is greatest; as soon as the flow is complete, it has reached its maximum.

Speaking generally, we have here a pseudo-algorithm which is very effective in most cases, but which does not always give the maximum flow and therefore must be used only to provide a first approximation.

The Problem of Minimum Flow

Given a transport network, *we wish to find the minimum value of a flow ϕ which satisfies:*

$$(3') \qquad \phi(u) \geqslant c(u) \qquad (u \in U)$$

This problem, which is very closely related to the preceding one, is also frequently encountered.

EXAMPLE. Minimum cover of a simple graph (X, Y, Γ). A *cover* of a graph is a set V of arcs such that every vertex of the graph is a vertex of at least one arc of V. We wish to find a cover V for which $|V|$ is as small as possible. The general solution to this problem will be given in Chapter 18; it is a problem of minimum flow if the graph is simple, that is to say it consists of two disjoint sets X and Y and a function Γ mapping X into Y.

Connect a source x_0 to every vertex $x_i \in X$ by an arc of capacity $+1$, and connect every vertex $y_j \in Y$ to a sink z by an arc of capacity $+1$, all other arcs having a capacity of 0; if we have a minimum flow ϕ, the arcs $u \in U$ for which $\phi(u) > 0$ constitute a minimum cover.

ALGORITHM FOR THE PROBLEM OF MINIMUM FLOW. We can use a method which is strictly analogous to that given for the problem of maximum flow; starting with any flow ϕ which satisfies (3′), we diminish its value by a labelling process:

1. the sink is labelled x_n;
2. if x_j is a labelled vertex, any unlabelled vertex x_i such that $(x_i, x_j) \in U$ and $\phi(x_i, x_j) > c(x_i, x_j)$ is given the label $+j$; similarly, any unlabelled vertex x_i such that $(x_j, x_i) \in U$ is given the label $-j$.

If, in this manner, the source x_0 can be labelled, the value of the flow can be decreased by one; if x_0 cannot be labelled, the flow ϕ is a minimum flow.

Problems of Flow when a Set of Values is Associated with each Arc

We consider here a transport network in which associated with every arc u is a set $C(u)$ of positive integers instead of the single number $c(u) \geqslant 0$. The following problems are now proposed:

PROBLEM 1. *Construct a flow $\phi(u)$ such that $\phi(u) \in C(u)$ for all u.*

PROBLEM 2. *Construct a maximum flow $\phi(u)$ such that $\phi(u) \in C(u)$ for all u.*

Problem 2 contains Problem 1; if we put:

$$C(u) = \{0, 1, 2, \ldots, c(u)\}$$

the problem of maximum flow is also included.

EXAMPLE 1 (R. A. FISHER). Consider the individuals (1), (2), ..., (n); what is the maximum number of triads which can be formed from them if no two individuals are to be together in more than p distinct triads? This problem reduces to finding the maximum flow of the network of Fig. 8.4,

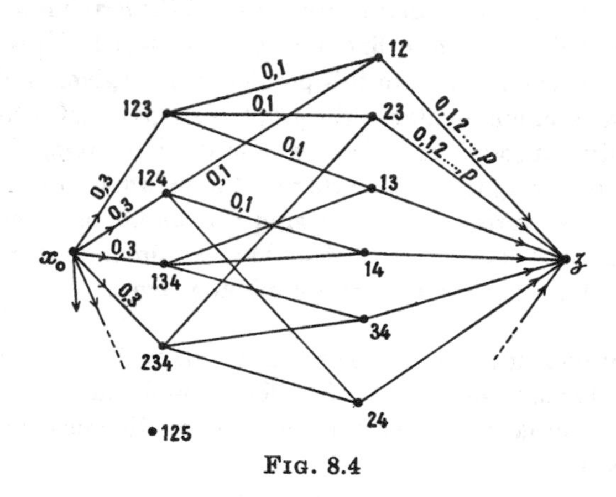

FIG. 8.4

where the elements of $C(u)$ are indicated on each arc u. We remark that this type of problem has been introduced into biology (under the name 'balanced incomplete block design') to test several treatments on different varieties of corn; it will not, however, be solved by the subsequent theory.

EXAMPLE 2 (HOFFMAN and KUHN [5]). Let X be a finite set, and let $S_1, S_2, \ldots, S_n$ be subsets of X; $T_1, T_2, \ldots, T_p$ be disjoint subsets of X; $b_1, b_2, \ldots, b_p, c_1, c_2, \ldots, c_p$ be integers (with $0 \leq b_i \leq c_i$ for all i). Find a set $Y = \{y_1, y_2, \ldots, y_n\}$ in X such that $y_k \in S_k$ for $k = 1, 2, \ldots, n$, and satisfying:

$$b_i \leq |Y \cap T_i| \leq c_i \qquad (i = 1, 2, \ldots, p)$$

This problem reduces to one of finding a maximum flow ϕ for the network of Fig. 8.5; the set Y consists of the y's for which $\phi(y, \bar{y}) > 0$. The subsequent theory provides a complete solution to this problem.

We shall first consider a special case of Problem 1, which can be formulated as follows: *given a transport network with a capacity $c(u) \geq 0$ associated* *with each arc u, does a flow $\phi^0(u)$ exist such that $\phi^0(u) \leqslant c(u)$, and such that the arcs leading to the sink z are saturated?*

To simplify the notation, put:

$$d(y) \begin{cases} = c(y, z) & \text{if} \quad y \in \Gamma^{-1}z \\ = 0 & \text{otherwise} \end{cases}$$

$d(y)$ is the 'demand' at y which we are trying to satisfy; similarly, if $A \subset X$, the quantity:

$$d(A) = \sum_{y \in A} d(y)$$

is called the 'demand' of the set A.

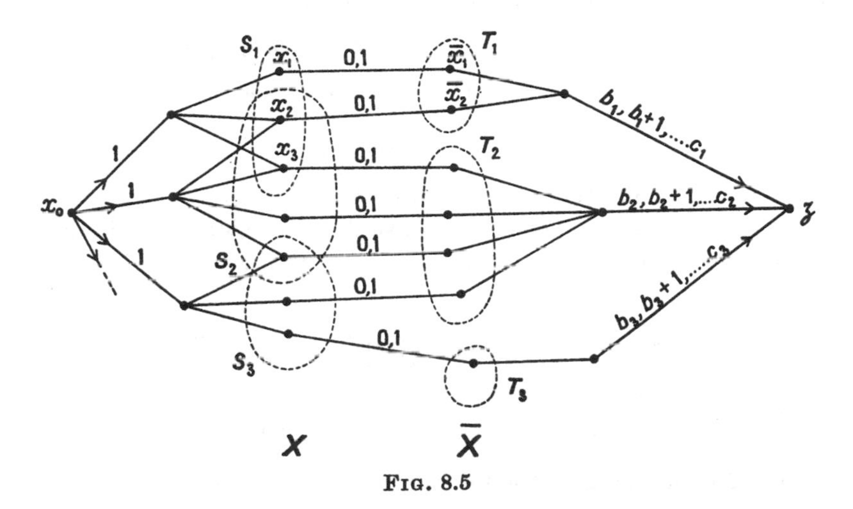

Fig. 8.5

As usual, we represent the set of outwardly directed arcs incident to A, by U_A^+, and the set of inwardly directed arcs incident to A by U_A^-; we denote by $F(A)$ the total amount of material which can be made to enter A (after deleting the arcs of U_A^+).

Finally, the total capacity of a set V of arcs is written:

$$c(V) = \sum_{u \in V} c(u)$$

Lemma. *Let $A_0 \subset \bar{X} = X - \{x_0, z\}$; if for every set A for which $A_0 \subset A \subset \bar{X}$, we have $c(U_A^-) \geqslant d(A)$, then we have*

$$F(A_0) \geqslant d(A_0)$$

Consider a set $A_0 \subset \bar{X}$, and reduce it to a sink z_0 by deleting the arcs of $U^+_{A_0}$ and identifying all points of A_0 with the vertex z_0; for this new network, using Ford and Fulkerson's result, the maximum value of the flow is:

$$\min_W c(W) = F(A_0)$$

Let W_0 be a minimum cut defined by a set A (with $x_0 \notin A$, $A \supset A_0$, $z \notin A$); we have $W_0 = U^-_A$, from which:

$$F(A_0) = c(W_0) = c(U^-_A) \geqslant d(A) \geqslant d(A_0)$$

THEOREM 1 (GALE [4]). *Given a transport network with a capacity $c(u) \geqslant 0$ associated with each arc u, a flow ϕ^0 such that $\phi^0(u) \leqslant c(u)$ and such that the arcs leading to z are saturated, exists if and only if:*

(1) $$c(U^-_A) \geqslant d(A) \qquad (A \subset \bar{X} = X - \{x_0, z\})$$

1. If a flow exists which saturates the arcs incident to z, then, for all $A \subset \bar{X}$, we have

$$c(U^-_A) \geqslant F(A) \geqslant d(A)$$

Therefore inequality (1) certainly holds.

2. Suppose that inequality (1) is satisfied, and consider an arbitrary cut defined by a set S for which $x_0 \notin S$, $z \in S$. Put $A = S - \{z\}$; then we have:

$$d(A) \leqslant c(U^-_A) = c(U^-_S) - \sum_{x \notin A} c(x, z) = c(W) - d(X - A)$$

Therefore, for any cut W, we have:

$$c(W) \geqslant d(A) + d(X - A) = d(X)$$

Therefore the maximum flow ϕ^0 satisfies:

$$\phi^0_z = \min_W c(W) \geqslant d(X)$$

This flow, therefore, must saturate the arcs leading to the sink z.

COROLLARY (saturation theorem). *A flow ϕ such that $\phi(u) \leqslant c(u)$ and such that the arcs leading to z are saturated, exists if and only if:*

(2) $$F(Y) \geqslant d(Y) \quad (Y \subset \Gamma^{-1} z) \cap \bar{X})$$

We need to show that inequality (1) of the preceding theorem is equivalent to inequality (2).

(1) $\Rightarrow$ (2), since we have, from the lemma,

$$F(A) \geqslant d(A) \quad (A \subset \bar{X})$$

and, in particular, this gives:

$$F(Y) \geqslant d(Y) \quad (Y \subset \Gamma^{-1} z \cap \bar{X})$$

(2) $\Rightarrow$ (1), since if we consider a set $A \subset \bar{X}$, and if we put $Y = A \cap \Gamma^{-1}z$, then it is certainly true that

$$d(A) = d(Y) \leqslant F(Y) \leqslant F(A)$$

The following result is an important generalization of Theorem 1:

THEOREM 2 (HOFFMAN)†. *Consider a transport network in which two integers $b(u)$ and $c(u)$ [with $0 \leqslant b(u) \leqslant c(u)$] are associated with every arc u. Let $\mathscr{A}$ be the family of subsets of X which contain neither x_0 nor z, or which contain both x_0 and z. A flow ϕ such that $b(u) \leqslant \phi(u) \leqslant c(u)$ for all u, exists if and only if*

$$c(U_A^-) \geqslant b(U_A^+) \qquad (A \in \mathscr{A})$$

Given a network G with two numbers $b(u)$ and $c(u)$ for each arc u, we can construct a corresponding network $\bar{G}$ with capacities $\bar{c}(u)$ as follows: add two points to the vertices of G, $\bar{x}_0$ and $\bar{z}$ which will be the source and sink of $\bar{G}$; give every arc $u = (x, y)$ of G a capacity $\bar{c}(u) = c(u) - b(u)$; connect the source $\bar{x}_0$ to the vertex y by an arc of capacity $\bar{c}(\bar{x}_0, y) = b(u)$, and connect the vertex x to the sink $\bar{z}$ by an arc of capacity $\bar{c}(x, \bar{z}) = b(u)$. Finally, join z to x_0 by an arc of capacity:

$$\bar{c}(z, x_0) = +\infty$$

1. We shall show that the two assertions below are equivalent.

(1): a flow ϕ exists for the network G such that

$$b(u) \leqslant \phi(u) \leqslant c(u)$$

(2): a flow $\bar{\phi}$ such that $\bar{\phi}(u) \leqslant \bar{c}(u)$ and such that the arcs leading to the sink $\bar{z}$ are saturated, exists for the network $\bar{G}$.

(1) $\Rightarrow$ (2), for if ϕ exists, we simply define a flow $\bar{\phi}$ in $\bar{G}$ by:

$$\begin{aligned} \bar{\phi}(u) &= \phi(u) - b(u) \quad \text{if} \quad u \in U \\ \bar{\phi}(x, \bar{z}) &= \sum_{y \in \Gamma x} b(x, y) \\ \bar{\phi}(\bar{x}_0, y) &= \sum_{x \in \Gamma^{-1} y} b(x, y) \\ \bar{\phi}(z, x_0) &= \phi_z \end{aligned}$$

† This result was proved by A. HOFFMAN using linear programming techniques, and related to the theory of networks by L. FORD.

$\bar{\phi}$ is a flow in $\bar{G}$ which saturates the arcs leading to the sink $\bar{z}$, for:

(1) $$\bar{\phi}(u) = \phi(u)-b(u) \geqslant b(u)-b(u) = 0$$

(2) $$\bar{\phi}(u) = \phi(u)-b(u) \leqslant c(u)-b(u) = \bar{c}(u)$$

(3) $$\sum_{u \in U_x^-} \bar{\phi}(u) - \sum_{u \in U_x^+} \bar{\phi}(u) = \sum_{u \in U_x^-} \phi(u) - \sum_{u \in U_x^+} \phi(u) = 0$$

(2) $\Rightarrow$ (1), for if $\bar{\phi}$ exists, we define a function ϕ by:

$$\phi(u) = \bar{\phi}(u)+b(u) \qquad (u \in U)$$

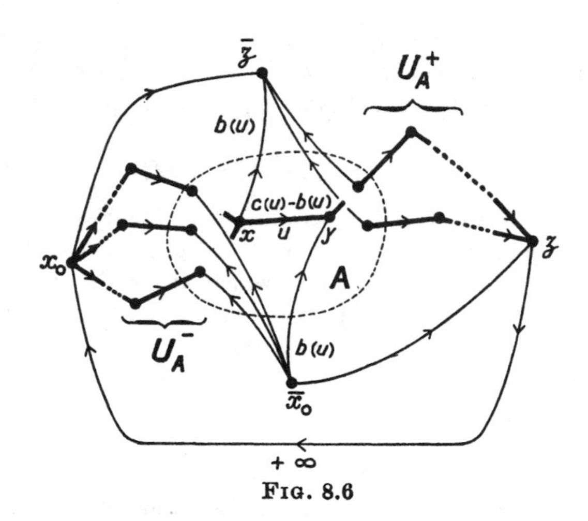

Fig. 8.6

ϕ is certainly a flow in G which satisfies the desired conditions, for:

(1) $$\phi(u) = \bar{\phi}(u)+b(u) \geqslant 0+b(u) = b(u)$$

(2) $$\phi(u) = \bar{\phi}(u)+b(u) \leqslant \bar{c}(u)+b(u) = c(u)$$

(3) $$\sum_{u \in U_x^-} \phi(u) - \sum_{u \in U_x^+} \phi(u) = 0$$

2. Thus we now wish to find a necessary and sufficient condition for (2) to hold. For all $A \subset X$, let V_A^- be the set of arcs of G incident into A; it has been shown above (Theorem 1) that condition (2) is equivalent to:

$$\bar{c}(V_A^-) \geqslant \bar{d}(A) \qquad (A \subset X)$$

If this inequality is satisfied for a set S containing neither x_0 nor z, then, *a fortiori*, it is satisfied for all sets $T = S \cup \{z\}$ since:

$$\bar{c}(V_T^-) \geqslant \bar{c}(V_S^-) \geqslant \bar{d}(S) = d(T)$$

The inequality always holds for all sets $R = S \cup \{z\}$, since $\bar{c}(V_R^-) = +\infty$. Therefore, if we limit ourselves to sets A of the family $\mathscr{A}$, the inequality becomes:

$$\sum_{u \in U_A^-} \bar{c}(u) + \sum_{x \in A} \bar{c}(\bar{x}_0, x) \geqslant \sum_{x \in A} \bar{c}(x, \bar{z}) \qquad (A \in \mathscr{A})$$

which gives:

$$\sum_{u \in U_A^-} (c(u) - b(u)) + \sum_{u \in U_A^-} b(u) + \sum_{\substack{u=(x,y) \\ x \in A \\ y \in A}} b(u) \geqslant \sum_{\substack{u=(x,y) \\ x \in A \\ y \in A}} b(u) + \sum_{u \in U_A^+} b(u) \quad (A \in \mathscr{A})$$

which gives finally:

$$c(U_A^-) \geqslant b(U_A^+) \qquad (A \in \mathscr{A})$$

ALGORITHM FOR PROBLEMS 1 AND 2 WHEN THE ALLOWED SET $C(u)$ IS AN INTERVAL. 1. Given a network G in which associated with each arc u are two numbers $b(u)$ and $c(u)$, with $0 \leqslant b(u) \leqslant c(u)$, we determine a flow ϕ such that $b(u) \leqslant \phi \leqslant c(u)$ by finding a maximum flow $\bar{\phi}$ for the network $\bar{G}$ (defined above) and putting as before:

$$\phi(u) = \bar{\phi}(u) + b(u)$$

2. The flow thus obtained is not necessarily a maximum flow; it may be given a maximum value by using a method analogous to that of Ford and Fulkerson:

(i) label the vertex x_0 with the coefficient: 0;
(ii) if x_i is a labelled vertex, any unlabelled vertex y such that:

$$(x_i, y) \in U \qquad \phi(x_i, y) < c(x_i, y)$$

is given the label $+i$;
(iii) if x_i is a labelled vertex, any unlabelled vertex y such that:

$$(y, x_i) \in U \qquad \phi(y, x_i) > b(y, x_i)$$

is given the label $-i$.

If, by these means, the vertex z can be labelled, the flow can be improved as in the Ford–Fulkerson algorithm.

If, for some flow ϕ^0, z cannot be labelled, let A be the set of vertices which cannot be labelled. We have $x_0 \notin A, z \in A$; the value of the flow ϕ^0 is:

$$\phi_z^0 = \sum_{u \in U_A^-} \phi^0(u) - \sum_{u \in U_A^+} \phi^0(u) = c(U_A^-) - b(U_A^+)$$

Since, on the other hand, the value of a flow cannot exceed $c(U_A^-)-b(U_A^+)$, the flow ϕ^0 must be a maximum flow.

From the preceding arguments, we deduce:

THEOREM 3. *If a transport network possesses a flow ϕ for which*

$$0 \leqslant b(u) \leqslant \phi(u) \leqslant c(u)$$

for all u, the maximum value of such a flow is:

$$\phi_z = \min_{\substack{A \ni z \\ A \not\ni x_0}} (c(U_A^-)-b(U_A^+))$$

Infinite Transport Networks

Consider a transport network defined by a graph (X, U), with capacity $c(u)$; in this paragraph we shall no longer suppose that $|X| < \infty$. We are going to extend the saturation theorem.

THEOREM 4. *In an infinite network, let $Y = \{y_1, y_2, \ldots, y_n, \ldots\}$ be the set (assumed to be denumerable) of vertices adjacent to the sink z; if the graph is regressively finite and Γ^{-1}-finite at every point except z, a necessary and sufficient condition for the existence of a flow ϕ^0 which saturates every arc leading to z is that:*

$$F(A) \geqslant d(A) \qquad (A \subset Y;\ |A| < \infty)$$

Put $Y_n = \{y_1, y_2, \ldots, y_n\}$, and denote by X_n the set of vertices belonging to at least one path going from x_0 to z via one of the points of Y_n.

By changing the orientation of each arc and deleting z, a graph is obtained which is Γ-finite and progressively finite, and therefore, from Theorem 2 (Chapter 3), we have

$$|(\hat{\Gamma}^{-1})y| < \infty \qquad (y \in Y)$$

Therefore, both the set:

$$X_n = \bigcup_{k=1}^{n} (\hat{\Gamma}^{-1})y_k \cup \{z\}$$

and the subnetwork G_n which it determines, are finite.

Let the set of flows ϕ_n which saturate the outgoing arcs be written Φ_n [if u is not an arc of G_n, we also put $\phi_n(u) = 0$]. Since G_n is a finite network, we have $|\Phi_n| < \infty$; using Gale's theorem, we also have $\Phi_n \neq \emptyset$; lastly, it is obvious that for any $\phi_n \in \Phi_n$, a corresponding flow $\phi_{n-1} \in \Phi_{n-1}$ can be formed such that $\phi_{n-1} \leqslant \phi_n$: then, from König's corollary (Chapter 3), a sequence $(\phi_1^0, \phi_2^0, \ldots, \phi_n^0, \ldots)$, exists, with $\phi_n^0 \in \Phi_n$ for all n, and such that:

$$\phi_1^0 \leqslant \phi_2^0 \leqslant \ldots \leqslant \phi_n^0 \leqslant \ldots$$

We shall show that by putting $\phi^0(u) = \sup \phi_n^0(u)$, a flow is obtained which is compatible with the network G.

Since the set U_a^- of arcs incident into a vertex a, with $a \neq z$, is finite, an integer p exists such that:

$$\phi_p^0(u) = \phi^0(u) \qquad (u \in U_a^-)$$

and, since ϕ_p^0 is a flow of the network G_p, we have:

$$\sum_{u \in U_a^-} \phi^0(u) = \sum_{u \in U_a^-} \phi_p^0(u) = \sum_{u \in U_a^+} \phi_p^0(u) = \sum_{u \in U_a^+} \phi^0(u)$$

Therefore ϕ^0 is a flow; further, since:

$$\phi^0(u) = \sup_n \phi_n^0(u) \leqslant c(u) \qquad (u \in U)$$

this flow is compatible with the capacities of the arcs of the network.

Finally, this flow obviously saturates all outgoing arcs; it must, therefore, be the required flow.

9. The Theorem of the Demi-Degrees

Inward or Outward Demi-Degree

To make our treatment more general, we shall consider here not a graph, but an s-graph: if s is an integer, the graph (X, U) is an *s-graph* if several distinct arcs of U can be drawn (in the same direction) connecting the same pair of vertices, and if the number of such arcs $\leqslant s$.

In an s-graph, the *outward demi-degree* of a vertex x is defined to be the number $d^+(x)$ of arcs having x as initial vertex; the *inward demi-degree* of the vertex x is the number $d^-(x)$ of arcs having x as terminal vertex. For a graph (X, Γ), we have

$$d^+(x) = |\Gamma x|$$
$$d^-(x) = |\Gamma^{-1} x|$$

If $d^+(x) = d^-(x) = 0$, the vertex x is said to be an *isolated point*; if $d^+(x) \neq 0$, $d^-(x) = 0$, the vertex x is said to be a *source*; if $d^+(x) = 0$, $d^-(x) \neq 0$, the vertex x is a *sink*.

We could ask if, for given integers $r_1, r_2, \ldots, r_n, s_1, s_2, \ldots, s_n$, an s-graph exists whose vertices $x_1, x_2, \ldots, x_n$ satisfy:

$$d^+(x_k) = r_k$$
$$d^-(x_k) = s_k \qquad (k = 1, 2, \ldots, n)$$

In this chapter we shall investigate this problem.

Theorem of the demi-degrees. *Consider n pairs of integers (r_1, s_1), $(r_2, s_2), \ldots, (r_n, s_n)$ in which, by altering the indices if necessary, we have*

$$s_1 \geqslant s_2 \geqslant \ldots \geqslant s_n \geqslant 0$$

The (r_k, s_k) constitute the demi-degrees of an s-graph if and only if we have:

$$(1) \qquad \sum_{i=1}^{n} \min\{sk, r_i\} \geqslant \sum_{j=1}^{k} s_j \qquad (k = 1, 2, \ldots, n)$$

$$(2) \qquad \sum_{i=1}^{n} r_i = \sum_{j=1}^{n} s_j$$

A transport network G consisting of vertices $x_1, x_2, \ldots, x_n, \bar{x}_1, \bar{x}_2, \ldots, \bar{x}_n$, a source x_0 and a sink z is built up to correspond to the set of pairs (r_k, s_k); vertices x_i and $\bar{x}_j$ are connected by an arc of capacity $c(x_i, \bar{x}_j) = s$; vertices x_0 and x_i are connected by an arc of capacity $c(x_0, x_i) = r_i$; vertices $\bar{x}_j$ and z are connected by an arc of capacity $c(\bar{x}_j, z) = s_j$.

Any flow which saturates the in- and out-going arcs of G defines an s-graph (X, U) with the (r_k, s_k) as its degrees; conversely, any s-graph with (r_k, s_k) as its degrees defines a flow G which saturates the ingoing and outgoing arcs. From the saturation theorem (Chapter 8), the necessary and sufficient condition that such a flow should exist is that (2) holds and that for all sets $A = \{\bar{x}_{i_1}, \bar{x}_{i_2}, \ldots, \bar{x}_{i_k}\}$ the maximum amount of material which can flow into A should not be less than the demand of A; that is:

$$F(\bar{x}_{i_1}, \bar{x}_{i_2}, \ldots, \bar{x}_{i_k}) \geqslant d(\bar{x}_{i_1}, \bar{x}_{i_2}, \ldots, \bar{x}_{i_k}) \qquad (k \leqslant n; i_1 < i_2 < \ldots < i_k)$$

which implies:

$$(1') \quad \sum_{i=1}^{n} \min\{r_i, sk\} \geqslant s_{i_1} + s_{i_2} + \ldots + s_{i_k} \qquad (k \leqslant n; i_1 < i_2 < \ldots < i_k)$$

We have $(1') \Rightarrow (1)$, and since the indices of the s_j are in decreasing order of magnitude, we have $(1) \Rightarrow (1')$; we have thus regained the stated conditions.

By putting $s = 1$, the theorem gives a necessary and sufficient condition for the pairs (r_k, s_k) to form the demi-degrees of a graph. It is possible to reformulate the result so that it can be tested more rapidly. Let:

$$\boldsymbol{r} = (r_1, r_2, \ldots)$$

be a finite sequence of (non-negative) integers (indexed in decreasing order), and let $\boldsymbol{r}^* = (r_1^*, r_2^*, \ldots)$ be the *dual* sequence to the sequence $\boldsymbol{r}$, where r_k^* stands for the number of the r_i greater than or equal to the integer k [the reader may verify that $(\boldsymbol{r}^*)^* = \boldsymbol{r}$]. We have then:

COROLLARY†. *In a given set of integer pairs (r_k, s_k) alter the indices of the r_k and of the s_k separately so that*

$$r_1 \geqslant r_2 \geqslant \ldots \geqslant r_n$$
$$s_1 \geqslant s_2 \geqslant \ldots \geqslant s_n$$

† This condition appears to be due to GALE [1]; note that, from the well-known theorem of HARDY–LITTLEWOOD–POLYA [2], these inequalities state that the vector $\boldsymbol{r}^*$ is the image of the vector $\boldsymbol{s}$ resulting from a linear doubly stochastic transformation (for a proof, cf. for example [E.T.F.M.], Chapter 8).

Next form the dual sequence $\mathbf{r}^* = (r_1^*, r_2^*, \ldots)$ *of* $\mathbf{r} = (r_1, r_2, \ldots)$, *and bring the sequences* $\mathbf{r}^*$ *and* $\mathbf{s}$ *to the same length m by the addition of appropriate zeros.*

A necessary and sufficient condition for the given pairs to constitute the degrees of a graph is that:

$$r_1^*+r_2^*+\ldots+r_k^* \geqslant s_1+s_2+\ldots+s_k \qquad (k < m)$$
$$r_1^*+r_2^*+\ldots+r_m^* = s_1+s_2+\ldots+s_m$$

This follows immediately from the preceding theorem if we show that for every integer k, we have

$$\sum_{i=1}^{n} \min\{k, r_i\} = \sum_{j=1}^{k} r_j^*$$

This equation is satisfied for $k = 1$; on the other hand, if it is satisfied for any integer k, it is also satisfied for $k+1$, since:

$$\begin{aligned}\sum_{i=1}^{n} \min\{(k+1), r_i\} &= \sum_{r_i \leqslant k} \min\{k, r_i\} + \Big(\sum_{r_i > k} \min\{k, r_i\} + r_{k+1}^*\Big)\\ &= \sum_{i=1}^{n} \min\{k, r_i\} + r_{k+1}^*\\ &= r_1^*+r_2^*+\ldots+r_k^*+r_{k+1}^*\end{aligned}$$

Q.E.D.

EXAMPLE (The excursion problem). m families go for a picnic in n cars; there are $r_1, r_2, \ldots, r_m$ people in the first, second, ..., m-th family, respectively, and $s_1, s_2, \ldots, s_n$ seats in the first, second, ..., n-th car. Can the seating be arranged so that no two members of the one family are in the same car? This problem reduces to constructing a simple graph with the given inward and outward demi-degrees; the graph is defined by the sets:

$$X = \{x_1, x_2, \ldots, x_m\} \quad \text{and} \quad Y = \{y_1, y_2, \ldots, y_n\}$$

The demi-degrees of x_k are $(r_k, 0)$, and the demi-degrees of y_k are $(0, s_k)$.

To be definite, let us take:

Seats available: $s_k = 5, 5, 4, 4, 4, 3, 2, 1, 1, 1,$

Families: $r_k = 8, 8, 5, 5, 4.$

The dual sequence consists of:

$$r_1^* = r_2^* = r_3^* = r_4^* = 5;\ r_5^* = 4;\ r_6^* = r_7^* = r_8^* = 2$$

Let us see if the problem is possible:

$$\begin{aligned}
& & 5 &\geqslant 5 & & \\
5+5 &= 10 &\geqslant& \;5+5 &= 10 \\
10+5 &= 15 &\geqslant& \;10+4 &= 14 \\
15+5 &= 20 &\geqslant& \;14+4 &= 18 \\
20+4 &= 24 &\geqslant& \;18+4 &= 22 \\
24+2 &= 26 &\geqslant& \;22+3 &= 25 \\
26+2 &= 28 &\geqslant& \;25+2 &= 27 \\
28+2 &= 30 &\geqslant& \;27+1 &= 28 \\
& & 30 &\geqslant 28+1 &= 29 \\
& & 30 &= 29+1 &
\end{aligned}$$

The problem therefore possesses a solution which may be found by solving a problem of maximum flow.

Suppose for a given system of demi-degrees, we have succeeded in constructing the graph $G = (X, U)$, and we now consider the vertices x, x', y, y', with:

$$x \neq x' \qquad y \neq y'$$
$$(x,y) \in U \qquad (x',y') \in U$$
$$(x,y') \notin U \qquad (x',y) \notin U$$

If we substitute the arcs (x,y') and (x',y) for the arcs (x,y) and (x',y'), we again have a graph with the same system of demi-degrees: we shall call such a substitution a *transfer*; its importance is due to the following result:

THEOREM 2. *If* $G = (X, U)$ *and* $H = (X, V)$ *are two graphs with the same demi-degrees at each vertex,* H *can be derived from* G *by means of a finite number of transfers.*

We can assume that G and H do not possess any arcs in common, for such arcs could be deleted without altering the theorem in any way. Let us associate a square matrix A of order n with the graph G, where the elements a_j^i of A satisfy:

$$a_j^i \begin{cases} = 0 & \text{if } (x_i, x_j) \notin U \\ = 1 & \text{if } (x_i, x_j) \in U \end{cases}$$

In the same way, a matrix $B = (b_j^i)$ corresponding to H may be formed; we have:

$$\sum_{j=1}^{n} a_j^i = \sum_{j=1}^{n} b_j^i = d^+(x_i)$$
$$\sum_{i=1}^{n} a_j^i = \sum_{i=1}^{n} b_j^i = d^-(x_j)$$

The matrix $C = A - B$, is defined to consist of elements: $c_j^i = a_j^i - b_j^i$, which are equal to $+1$, 0 or -1, and the sum of the elements of any row or column is zero.

If $\sum_{i,j} |c_j^i| = 0$, the graphs G and H coincide; if $\sum_{i,j} |c_j^i| > 0$, we shall show that by a succession of transfers on G, we can derive a matrix

$$\bar{C} = \bar{A} - B = (\bar{c}_j^i)$$

with

$$\sum_{i,j} |\bar{c}_j^i| < \sum_{i,j} |c_j^i|$$

By repeating this operation we shall eventually reach a graph $\bar{G}$ which coincides with H.

Let c_j^i be a coefficient equal to $+1$; since each column contains as many positive as negative elements, column j must also contain an element $c_j^k = -1$. Likewise, in row k there must be some element $c_l^k = +1$, and in column l there must be $c_l^m = -1$, etc. In this way we can define a certain path $c_j^i, c_j^k, c_l^k, c_l^m, \ldots, c_r^s, c_r^i$. If at some stage, we re-enter column q by means of the element $c_q^p = +1$ which has not been used previously, we may leave it by selecting an element $c_q^h = -1$ which has not yet been used (this is always possible, since the number of positive and negative elements in column q is equal); the matrix being finite, we must sooner or later return to the initial point c_j^i, thus completing an elementary circuit.

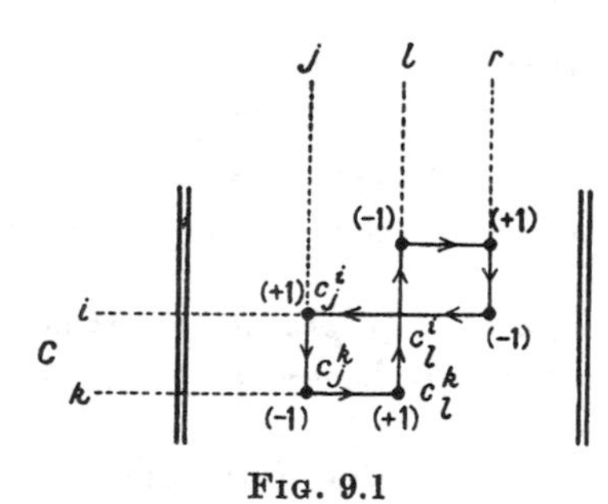

Fig. 9.1

Consider the square: c_j^i, c_j^k, c_l^k, c_l^i; if $c_l^i = -1$, we make the transfer:

$$(x_i, x_j) \rightarrow (x_i, x_l)$$
$$(x_k, x_l) \rightarrow (x_k, x_j)$$

In this way we get a matrix $\bar{C}$ with $\bar{c}_j^i = \bar{c}_j^k = \bar{c}_l^k = \bar{c}_l^i = 0$ (the other elements remaining unchanged), which achieves the desired result. If, however, $c_l^i = 0$, we have $(x_i, x_l) \notin U$ (for we have assumed that arcs common to G and H have been eliminated), and the same transfer gives $\bar{c}_j^i = \bar{c}_l^k = \bar{c}_j^k = 0$, $\bar{c}_l^i = 1$, which again achieves the desired result. Only the case $c_l^i = +1$ can hold us up, but in this case we can replace the elementary circuit by a shorter elementary circuit: $c_l^i, c_l^m, \ldots, c_r^i, c_l^i$.

If the theorem still cannot be used, we shorten the circuit step by step until finally we obtain a unique square which has as vertices $+1$, -1, $+1$, -1; an obvious transfer then gives the required result.

Q.E.D.

Consider the case in which:

$$X = \{x_1, x_2, \ldots, x_n, y_1, y_2, \ldots, y_n\}$$

$$d^+(x_i) = +1, \qquad d^-(x_i) = 0$$

$$d^+(y_j) = 0, \qquad d^-(y_j) = +1$$

The graph required corresponds to a permutation of the indices 1, 2, ..., n, and Theorem 2 becomes a well-known result:

If $(i_1, i_2, \ldots, i_n)$ *is a permutation, then by interchanging the positions of pairs of indices a certain number of times, we can obtain any other permutation which has been chosen in advance.*

10. Matching of a Simple Graph

The Maximum Matching Problem

The discussion in this chapter deals only with a *simple* graph, that is, one defined by two disjoint sets X and Y and by a multivalued function Γ mapping X into Y; later on (Chapter 18) we shall consider the same problems applied to any arbitrary graph, but very different techniques will be employed. We assume to begin with that the graph is finite.

A *matching* of a simple graph (X, Y, Γ) is a set of arcs W such that no two arcs of W are adjacent; if a matching defines a one–one correspondence between a set $A \subset X$ and a set $B \subset Y$, we can also say that we have *matched* the set A *on to* the set B, or that we have *matched A into Y*. We are interested in finding the maximum number of arcs in a matching.

EXAMPLE 1. *The personnel assignment problem.* In a certain firm, p workmen $x_1, x_2, \ldots, x_p$ are available to fill p positions $y_1, y_2, \ldots, y_p$, each workman being qualified for one or more of these jobs. Can each man be assigned to a position for which he is qualified? If we denote by Γx_i the set of positions for which workman x_i is qualified, the problem becomes one of matching X into Y with respect to the simple graph (X, Y, Γ).

EXAMPLE 2. *The end-of-term dance.* Every girl at an American co-educational college has m boy-friends, and every boy has m girl-friends; is it possible simultaneously for each girl to dance with one of her boy-friends, and for each boy to dance with one of his girl-friends? It will be shown shortly (Corollary 2) that this is always possible.

THEOREM OF KÖNIG AND HALL. *X can be matched into Y if and only if $|\Gamma A| \geqslant |A|$ for all sets $A \subset X$.*

Consider the transport network defined by the points of X, those of Y, a source x_0 and a sink z; join x_0 to every $y_j \in Y$ by an arc of capacity $c(x_0, y_j) = 1$, join every $x_i \in X$ to z by an arc of capacity $c(x_i, z) = 1$, and finally join y_j to x_i by an arc of infinite capacity whenever $y_j \in \Gamma x_i$. If $A \subset X$, the total demand of the set A is $d(A) = |A|$; the maximum amount of flow which can enter A is $F(A) = |\Gamma A|$.

Any flow through the network defines a matching of the simple graph,

the points x_i and y_j corresponding to one another whenever unit amount flows along the arc (y_j, x_i); conversely, any matching defines a flow. X can

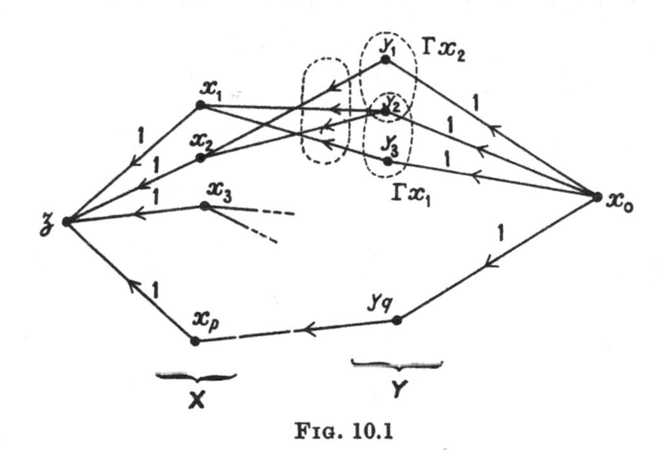

Fig. 10.1

be matched into Y if and only if the maximum flow saturates the terminal arcs, that is, from the saturation theorem (Chapter 8), if:

$$F(A) \geqslant d(A) \quad (A \subset X)$$

or

$$|\Gamma A| \geqslant |A| \quad (A \subset X)$$

Q.E.D.

Naturally, finding the maximum matching is a problem of maximum flow, for which we can use Ford and Fulkerson's algorithm (Chapter 8).

Corollary 1. ***In a simple graph $G = (X, Y, \Gamma)$, with $|X| = p$, $|Y| = q$, index the vertices $x_i \in X$ and $y_j \in Y$ so that:***

$$|\Gamma x_1| \leqslant |\Gamma x_2| \leqslant \ldots \leqslant |\Gamma x_p|$$
$$|\Gamma^{-1} y_1| \geqslant |\Gamma^{-1} y_2| \geqslant \ldots \geqslant |\Gamma^{-1} y_q|$$

A sufficient condition to enable X to be matched into Y is that:

$$(1) \quad |\Gamma x_1| + |\Gamma x_2| + \ldots + |\Gamma x_k| > |\Gamma^{-1} y_1| + |\Gamma^{-1} y_2| + \ldots + |\Gamma^{-1} y_{k-1}| \quad (k = 1, 2, \ldots p)$$

Consider two sets:

$$A = \{x_{i_1}, x_{i_2}, \ldots, x_{i_k}\} \quad \text{and} \quad B = \{y_{j_1}, y_{j_2}, \ldots, y_{j_{k-1}}\}$$

containing k and $k-1$ elements respectively; from (1), we have:

$$|\Gamma x_{i_1}| + |\Gamma x_{i_2}| + \ldots + |\Gamma x_{i_k}| > |\Gamma^{-1} y_{j_1}| + |\Gamma^{-1} y_{j_2}| + \ldots + |\Gamma^{-1} y_{j_{k-1}}|$$

The number of arcs leading out of A is strictly greater than the number of arcs incident into B; consequently $\Gamma A \not\subset B$, and this holds for any set B of $k-1$ elements; hence we may say $|\Gamma A| > k-1 = |A|-1$, or, finally:

$$|\Gamma A| \geqslant |A| \quad (A \subset X)$$

Therefore we can match X into Y.

COROLLARY 2. *If for a simple graph (X, Y, Γ), an integer m exists such that $|\Gamma x| \geqslant m$ for all $x \in X$ and $|\Gamma^{-1} y| \leqslant m$ for all $y \in Y$, X can be matched into Y.*

The condition of this corollary in fact satisfies condition (1) of the preceding corollary.

COROLLARY 3. *In a simple graph (X, Y, Γ), put:*

$$m = \max \{|\Gamma x|, |\Gamma^{-1} y| \,\big|\, x \in X, y \in Y\}$$

A matching exists which uses all vertices x for which $|\Gamma x| = m$ and all vertices y for which $|\Gamma^{-1} y| = m$.

1. We shall first show that we can construct a simple graph (X', Y', Γ') in which all vertices are of degree m and which contains the graph (X, Y, Γ) as a partial subgraph.

To begin, make m identical copies of the graph (X, Y, Γ):

$$(X^1, Y^1, \Gamma), (X^2, Y^2, \Gamma), \ldots, (X^m, Y^m, \Gamma)$$

If, for a vertex $x \in X$, $m - |\Gamma x| = k > 0$, each of the corresponding points $x^1 \in X^1, x^2 \in X^2, \ldots, x^m \in X^m$ is joined to all the vertices of a set:

$$Y(x) = \{y(1, x), y(2, x), \ldots, y(k, x)\}$$

consisting of k points added as required. When this procedure has been gone through for all points $x \in X$, we turn our attention to a point $y \in Y$ and in like fashion connect the points $y^1 \in Y^1, y^2 \in Y^2, \ldots, y^m \in Y^m$ to a set $X(y)$ of supplementary vertices, etc.

Put:

$$X' = \bigcup_{i \leqslant m} X^i \cup \bigcup_{y \in Y} X(y)$$

$$Y' = \bigcup_{j \leqslant m} Y^j \cup \bigcup_{x \in X} Y(x)$$

The simple graph defined by X', Y', all the arcs defined by Γ, and the arcs associated with the sets $X(y)$ and $Y(x)$, satisfies the proposed conditions and contains (X^1, Y^1, Γ) as a partial subgraph.

We note that $|X'| = |Y'|$.

2. In this new graph X' can be matched into Y' (Corollary 2); to every vertex $x^1 \in X^1$ for which $|\Gamma x^1| = m$, there corresponds a vertex in Y^1, and to every vertex $y^1 \in Y^1$ for which $|\Gamma^{-1} y^1| = m$, there corresponds a vertex in X^1: therefore in the graph (X^1, Y^1, Γ) a matching must exist which uses all these vertices.

COROLLARY 4. *The chromatic index of a simple graph* (X, Y, Γ) *is*

$$m = \max \{|\Gamma x|, |\Gamma^{-1} y| \,\big|\, x \in X, y \in Y\}$$

We have to show that by using m different colours, we can colour the arcs of the graph in such a way that no two adjacent arcs are of the same colour. The first colour is applied to the arcs of a matching which includes all vertices with a demi-degree equal to m; the second colour to the arcs of a matching using all the vertices of maximum demi-degree in the graph remaining, etc.... The preceding corollary shows that all these matchings exist, and at each stage the maximum demi-degree of the graph is diminished by one. Therefore m colours are sufficient to colour all the arcs of the graph (X, Y, Γ).

EXAMPLE (Latin squares). Corollary 4 above has a curious application to the theory of Latin squares invented by Euler. We say that a Latin square of order n is formed if the spaces of a square of size $n \times n$ are filled with the letters $y_1, y_2, \ldots, y_n$ in such a way that the same letter does not appear twice in any row or column. In order to recognize if a rectangle of letters forms part of a Latin square, we have the following theorem:

Let T *be a* $p \times q$ *rectangle filled with the letters* $y_1, y_2, \ldots, y_n$ *such that no letter appears twice in the same row or column, and let* $m(y)$ *be the number of times that the letter* y *appears in* T; *a necessary and sufficient condition for the formation of a Latin square of order* n *by the addition of* $n-p$ *rows and* $n-q$ *columns is that:*

$$m(y_k) \geqslant p+q-n \qquad (k = 1, 2, \ldots, n)$$

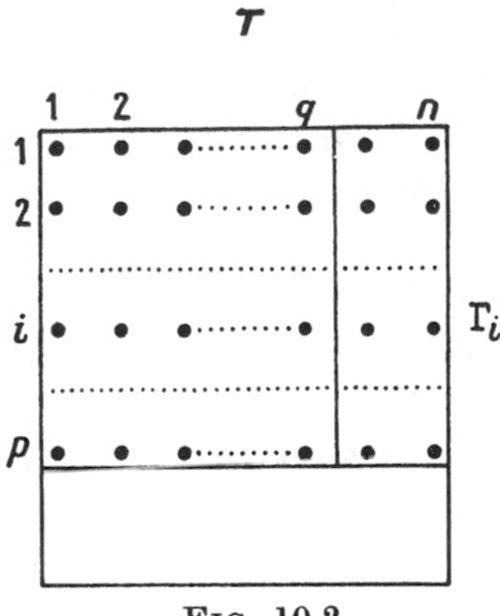

FIG. 10.2

1. Necessity of the condition: let Γ_i be the set of $n-q$ letters which do not appear in the i-th row of T, and let $m'(y)$ be the number of times y appears in the sets $\Gamma_1, \Gamma_2, \ldots, \Gamma_p$; if a Latin square can be formed, we must have $m'(y) \leqslant n-q$, from which:

$$m(y) = p - m'(y) \geqslant p-n+q$$

2. Sufficiency of the condition: put $P = \{1,2,\ldots,p\}$, $Y = \{y_1, y_2, \ldots, y_n\}$, define Γ_i as before, and consider the simple graph (P, Y, Γ); we have

$$|\Gamma_i| = n-q \qquad (i \in P)$$
$$|\Gamma^{-1} y| = m'(y) \leqslant n-q \qquad (y \in Y)$$

Therefore, from Corollary 4, we can colour the arcs of the graph with $n-q$ colours: blue, red, etc. We can now start completing T by adding $n-q$ columns: the first consists of the terminal vertices of the blue arcs leaving P, the second of the terminal vertices of the red arcs leaving P, etc. We now have a $p \times n$ rectangle $\overline{T}$, which still satisfies the conditions of the theorem since $\overline{m}(y) = p = (p+n)-n$; and the same argument can be used again, attention being given now to the columns instead of to the rows.

Deficiency of a Simple Graph

Given a simple graph $G = (X, Y, \Gamma)$, the *deficiency* of the graph G is defined to be the number:

$$\delta_0 = \max_{A \subset X} (|A| - |\Gamma A|)$$

We always have $\delta_0 \geqslant 0$ (since $|\varnothing| - |\Gamma\varnothing| = 0$), and therefore the theorem of König and Hall may be restated: a necessary and sufficient condition to enable X to be matched into Y is that $\delta_0 = 0$. We intend here to elaborate this result and, more precisely, to determine the maximum number of arcs in a matching.

First, a *support* of a simple graph $G = (X, Y, \Gamma)$ is defined to be any set $C \subset X \cup Y$ such that: at least one of the vertices of any arbitrary arc is in C. This is an important concept in itself, and the following results are derived from it:

THEOREM 2. *In a simple graph $G = (X, Y, \Gamma)$, the minimum number of elements of a support C is $|X| - \delta_0$.*

In fact, we have:

$$|X| - \delta_0 = |X| - \max_{A \subset X} (|A| - |\Gamma A|) = \min_{A \subset X} (|X| - |A| + |\Gamma A|)$$
$$= \min_{B \subset X} (|B| + |\Gamma(X-B)|)$$

(where $B = X - A$).

Now the set $B \cup \Gamma(X-B)$ is a support, and since any minimum support will be of this form, the theorem has been proved.

THEOREM 3. *The coefficient of internal stability of a simple graph* $G = (X, Y, \Gamma)$ *is* $\alpha(G) = |Y| + \delta_0$.

In fact, if C is a support, its complement $S = (X \cup Y) - C$ is an internally stable set (since no arc can have both its vertices in S); conversely, if S is an internally stable set, its complement is a support. Consequently:

$$\alpha(G) = \max_S |S| = \max_C (|X| + |Y| - |C|) = |X| + |Y| - \min_C |C|$$
$$= |X| + |Y| - (|X| - \delta_0) = |Y| + \delta_0$$

Remark. Theorem 3 may be stated in another form. In any arbitrary graph, let us denote by $\Delta x = \{y | y > x\}$ the set of descendants of a vertex x (cf. Chapter 2), and say that two vertices x_1 and x_2 are *unrelated* if we have neither $x_1 > x_2$ nor $x_2 > x_1$. In the case of a simple graph (X, Y, Γ), we have therefore:

$$\Delta x = \Gamma x \quad \text{if } x \in X$$
$$\Delta x = \emptyset \quad \text{if } x \in Y$$

Therefore we can rephrase Theorem 3 thus: *the maximum number of vertices which are pairwise unrelated in a simple graph is*

$$\max_{A \subset X \cup Y} (|A| - |\Delta A|)$$

It is a remarkable fact that this statement can be shown to hold for any graph whatever†.

† This may be proved as follows: if (X, Γ) is a given finite graph, construct the simple graph $(X, \bar{X}, \Delta)$ in which the descendants of each vertex are marked, and consider a minimum support $C = \{x_{i_1}, x_{i_2}, \ldots; \bar{x}_{j_1}, \bar{x}_{j_2}, \ldots\}$.

1. We show first that these vertices all have different indices. If, for example, $i_1 = j_1$, an index $k \neq i_1, i_2, \ldots$, exists for which:

$$\bar{x}_{i_1} \in \Delta x_k$$

(since $C - \{\bar{x}_{i_1}\}$, which possesses one element less than C, cannot therefore be a support); similarly, an index $h \neq j_1, j_2, \ldots$, exists for which:

$$\bar{x}_h \in \Delta x_{i_1}$$

and hence:

$$x_k < x_{i_1}, x_{i_1} < x_h \Rightarrow x_k < x_h$$

which contradicts the fact that neither of the vertices of the arc $(x_k, \bar{x}_h)$ are in the set C.

2. In the graph (X, Γ), consider the set $S = X - \{x_{i_1}, x_{i_2}, \ldots; x_{j_1}, x_{j_2}, \ldots\}$; its elements are pairwise unrelated, it is a maximum set with this property, and from Theorem 2 we have:

$$|S| = |X| - |C| = |X| - [|X| - \max_{A \subset X} (|A| - |\Delta A|)] = \max_{A \subset X} (|A| - |\dot{\Delta} A|)$$

Q.E.D.

Let us now examine the connection between the deficiency of a simple graph and the maximum number of arcs in a matching.

LEMMA 1. *If for $A \subset X$, we put $\delta(A) = |A| - |\Gamma A|$, we have:*

$$\delta(A_1 \cup A_2) + \delta(A_1 \cap A_2) \geqslant \delta(A_1) + \delta(A_2)$$

Consider the inequalities:

$$|A_1 \cup A_2| + |A_1 \cap A_2| = |A_1| + |A_2|$$

$$|\Gamma(A_1 \cup A_2)| + |\Gamma(A_1 \cap A_2)| \leqslant |\Gamma A_1 \cup \Gamma A_2| + |\Gamma A_1 \cap \Gamma A_2| = |\Gamma A_1| + |\Gamma A_2|$$

By subtracting the second inequality from the first, we obtain the desired result.

LEMMA 2. *The family $\mathscr{A} = \{A \mid \delta(A) = \delta_0\}$ is closed with respect to the operations $\cup$ and $\cap$; that is to say:*

$$A_1, A_2 \in \mathscr{A} \Rightarrow A_1 \cup A_2 \in \mathscr{A},\ A_1 \cap A_2 \in \mathscr{A}$$

In fact, if A_1, $A_2 \in \mathscr{A}$, then from Lemma 1

$$\delta(A_1 \cup A_2) + \delta(A_1 \cap A_2) \geqslant \delta(A_1) + \delta(A_2) = 2\delta_0$$

Since neither of the terms $\delta(A_1 \cup A_2)$ and $\delta(A_1 \cap A_2)$ can exceed δ_0, we must have $\delta(A_1 \cup A_2) = \delta_0$, $\delta(A_1 \cap A_2) = \delta_0$, which establishes the proposition.

LEMMA 3. *If $\delta_0 > 0$, the intersection A_0 of (the sets of) $\mathscr{A}$, is itself a set of $\mathscr{A}$ and is non-empty.*

Since the graph is finite, A_0 is an intersection of a finite number of sets of $\mathscr{A}$, and therefore from the preceding lemma $A_0 \in \mathscr{A}$; if, however, $A_0 = \emptyset$, we would have

$$\delta_0 = \delta(A_0) = \delta(\emptyset) = 0$$

which contradicts the hypothesis that $\delta_0 > 0$.

THEOREM 4 (KÖNIG–ORE). *In a simple graph, the maximum number of arcs in a matching is $|X| - \delta_0$.*

If $\delta_0 = 0$, the theorem is proved (from the theorem of König and Hall); let us assume therefore that $\delta_0 > 0$ and consider the (non-empty) intersection A_0 of the sets A such that $\delta(A) = \delta_0$.

1. *If we eliminate a vertex $x_0 \notin A_0$ from the graph, the deficiency is unaltered.* The deficiency δ_0' of the new graph must be less than or equal to δ_0; since the new graph contains a set A_0 with $\delta(A_0) = \delta_0$, we must have $\delta_0' = \delta_0$.

2. *If we eliminate a vertex $x_0 \in A_0$ from the graph, the deficiency of the graph is decreased by exactly one unit.* In fact, since we have eliminated all sets A for which $\delta(A) = \delta_0$, the deficiency δ_0' of the new graph must be strictly less than δ_0; the set $A_0' = A_0 - \{x_0\}$ satisfies:

$$\delta(A_0') = |A_0| - 1 - |\Gamma A_0'| < \delta_0 = |A_0| - |\Gamma A_0|$$

and hence

$$|\Gamma A_0'| \geqslant |\Gamma A_0|$$

Since $A_0' \subset A_0$, the reverse inequality also holds, from which:

$$|\Gamma A_0'| = |\Gamma A_0|$$

which gives finally:

$$\delta(A_0') = |A_0'| - |\Gamma A_0'| = (|A_0| - 1) - |\Gamma A_0| = \delta_0 - 1$$

Since $\delta_0' < \delta_0$, this shows that $\delta_0' = \delta_0 - 1$.

3. From the preceding, we deduce that δ_0 is the minimum number of vertices which we must eliminate from X in order to obtain a graph whose deficiency $\delta_0' = 0$, that is, a graph in which X can be matched into Y; this gives the required result.

Corollary (König's theorem). *In a simple graph, the minimum number of vertices in a support is equal to the maximum number of arcs of a matching.* (This follows immediately on comparing Theorems 2 and 4.)

This last result allows us to set up a new algorithm for finding a maximum matching; this algorithm, which is more instructive than that for maximum flow in so far as it concerns the theory of graphs, is based on work of Egerváry [2] and of Kuhn [5], and is generally known under the name of 'the Hungarian method'.

The Hungarian Method

Consider a simple graph (X, Y, Γ) and a matching W, the arcs of which are represented by heavy lines (the other arcs being shown by light lines);

we assume that the matching W is *complete*: any arc which is not an element of W is adjacent to at least one element of W.

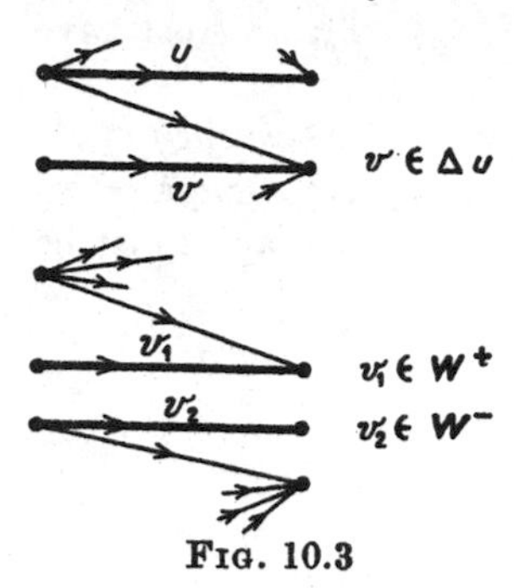

FIG. 10.3

A vertex z which is not a vertex of any arc in W is said to be *unsaturated*. If $w \in W$, we shall denote by Δw the set of arcs of W whose terminal vertices are adjacent to the initial vertex of w; we shall denote by W^+ the set of arcs of W whose terminal vertices are adjacent to an unsaturated point, and by W^- the set of arcs of W whose initial vertices are adjacent to an unsaturated point.

We can now state:

THEOREM 5. *The following conditions are equivalent:*

(1) *W is a maximum matching of the simple graph (X, Y, Γ);*

(2) *the simple graph (X, Y, Γ) contains no alternating chains (of links represented by light and heavy lines alternately) connecting two distinct unsaturated points;*

(3) *the graph (W, Δ) contains no path going from a point of W^+ to a point of W^-;*

(4) *in the graph (W, Δ), each vertex may be labelled with the sign $+$ or the sign $-$ in such a way that each point of W^+ is labelled $+$, each point of W^- is labelled $-$, and there is no oriented arc going from a $+$ to a $-$.*

We shall show that (1) $\Rightarrow$ (2) $\Rightarrow$ (3) $\Rightarrow$ (4) $\Rightarrow$ (1).

(1) $\Rightarrow$ (2), for if an alternating chain U_0 connecting two distinct unsaturated vertices existed, the matching $W' = (W - U_0) \cup (U_0 - W)$ would have one arc more than W, and therefore W would not be a maximum.

(2) $\Rightarrow$ (3), for any path of (W, Δ) going from W^+ to W^- corresponds to an alternating chain connecting two unsaturated points $x_0 \in X$ and $y_0 \in Y$.

(3) $\Rightarrow$ (4). Label all the vertices of $\hat{\Delta} W^+$ with a $+$, and all those of $W - \hat{\Delta} W^+$ with a $-$; $W^+ \subset \hat{\Delta} W^+$, and $W^- \subset W - \hat{\Delta} W^+$ (since no path exists going from W^+ to W^-); condition (4) is therefore satisfied.

(4) $\Rightarrow$ (1). In the graph (X, Y, Γ) with the matching W, label the arcs of W with a $+$ or a $-$ according to (4), define X^- to be the set of initial vertices of the arcs of W marked with a $-$, and Y^+ to be the set of terminal vertices of the arcs of W marked with a $+$.

If an arc $u_0 = (x_0, y_0)$ satisfies $x_0 \notin X^-$, $y_0 \notin Y^+$, we can imagine four configurations:

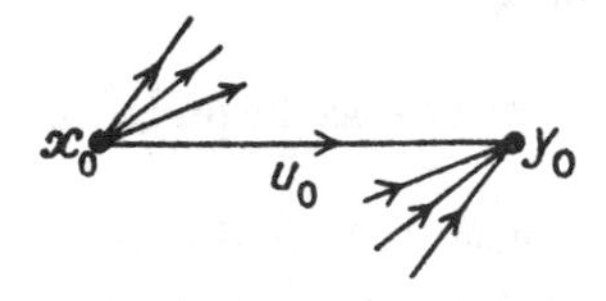

Case 1. x_0, y_0 unsaturated: impossible, since W is a complete matching.

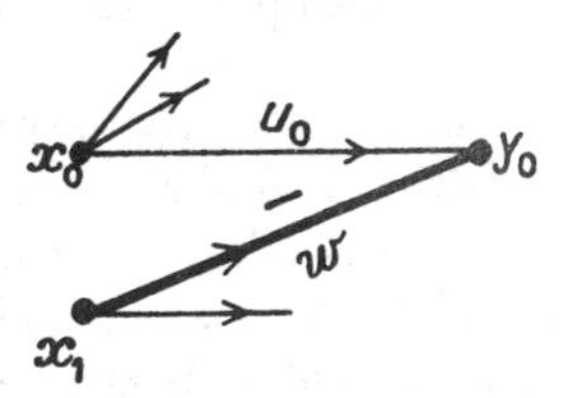

Case 2. x_0 unsaturated: impossible, because $w \in W^+$ and should therefore be labelled $+$.

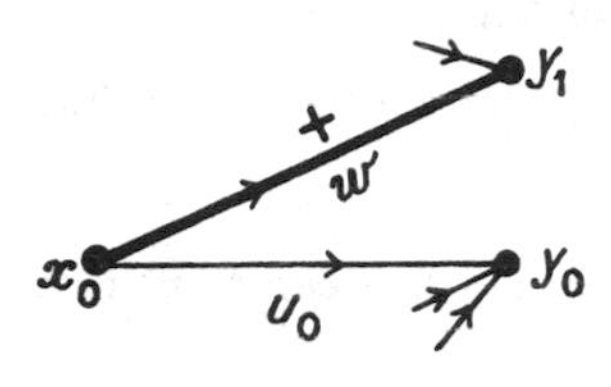

Case 3. y_0 unsaturated: impossible, because $w \in W^-$ and should therefore be labelled $-$.

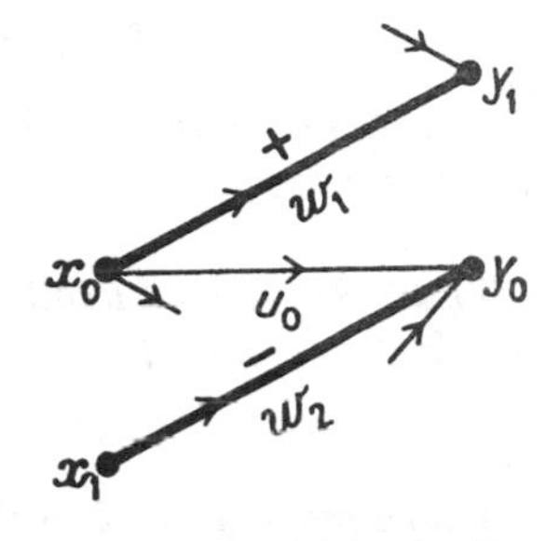

Case 4. x_0, y_0 saturated: impossible, because the graph (W, Δ) cannot contain an oriented arc going from w_1 (labelled $+$) to w_2 (labelled $-$).

FIG. 10.4

Since none of the four cases of Fig. 10.4 is possible, no arc (x_0, y_0) with $x_0 \notin X^-$ and $y_0 \notin Y^+$ exists, and the set $C = X^- \cup Y^+$ is therefore a support.

From König's theorem it is known that for any matching W' and any support C', $|W'| \leqslant |C'|$; since we have $|W| = |C|$ here, W is a maximum matching and C is a minimum support.

ALGORITHM FOR CONSTRUCTING A MAXIMUM MATCHING. We start by trying to match each vertex of X in turn into Y; as soon as no more vertices can be matched into Y we have a complete matching W.

Form the graph (W, Δ) and look for a path going from W^+ to W^- (using one of the methods of Chapter 7). If such a path exists, replace the matching

W by a matching W' which contains one element more than W, and start again.

If no such path exists, W is a maximum matching.

ALGORITHM FOR CONSTRUCTING A MINIMUM SUPPORT. We use the preceding algorithm to find a maximum matching.

In the graph (X, Y, Γ), label the arcs of $\hat{\Delta} W^+$ with a $+$, and those of $W - \hat{\Delta} W^+$ with a $-$; a minimum support C consists of the initial vertices of the arcs of W labelled $-$, and of the terminal vertices of the arcs of W labelled $+$.

It will be observed that by this construction we also get a maximum internally stable set (by taking the complement of C).

Extensions to the Infinite Case

It is possible to extend some of the preceding results to locally finite simple graphs.

THEOREM 6. *In a locally finite simple graph* (X, Y, Γ), *X can be matched into Y if and only if:*

$$|\Gamma A| \geqslant |A| \quad (A \subset X;\ |A| < \infty)$$

This may be proved in the same way as the theorem of König and Hall by using the results given for infinite transport networks in Chapter 8.

THEOREM 7. *In a locally finite simple graph* (X, Y, Γ), *the minimum number of vertices which must be eliminated from X in order to be able to match X into Y is:*

$$\delta_0 = \max_{\substack{A \subset X \\ |A| < \infty}} (|A| - |\Gamma A|)$$

This is proved in the same way as the theorem of König and Ore by considering the intersection A_0 of the family:

$$\mathscr{A} = \{A \mid |A| < \infty;\ \delta(A) = \delta_0\}$$

(note that A_0 is the intersection of a *finite number* of the sets of $\mathscr{A}$)†.

† It has been shown elsewhere that this statement can be extended to all transport networks (BERGE [1]); it then becomes: *if the graph of a transport network is regressively finite and Γ^{-1}-finite, the minimum total amount by which the demands must be reduced in order to be able to satisfy them all is:*

$$\delta_0 = \sup_{\substack{A \subset \Gamma^{-1} z \\ |A| < \infty}} (d(A) - F(A))$$

We shall now state an important theorem which was originally used in the theory of transfinite numbers and which has since been applied in various forms for various purposes.

BERNSTEIN'S THEOREM. *Given a simple graph $(X, \overline{Y}, \Gamma)$, finite or infinite, and a set $Y \subset \overline{Y}$, the necessary and sufficient condition for the existence of a matching W_0 which uses both the vertices of X and those of Y is that it should be possible on the one hand to match X into $\overline{Y}$, and on the other to match Y into X.*

We need to show that the condition is sufficient; consider a matching of X into $\overline{Y}$, such that to every $x \in X$ there corresponds a vertex $y = \sigma x$ of $\overline{Y}$. Consider a matching of Y into X such that to every $y \in Y$ there corresponds a vertex $x = \tau y$ of X. From these definitions, σ and τ are single-valued functions, as are their inverses σ^{-1} and τ^{-1}; since τ is only defined on Y, we agree to put $\tau y = \emptyset$ for $y \in \overline{Y} - Y$. Put

$$A = X - \tau Y$$

Form the family $\mathscr{A}$ of sets $A_i \subset X$ such that:

$$A_i \supset A \tag{1}$$

$$\tau(\sigma A_i) \subset A_i \tag{2}$$

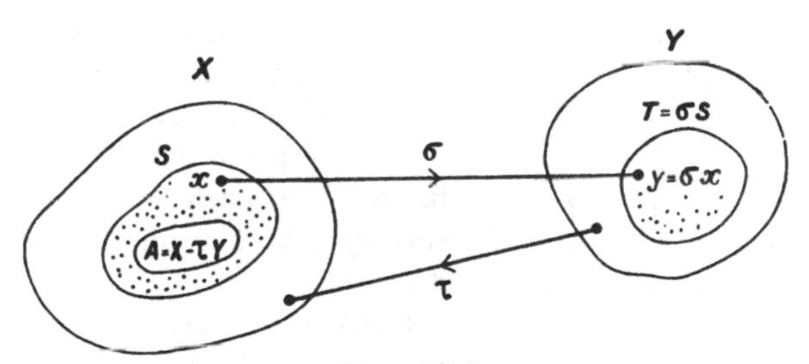

FIG. 10.5

This family $\mathscr{A}$ is non-empty since $X \in \mathscr{A}$; if we put $S = \bigcap_{i \in I} A_i$, we have:

$$\begin{cases} S \supset A \\ \tau(\sigma S) = \tau\sigma \bigcap_{i \in I} A_i \subset \bigcap_{i \in I} \tau\sigma A_i \subset \bigcap_{i \in I} A_i = S \end{cases}$$

Therefore $S \in \mathscr{A}$, as is $S_0 = A \cup \tau(\sigma S)$, since:

$$\begin{cases} S_0 \supset A \\ \tau(\sigma S_0) \subset \tau(\sigma S) \subset A \cup \tau\sigma S = S_0 \end{cases}$$

Therefore we have $S_0 \supset S$ (since $S_0 \in \mathscr{A}$), and on the other hand:

$$S_0 = A \cup \tau(\sigma S) \subset A \cup S = S$$

which gives finally:

$$A \cup \tau(\sigma S) = S$$

Consider now the set W_0 of arcs (x, y) such that:

$$y = \sigma x \quad \text{if } x \in S$$
$$y = \tau^{-1} x \quad \text{if } x \notin S$$

To show that this is a matching of X on to Y, it is sufficient to show that on putting $\sigma S = T$ we have:

$$\tau(Y - T) = X - S$$

But this follows immediately, for we can write:

$$\tau(Y - T) = \tau Y - A - \tau T = \tau Y - (A \cup \tau\sigma S) = \tau Y - S = X - S$$

Q.E.D.

EXAMPLE (*Plane geometry*). Consider a plane curve and let X and Y be its projections on to two rectangular axes. Let us assume that for all $x \in X$, there are two points (x, y_1) and (x, y_2) of the curve which project on to x, and put $\Gamma x = \{y_1, y_2\}$; likewise assume that for all $y \in Y$, there are two points of the curve which project on to y.

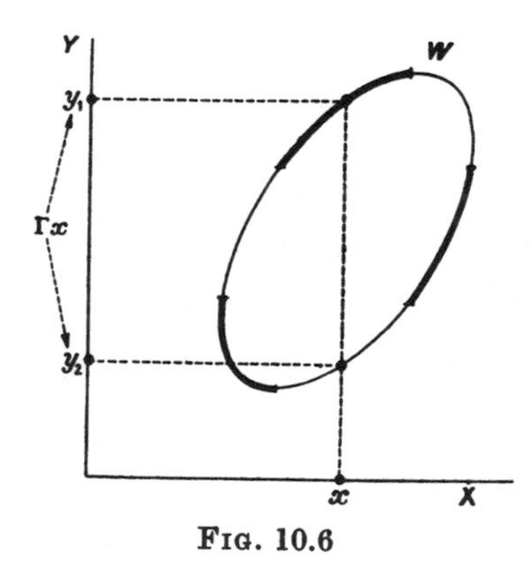

FIG. 10.6

The simple graph (X, Y, Γ) is locally finite, and each arc of this graph corresponds to a point on the curve of Fig. 10.6.

Further, we have:

$$|\Gamma x| = 2 \qquad (x \in X)$$
$$|\Gamma^{-1} y| = 2 \qquad (y \in Y)$$

It follows then from Theorem 6 that we can match X into Y, and Y into X. Therefore from Bernstein's theorem, a matching W exists which uses all the vertices of X and all those of Y; therefore there exists a part of the curve (which has been heavily outlined in the figure) which has the following property: a line drawn parallel to either of the axes meets this part of the curve at one point only. A sensitive existence theorem which is of particular value in logic, has thus been proved.

Application to the Theory of Matrices

A square matrix:

$$P = (p_j^i) = \begin{pmatrix} p_1^1 & p_2^1 & \dots & p_j^1 & \dots & p_n^1 \\ p_1^2 & p_2^2 & \dots & p_j^2 & \dots & p_n^2 \\ \dots & \dots & \dots & \dots & \dots & \dots \\ p_1^n & p_2^n & \dots & p_j^n & \dots & p_n^n \end{pmatrix}$$

is said to be *doubly stochastic* if:

$$p_j^i \geqslant 0 \qquad (i \leqslant n, j \leqslant n)$$

$$\sum_i p_j^i = 1 \qquad (j \leqslant n)$$

$$\sum_j p_j^i = 1 \qquad (i \leqslant n)$$

Doubly stochastic matrices play a large part in different generalizations of convex functions (cf. [E.T.F.M.], Chapter 8); G. Birkhoff [2] has shown that the principal properties of these matrices may be developed from the theorem of König and Hall.

THEOREM 9. *In the development of the determinant of a square matrix $A = (a_j^i)$ of order n, each monomial will be zero if and only if a submatrix of dimensions $p \times (n-p+1)$ (with $p \leqslant n$) which consists entirely of zeros, can be extracted from A.*

Put $N = \{1, 2, \dots, n\}$, and, for $i \in N$, put:

$$\Gamma i = \{j \mid a_j^i \neq 0\}$$

In this way a simple graph $(N, \bar{N}, \Gamma)$ is formed.

The mononomials in the development of the determinant are of the form:

$$\pm a_1^{i_1} a_2^{i_2} \dots a_n^{i_n}$$

and they will all be zero so long as no matching of N into $\bar{N}$ exists; from the theorem of König and Hall no such matching exists if and only if a set $I \subset N$ exists for which:

$$|I| > |\Gamma I|$$

In this case, we can then extract from A a submatrix of zeros of dimensions:

$$|I| \times (n - |I| + 1)$$

THEOREM 10. *If $P = (p_j^i)$ is a doubly stochastic matrix, at least one of the monomials in its development is non-null.*

Suppose, in fact, that this were not the case and that, in consequence, the matrix P can be divided into four submatrices $R = (r_j^i)$, $S = (s_j^i)$, $T = (t_j^i)$ and the null matrix 0, as in Fig. 10.7. We have:

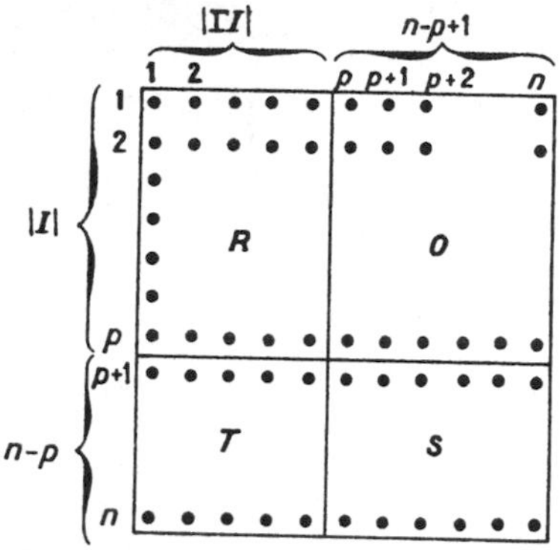

FIG. 10.7

$$\sum_i \sum_j s_j^i = n-p+1$$

Therefore:

$$\sum_i \sum_j t_j^i = (n-p)-(n-p+1) = -1$$

This is impossible since it contradicts the hypothesis that $t_j^i \geqslant 0$.

THEOREM 11 (BIRKHOFF and VON NEUMANN). *Every doubly stochastic matrix is the centre of gravity of some set of permutation matrices.*

What we must show is that if P is a doubly stochastic matrix, then there exist permutation matrices P_1, P_2, ..., P_k (having exactly one element equal to one in each row and column, all other elements being zero), and numbers $\lambda_1, \lambda_2, \ldots, \lambda_k$ such that:

$$\lambda_1, \lambda_2, \ldots, \lambda_k \geqslant 0$$

$$\lambda_1 + \lambda_2 + \ldots + \lambda_k = 1$$

$$\lambda_1 P_1 + \lambda_2 P_2 + \ldots + \lambda_k P_k = P$$

If not all the terms of the determinant of $P = (a_j^i)$ are zero, consider a term $a_1^{i_1} a_2^{i_2} \ldots a_n^{i_n} \neq 0$, put $\lambda_1 = \min\{a_1^{i_1}, a_2^{i_2} \ldots, a_n^{i_n}\}$, and denote by P_1 the permutation matrix $(i_1, i_2, \ldots, i_n)$; now put:

$$P = \lambda_1 P_1 + R$$

The matrix R contains more zero elements than P; if all the terms of the determinant of R are zero, we stop, and if not we decompose R in the same way; finally, we obtain:

$$P = \lambda_1 P_1 + \lambda_2 P_2 + \ldots + \lambda_k P_k + R$$

where all terms in the determinant of R are zero. It follows that $R = 0$ (for if this were not so, then for some $\lambda > 0$, the matrix λR would be doubly stochastic, and since all terms in its determinant are zero this would contradict Theorem 10). On the other hand:

$$\sum_{i=1}^{n} a_j^i = 1 = \lambda_1 + \lambda_2 + \ldots + \lambda_k$$

This completes the proof of the theorem.

This proof illustrates the value of the tool provided by the theory of graphs; the direct proof of the theorem of Birkhoff and von Neumann is very much longer.

11. Factors

Hamiltonian Paths and Circuits

In a graph (X, Γ), a path $\mu = [x_1, x_2, \ldots, x_n]$ is said to be *Hamiltonian* if it passes once and only once through every vertex of the graph; likewise a circuit $\mu = [x_0, x_1, \ldots, x_n]$ is said to be *Hamiltonian* if it passes once and only once through every vertex of the graph (except that $x_0 = x_n$).

It is often important to know if a given graph possesses a Hamiltonian path or circuit.

EXAMPLE 1. *Round tour of the world* (HAMILTON). Let us select 20 towns: $a, b, c, \ldots, t$, on the terrestrial globe; for simplicity, we assume that these towns are at the vertices of a regular dodecahedron (a polyhedron with 12 pentagonal faces and 20 vertices) which represents the earth. We propose to visit each of these towns once and once only and to return to our point of departure, using only the edges of the dodecahedron. This problem is the same as finding a Hamiltonian circuit for the symmetric graph of Fig. 11.1.

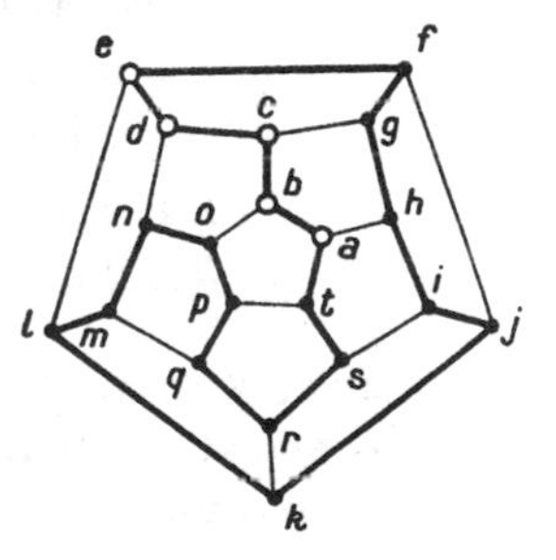

FIG. 11.1

Hamilton's own method is worthy of note; when the traveller reaches the terminal vertex of an edge he has to choose between: taking the edge on his right, an operation which we shall denote by R; taking the edge on his left, an operation which we shall denote by L; or staying where he is, an operation which we shall denote by 1. Next, we define a *product* of these operations: $R\,L$ is the operation consisting of going to the right and then to the left; $L\,L\,R$ or $L^2\,R$ is the operation of going twice to the left and then once to the right, etc.... Finally, two operations are *equal* if, starting from the same point of departure, they have the same terminus. The product is not commutative (for example: $R\,L \neq L\,R$) but it is associative [for example: $(L\,L)\,R = L(L\,R)$].

The following results hold:

$$R^5 = L^5 = 1$$
$$R\,L^2\,R = L\,R\,L$$
$$L\,R^2\,L = R\,L\,R$$
$$R\,L^3\,R = L^2$$
$$L\,R^3\,L = R^2$$

We can write:

$$\begin{aligned} 1 &= R^5 = R^2\,R^3 = (L\,R^3\,L)\,R^3 = (L\,R^3)^2 \\ &= [L(L\,R^3\,L)\,R]^2 = (L^2\,R^3\,L\,R)^2 \\ &= [L^2(L\,R^3\,L)\,R\,L\,R]^2 = [L^3\,R^3\,L\,R\,L\,R]^2 \\ &= L\,L\,L\,R\,R\,R\,L\,R\,L\,R\,L\,L\,L\,R\,R\,R\,L\,R\,L\,R \end{aligned}$$

This sequence contains 20 letters, and we cannot isolate from it any partial sequence which is also equal to 1: therefore it represents a Hamiltonian circuit. Another circuit is found by reading the sequence from right to left, and a simple argument may be used to show that no others exist.

It should be noted, however, that the journey can start from any one of the points on the round trip shown. In the solutions given here, every tour is assumed to start with the edge $[a, b]$. Hamilton, who had fixed in advance the first five towns of the voyage, a, b, c, d, e for example, kept only the voyages which began with $R\,L\,R$, and found four solutions:

RLRLRLLLRRRLRLRLLLRR
RLRLLLRRRLRLRLLLRRRL
RLRLRRRLLLRLRLRRRLLL
RLRRRLLLRLRLRRRLLLRL

EXAMPLE 2. *Tour of the world.* If, in the preceding problem, we do not impose the condition that one must return to one's point of departure, the number of possible voyages is very much greater: we then have to find all the Hamiltonian paths of the graph of Fig. 11.1.

We shall only record here the Hamiltonian paths which begin with R, since all the others can be deduced from these by interchanging the letters R and L.

RRRLRLRLLLRRRLRLRL
RLRLLLRRRLRLRLLLRR *RLRRRLLLRLRRRLLLRL*
RLLLRRRLRLRLLLRRRL *RLRLLLRRRLRLLLRRLRL*
RRRLLLRLRLRRRLLLRL
RLRRRLLLRLRLRRRLLL
RLRLRRRLLLRLRLRRRL *RRRLRRLRLRLRLLRLLL*

RRLRLRLLLRRRLRLRLL	*RRRLLRLRRRLLLRLRLR*
RLRLRLLLRRRLRLRLLL	*RRRLRRLRLRLLLRRRLR*
RLLLRLRLRRRLLLRLRL	
RRLLLRLRLRRRLLLRLR	*RLRLRLLLRRRLRLLRRR*
	RLLLRLRLLRRLRLRRRL
RRRLLLRLRLRLRRLRRR	
RLRRRLLLRLRLRRLRRR	*RRRLRRLRLRLRLLLRRR*
RLRLLLRRLRLLRRRLRL	*RLRLRRRLLLRRLRLLLR*
RLLLRLRRLLLRRRLRLR	
	RRRLLLRLRLRLRRRLLL
RLLLRLRLLRRRLRLLLR	*RLLLRLRRRLLRLRLLLR*
RLLLRLRRRLLLRLRRRL	

(sequences corresponding to equal operations have been grouped together).

EXAMPLE 3. *The knight's tour* (EULER). The problem consists of moving a knight on a chessboard in such a way that he goes once and once only through every square on the board. This is equivalent to finding a Hamiltonian path of a symmetric graph, and the problem has aroused the interest of many mathematicians, notably Euler, de Moivre, Vandermonde, etc. Innumerable methods have been used, and we cannot pretend to give all the solutions here. One rule which has been found to work well in practice, but has not yet been justified by theory, is: at each stage, move the knight to the position where he dominates the smallest number of squares which have not yet been visited.

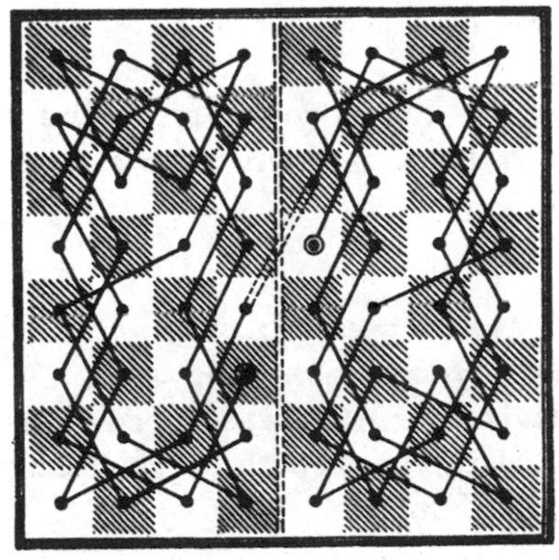

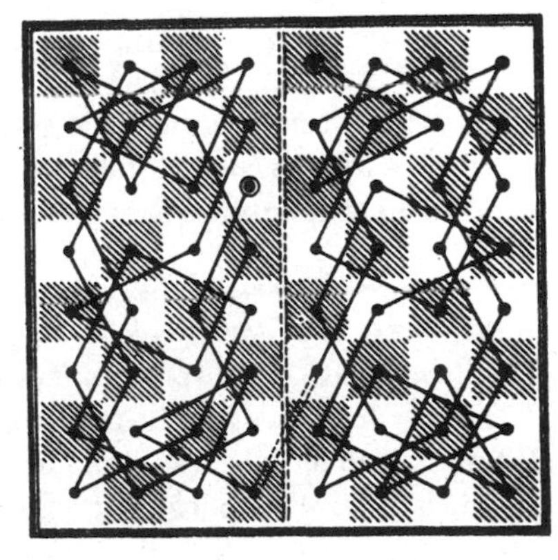

FIG. 11.2

Another method consists of finding a path contained in half a chessboard, and doubling it by symmetry to yield a solution (Fig. 11.2). This method depends on the special nature of the chessboard and cannot be extended to a more general graph. The reader intrigued by these solutions can find more about them in [1] and [6].

The key to many scheduling problems of operational research is the determination of a Hamiltonian path: more precisely, if we are trying to find the optimum order in which a certain number of operations should be carried out, we represent these operations by points $x_1, x_2, \ldots, x_n$, and join x_j to x_i by an arc (x_i, x_j) whenever no loss results from carrying out x_i before x_j; the required scheduling will generally be given by one of the Hamiltonian paths of the graph.

EXAMPLE. *The bookbinding problem* (S. JOHNSON). A printer with n books to publish, has two machines at his disposal: one for printing and the other for binding. Let a_k be the time required to print the k-th book, and b_k the time required to bind it. Naturally, a book must be printed before it is bound, and consequently time will be lost whenever the bookbinding machine runs out of material. What order of printing should be followed in order to finish as soon as possible?

It is easily seen that nothing is lost if the books are handled in the same order on each machine; therefore an optimum order will be defined by a permutation $(i_1, i_2, \ldots, i_n)$, or by a sequence of operations: $x_{i_1}, x_{i_2}, \ldots, x_{i_n}$. On the other hand, it can be shown that in a certain optimum order, book (i) should precede book (j) whenever:

$$\min\{a_i, b_j\} \leqslant \min\{a_j, b_i\}$$

Each time this inequality is satisfied, the corresponding arc (x_i, x_j) is drawn; if a unique Hamiltonian path can be found in the graph so obtained, it indicates the required optimum order.

THEOREM 1. *If the graph is complete (that is, if every pair of vertices is joined in at least one direction), it contains a Hamiltonian path.*

Let $\mu = [a_1, a_2, \ldots, a_p]$ be a path of length $p-1$, all of whose vertices are different, and let x be a vertex not yet included in the path; we shall show that it is possible to construct a path of the form

$$\mu_k = [a_1, a_2, \ldots, a_k, x, a_{k+1}, \ldots, a_p]$$

Suppose, in fact, that this cannot be done: then there exists no integer k between 1 and p such that:

$$(a_k, x) \in U \qquad (x, a_{k+1}) \in U$$

Therefore we have, for $1 \leqslant k \leqslant p$,

$$(a_k, x) \in U \quad \Rightarrow \quad (x, a_{k+1}) \notin U \quad \Rightarrow \quad (a_{k+1}, x) \in U$$

If the path $\mu_0 = [x, a_1, a_2, \ldots, a_p]$ does not exist, we must have $(a_1, x) \in U$, and therefore $(a_p, x) \in U$ and the path $\mu_p = [a_1, a_2, \ldots, a_p, x]$ exists; the supposition that x cannot be included in the path has thus been contradicted.

In this way, a path using all the vertices of the graph can gradually be built up†.

We deduce from this result that the competitors in a tournament can always be arranged in such a way that each one beats the player immediately following him.

To investigate the existence of a Hamiltonian path in more general cases, we make use of the theory of transport networks, as will be shown in the following paragraph.

Factors and Methods for Determining Them

A *factor* of a graph (X, Γ) is a partial graph (X, B) in which the inward and outward demi-degree of every vertex is equal to 1; a Hamiltonian circuit is a factor but the converse is not necessarily true: some factors consist of several disjoint circuits.

THEOREM 2. *The necessary and sufficient condition for the graph (X, Γ) to possess a factor is that $|S| \leqslant |\Gamma S|$ for all $S \subset X$.*

This is, in fact, König and Hall's theorem for matching X on to $\bar{X}$ in the simple graph $(X, \bar{X}, \Gamma)$.

COROLLARY. *If $|\Gamma x| = |\Gamma^{-1} x| = m$ for every vertex x of a graph (X, Γ), the arcs can be partitioned into m disjoint sets W_1, W_2, …, W_m, each of which constitutes a factor.*

In the simple graph $(X, \bar{X}, \Gamma)$, we have $|\Gamma x| = m$, $|\Gamma^{-1} \bar{x}| = m$; from the corollary to König and Hall's theorem, we can therefore match X into $\bar{X}$,

† This result is related to the following theorem which is due to DIRAC [2]; *If (X, Γ) is a connected, symmetric graph with no loops, and if $|\Gamma x| \geqslant \frac{1}{2}|X|$ for all $x \in X$, the graph possesses a Hamiltonian path.* In addition, if a few minor changes are made in the proof of Theorem 1, we can obtain another interesting result: if a graph is complete and strongly connected, it contains a Hamiltonian circuit.

thus determining W_1; in the remaining partial graph, we can, by the same argument, match X into $\overline{X}$, thus determining W_2, etc...†.

It is quite easy to find a factor, since this is a maximum flow problem; one method for finding a Hamiltonian circuit consists of finding all the factors of the graph, and retaining only those consisting of a single circuit. More often, we shall try to 'fuse' the different circuits of a factor.

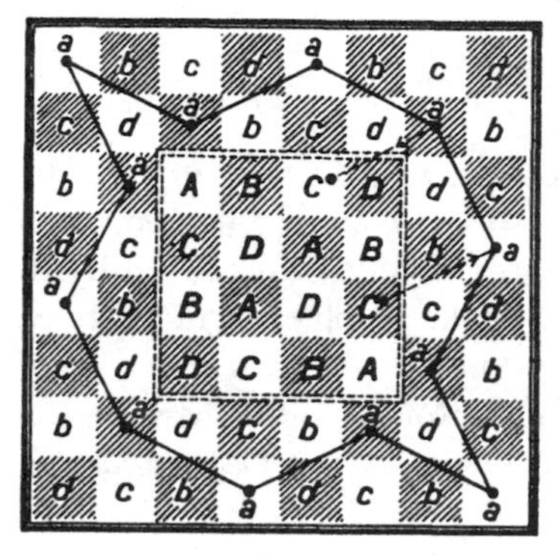

FIG. 11.3

EXAMPLE. Consider again the problem of the knight's tour of the chessboard in which we require a route which is a Hamiltonian circuit. Because of the symmetry of the chessboard, the factor of Fig. 11.3 (in which the squares marked 'a' form one circuit, those marked 'b' another, etc...) can be determined directly. Two circuits can be fused whenever two consecutive vertices of the first are respectively adjacent to two consecutive vertices of the second. Thus, the circuits 'a' and 'C' can be fused (by means of the arcs shown by dotted lines in the figure), likewise the circuits 'a' and 'B', etc.... The auxiliary graph of Fig. 11.4 shows the various possible fusions, the links being $[a,C]$, $[a,B]$, etc...; since an obvious Hamiltonian path of this auxiliary graph is $aDbCdAcB$, the different circuits are successively fused in this order, and we see that we must obtain the required Hamiltonian circuit.

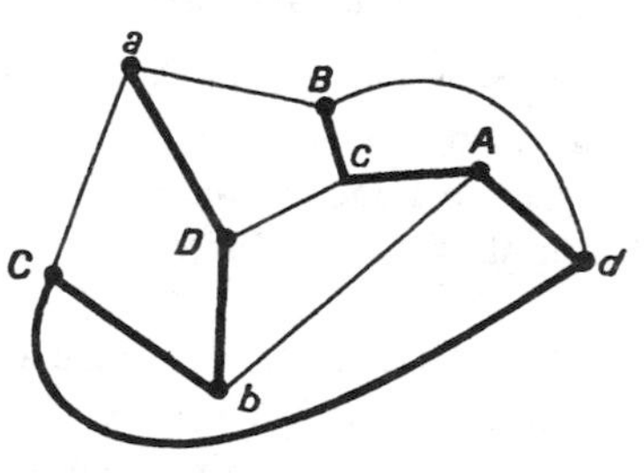

FIG. 11.4

The preceding results may be extended in a remarkable way; for a given graph (X,Γ), we define a *dissection* to be a collection of paths such that:

1. two paths of the dissection have no vertices in common;
2. every vertex of the graph is on one of the paths of the dissection.

Amongst the different paths of the dissection, we distinguish those which are elementary circuits, which we write $\alpha^i = [a_0^i, a_1^i, \ldots, a_p^i = a_0^i]$; those which are elementary paths (of length $\neq 0$), which we write: $\beta^j = [b_0^j, b_1^j, \ldots, b_q^j]$; and those which are paths of length 0 are written $\gamma^k = \{c^k\}$. Let A^i be the

† The existence of a factor in the problem of the 'tour round the world' follows from this corollary.

set of vertices of the circuit α^i, B^j be the set of vertices of the path β^j, and C^k be the set $\{c^k\}$. Finally, we shall put:

$$n(\alpha^i) = 0$$
$$n(\beta^j) = 1$$
$$n(\gamma^k) = 1$$

By definition, the *value* of a dissection

$$\mathscr{A} = (\alpha^1, \alpha^2, \ldots; \beta^1, \beta^2, \ldots; \gamma^1, \gamma^2, \ldots)$$

is:

$$n(\mathscr{A}) = \sum_i n(\alpha^i) + \sum_j n(\beta^j) + \sum_k n(\gamma^k)$$

A factor is a dissection of value 0. We now intend to find the minimum value of the dissections of a given graph.

THEOREM 3. *In a graph* (X, Γ), *the minimum value of all possible dissections is equal to the deficiency of the graph, and is given by:*

$$\delta_0 = \max_{S \subset X} (|S| - |\Gamma S|)$$

Consider a dissection $\mathscr{A} = (\alpha^1, \alpha^2, \ldots;\ \beta^1, \beta^2, \ldots;\ \gamma^1, \gamma^2, \ldots)$. Write the set X down twice (X and $\bar{X}$) and form the simple graph $(X, \bar{X}, \Gamma)$, where $\bar{x}_j \in \Gamma x_i$ if and only if $x_j \in \Gamma x_i$.

In this simple graph a matching of $|A^i|$ arcs corresponds to a circuit α^i of the dissection $\mathscr{A}$; similarly, a matching of $|B^j| - 1$ arcs corresponds to a path β^j. All these arcs taken together form a matching W in which the number of arcs is:

$$|W| = \sum_i |A^i| + \sum_j (|B^j| - 1)$$

We have:

$$\begin{aligned} n = |X| &= \sum_i |A^i| + \sum_j |B^j| + \sum_k |C^k| \\ &= \sum_i |A^i| + \sum_j (|B^j| - 1) + n(\mathscr{A}) \\ &= |W| + n(\mathscr{A}) \end{aligned}$$

There is a one–one correspondence between matchings and dissections, and therefore, from Theorem 4 (Chapter 10), we have:

$$\min n(\mathscr{A}) = n - \max |W| = n - (n - \delta_0) = \delta_0$$

COROLLARY 1. *If a Hamiltonian circuit exists, we have* $\delta_0 = 0$.

COROLLARY 2. *If a Hamiltonian path exists, we have* $0 \leqslant \delta_0 \leqslant 1$.

COROLLARY 3. *If the graph* (X, Γ) *contains no circuits, its deficiency* δ_0 *is equal to the smallest number of elementary disjoint paths in the graph (a single vertex here being considered as a path of length* 0).

COROLLARY 4. *If the graph contains no circuits, the necessary and sufficient condition for the existence of a Hamiltonian path is that* $\delta_0 = 1$.

All these corollaries follow immediately from the main theorem.

EXAMPLE. In the graph of Fig. 11.5 we shall determine a Hamiltonian path (using three different methods).

1. Since the graph (X, Γ) is complete, a Hamiltonian path certainly exists (Theorem 1); it may be found by following the method of Theorem 1: for example, having found the path *eacb*, we incorporate the vertex *d*, giving the solution *eadcb*.

2. The graph (X, Γ) clearly contains no circuits (because of the point *b*), and its deficiency is:

$$\delta_0 = \max_{S \subset X} (|S| - |\Gamma S|) = |\{b\}| - |\Gamma b| = 1$$

Therefore from Corollary 4, a Hamiltonian path exists, and this may be determined by finding a maximum matching for the graph $(X, \overline{X}, \Gamma)$ of Fig. 11.6, giving the solution: *eadcb*.

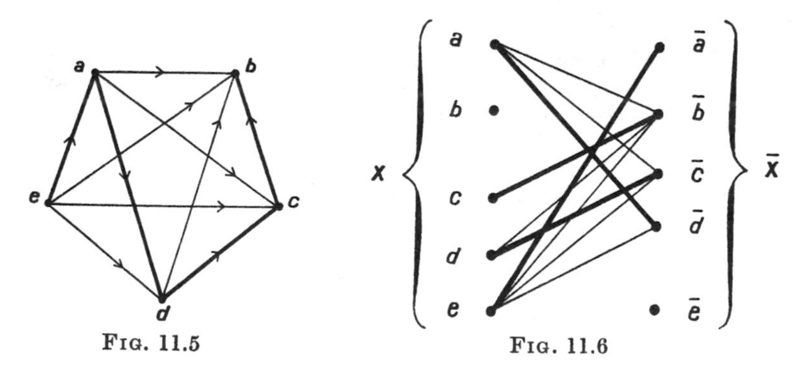

FIG. 11.5 FIG. 11.6

3. The graph (X, Γ) is transitive, that is:

$$z \in \Gamma y, y \in \Gamma x \quad \Rightarrow \quad z \in \Gamma x$$

Therefore the outward demi-degrees of the successive vertices of a Hamiltonian path form a decreasing sequence. These demi-degrees are:

$$|\Gamma a| = 3,\ |\Gamma b| = 0,\ |\Gamma c| = 1,\ |\Gamma d| = 2,\ |\Gamma e| = 4$$

This gives the path *eadcb*.

ALGORITHM FOR FINDING A HAMILTONIAN CIRCUIT (R. FORTET). The only general way of determining Hamiltonian circuits is to enumerate *all* the factors of the graph; recently, B. Roy† suggested a graph method for the solution of this problem. The method given here, which is based on Boolean equations, was formulated by R. Fortet‡, and has the advantage of being in a form suitable for machine use.

Let us consider the graph of Fig. 11.7:

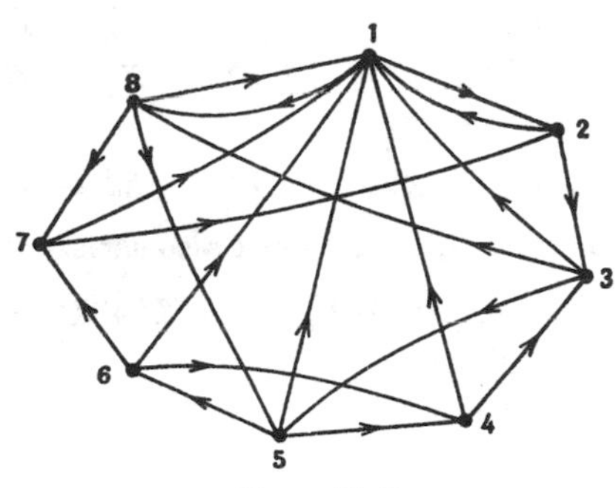

FIG. 11.7

We define a set of variables X_j^i which are associated with the required Hamiltonian circuit μ as follows:

$$X_j^i = \begin{cases} 1 & \text{if the circuit uses the arc } (i,j) \\ 0 & \text{otherwise.} \end{cases}$$

Since $\sum_j X_j^i = 1$ for all i, these variables must satisfy the system (1):

$$(1)\quad \begin{cases} X_2^1 + X_8^1 = 1 \\ X_1^2 + X_3^2 = 1 \\ X_1^3 + X_5^3 + X_8^3 = 1 \\ X_1^4 + X_3^4 = 1 \\ X_1^5 + X_4^5 + X_6^5 = 1 \\ X_1^6 + X_4^6 + X_7^6 = 1 \\ X_1^7 + X_2^7 = 1 \\ X_1^8 + X_5^8 + X_7^8 = 1 \end{cases}$$

† B. ROY, Recherche des circuits élémentaires et des circuits hamiltoniens dans un graphe quelconque, miméographe, *Société de Mathématiques Appliquées*, 1959.

‡ R. FORTET, L'algèbre de Boole et ses applications en recherche opérationnelle, miméographe, *Société de Mathématiques Appliquées*, 1959.

We now solve for the X_j^i in terms of certain auxiliary Boolean variables Y^i, Z^i, ...; these latter variables will be regarded as independent until otherwise stated.

$$(1')\quad \begin{cases} X_2^1 = Y^1 & X_8^1 = Y^{1'} & \\ X_1^2 = Y^2 & X_3^2 = Y^{2'} & \\ X_1^3 = Y^3 & X_5^3 = Y^{3'}Z^3 & X_8^3 = Y^{3'}Z^{3'} \\ X_1^4 = Y^4 & X_3^4 = Y^{4'} & \\ X_1^5 = Y^5 & X_4^5 = Y^{5'}Z^5 & X_6^5 = Y^{5'}Z^{5'} \\ X_1^6 = Y^6 & X_4^6 = Y^{6'}Z^6 & X_7^6 = Y^{6'}Z^{6'} \\ X_1^7 = Y^7 & X_2^7 = Y^{7'} & \\ X_1^8 = Y^8 & X_5^8 = Y^{8'}Z^8 & X_7^8 = Y^{8'}Z^{8'} \end{cases}$$

Since $\sum_i X_j^i = 1$ for every j, the X_j^i must also satisfy the system (2):

$$(2)\quad \begin{cases} X_1^2+X_1^3+X_1^4+X_1^5+X_1^6+X_1^7+X_1^8 = 1 \\ X_2^1+X_2^7 = 1 \\ X_3^2+X_3^4 = 1 \\ X_4^5+X_4^6 = 1 \\ X_5^3+X_5^8 = 1 \\ X_6^5 = 1 \\ X_7^6+X_7^8 = 1 \\ X_8^1+X_8^3 = 1 \end{cases}$$

If we substitute the solutions given in (1′) in (2), we obtain immediately:

$$Y^1 = Y^3 = Y^5 = Y^6 = Y^7 = Y^8 = 0$$
$$Z^5 = Z^8 = 0$$
$$Z^3 = Z^6 = 1$$
$$Y^4 = Y^{2'}$$

The remaining variables can all be expressed in terms of one independent variable, say Y^2. The system (1′) now becomes:

$$(1'')\quad \begin{cases} X_2^1 = 0 & X_8^1 = 1 & \\ X_1^2 = Y^2 & X_3^2 = Y^{2'} & \\ X_1^3 = 0 & X_5^3 = 1 & X_8^3 = 0 \\ X_1^4 = Y^{2'} & X_3^4 = Y^2 & \\ X_1^5 = 0 & X_4^5 = 0 & X_6^5 = 1 \\ X_1^6 = 0 & X_4^6 = 1 & X_7^6 = 0 \\ X_1^7 = 0 & X_2^7 = 1 & \\ X_1^8 = 0 & X_5^8 = 0 & X_7^8 = 1 \end{cases}$$

If we put $Y^2 = 0$, we obtain the Hamiltonian circuit [1,8,7,2,3,5,6,4,1]. If we put $Y^2 = 1$, we obtain a factor which consists of the two disjoint circuits [1,8,7,2,1] and [3,5,6,4,3].

The Problem of Finding a Partial Graph with Given Demi-Degrees

The following problem is an immediate generalization of that of finding a factor: given a graph (X, Γ), construct a partial graph with given demi-degrees. If $A \subset X$, $B \subset X$, we denote by $m(A, B)$ the number of arcs which originate from A and terminate in B.

THEOREM 4. *Given a graph (X, Γ), and integers $r(x)$ and $s(x)$ (for $x \in X$), the necessary and sufficient condition for the existence of a partial graph (X, Δ) with $|\Delta x| = r(x)$, $|\Delta^{-1} x| = s(x)$, for all $x \in X$, is that:*

(1) $$\sum_{x \in X} r(x) = \sum_{x \in X} s(x)$$

(2) $$\sum_{x \in X} \min\{r(x), |\Gamma x \cap S|\} \geqslant \sum_{x \in S} s(x) \qquad (S \subset X)$$

Condition (2) requires that a flow should exist which saturates all the terminal arcs of the transport network of Fig. 11.8, that is to say: $F(S) \geqslant d(S)$ for all $S \subset X$.

By putting $r(x) = s(x) = 1$, we regain the existence condition for a factor; by putting $\Gamma x = X$, we regain the theorem of the demi-degrees.

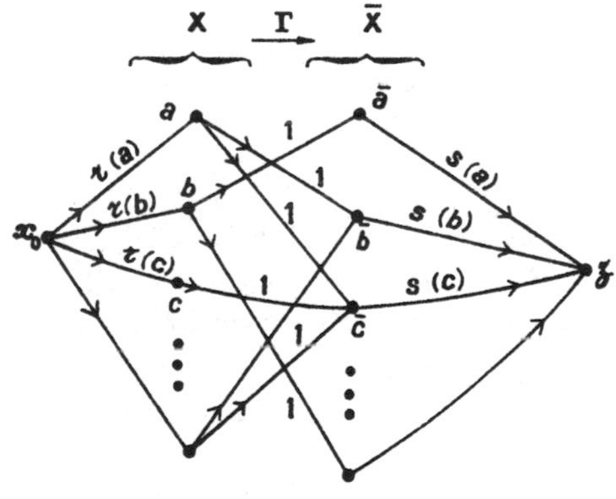

FIG. 11.8

GENERALIZATION 1. *Given two partitions $(A_1, A_2, \ldots, A_p)$ and $(B_1, B_2, \ldots, B_q)$ of the vertices of a graph (X, Γ), and the integers $r_1, r_2, \ldots, r_p$, $s_1, s_2, \ldots, s_q$, a partial graph (X, Δ) with*

$$\sum_{x \in A_k} |\Delta x| = r_k \qquad \sum_{x \in B_h} |\Delta^{-1} x| = s_h$$

$$(k \leqslant p, h \leqslant q)$$

exists if and only if:

(1) $$\sum_{k=1}^{p} r_k = \sum_{h=1}^{q} s_h$$

(2) $m(X-A,B) \geqslant \sum_{h \in J} s_h - \sum_{k \in I} r_k$

$$\left(I \subset \{1,2,\ldots,p\}; J \subset \{1,2,\ldots,q\}; A \subset \bigcup_{i \in I} A_i; B \supset \bigcup_{j \in J} B_j\right)$$

In the transport network of Fig. 11.9, a flow saturating the terminal arcs exists if, for any set $S = I \cup A \cup B \cup J$, with $I \subset \{1,2,\ldots,p\}, J \subset \{1,2,\ldots,q\}$, $A \subset X$, $B \subset \bar{X}$, we have

$$c(U_S^-) \geqslant d(S)$$

This is precisely condition (2).

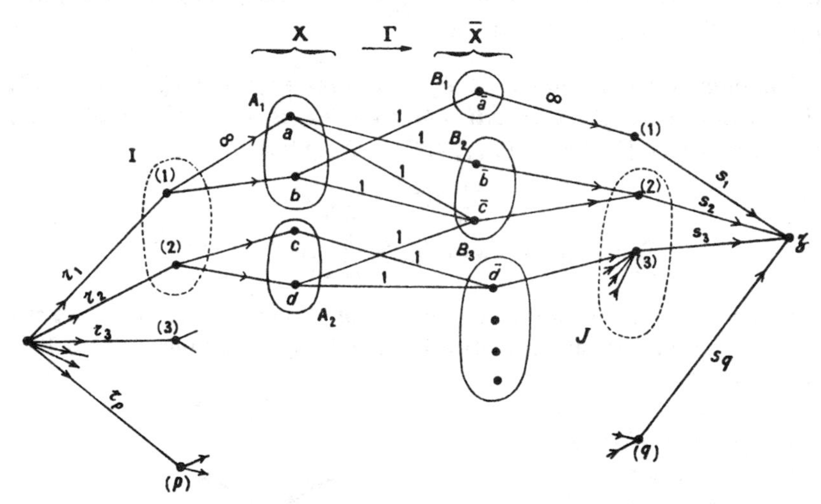

Fig. 11.9

Generalization 2. *With every vertex x of a graph (X, Γ), let there be associated integers $r(x)$, $r'(x)$, $s(x)$, $s'(x)$, with $0 \leqslant r(x) \leqslant r'(x)$, $0 \leqslant s(x) \leqslant s'(x)$; a necessary and sufficient condition for the existence of a partial graph (X, Δ), with $r(x) \leqslant |\Delta x| \leqslant r'(x)$, $s(x) \leqslant |\Delta^{-1}x| \leqslant s'(x)$, is that:*

(1) $$m(X-A,B) + \sum_{a \in A} r'(a) \geqslant \sum_{b \in B} s(b)$$

(2) $$m(X-A,B) + \sum_{y \notin B} s'(y) \geqslant \sum_{x \notin A} r(x)$$

We need to discover if a flow exists for the network of Fig. 11.8, the capacities being replaced by the allowed sets $[r(x), r'(x)]$, $[0,1]$, $[s(x), s'(x)]$. By using Theorem 2 (Chapter 8):

$$c(U_S^-) \geqslant b(U_S^+) \qquad (S \in \mathscr{A})$$

By taking the sets $S = A \cup B$, with $A \subset X$, $B \subset \bar{X}$, we get condition (1); by taking the sets $S = A \cup B \cup \{x_0\} \cup \{z\}$, we get condition (2).

12. Centres of a Graph

Centres

Let (X, Γ) be a graph (finite or otherwise); given two vertices x and y, the *(directed) distance* $d(x,y)$ from x to y is defined to be the length of the shortest path going from x to y.

If $x = y$, then by convention we put $d(x,x) = 0$; if $y \notin \hat{\Gamma}x$, we put

$$d(x,y) = \infty$$

THEOREM 1. *$d(x,y)$ satisfies*

$$d(x,x) = 0 \tag{1}$$

$$d(x,y)+d(y,z) \geqslant d(x,z) \tag{2}$$

In addition, if the graph is symmetric, we have

$$d(x,y) = d(y,x) \tag{3}$$

The proof is immediate; in the case of a symmetric graph, the function $d(x,y)$ which satisfies (1), (2) and (3), is then a *metric* in the topological sense of the word.

The *associated number* of a vertex x is the number

$$\dot{e}(x) = \max_{y \in X} d(x,y)$$

If the minimum associated number taken over all $x \in X$ is a finite number, any point having this value as associated number is called a *centre* of the graph; a point for which the associated number attains its maximum value is called a *peripheral point* of the graph; a graph may have several centres, or it may have none at all.

EXAMPLE. Fig. 12.1 shows the first network for communication by means of carrier pigeons (1870) in the form of a graph.

In this case a centre is a town which can communicate with the others without having to make a lot of retransmissions (and therefore taking a minimum length of time); Paris and Lyons are centres (with two as associated number), Grenoble and Nice are peripheral points (with infinity as associated number).

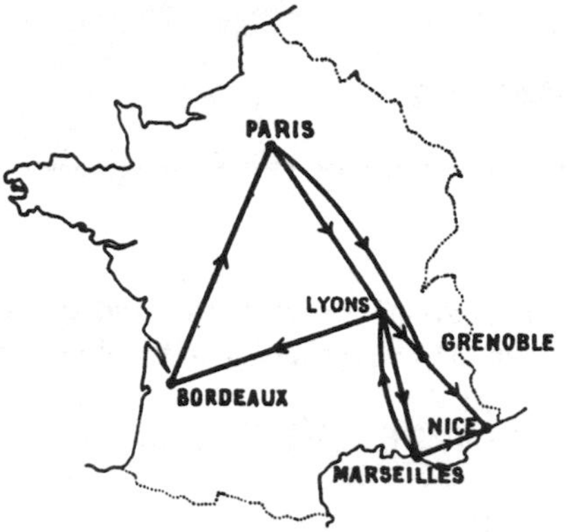

Fig. 12.1

Under what conditions does a graph possess a centre? A graph is said to be *demi-total* (upward) if for every pair x, y of vertices, there exist:

(1) a vertex z,

(2) a path (possibly of length zero) going from z to x,

(3) a path (possibly of length zero) going from z to y.

According to the definition of Chapter 2, a graph is *total* if every pair of vertices is connected by a path in at least one direction; hence a total graph is also demi-total. We now have:

Theorem 2 (B. Roy). *A graph G possesses a centre if and only if it is demi-total.*

It is obvious that if G possesses a centre x_0, and if x, y are two vertices of G, there exists a path going from x_0 to x, and from x_0 to y. Hence the graph is demi-total.

Conversely, if G is demi-total, let us enumerate its vertices $x_1, x_2, \ldots, x_n$. We define z_2 to be a vertex from which it is possible to go to both x_1 and x_2. Next, we define z_3 to be a vertex from which it is possible to go to both z_2 and x_3, etc. Then it must be possible to go to every vertex of the graph from z_n. Therefore G possesses a centre.

It follows from this theorem that *if a graph is total, it possesses a centre.* Clearly, also, if every pair of vertices is connected by an arc in at least one direction, that is, the graph is *complete*, then it possesses a centre. We shall also quote the result that a graph (X, Γ) possesses a centre if and only if any basis of the inverse graph (X, Γ^{-1}) has only one element (Chapter 2). Thus the problem of finding a centre is closely related to that of finding a basis.

Radius

If a graph has a centre x_0, we define the *radius* of the graph to be the associated number $e(x_0)$ of the vertex x_0; the formula for the radius is therefore:

$$\rho = \min_{x \in X} e(x)$$

Should the graph have no centre, we put $\rho = \infty$.

Theorem 3. *If, in a finite graph (X, Γ), we put $|X| = n$, $\max |\Gamma x| = p$, and if $1 < p < \infty$, the radius of the graph satisfies*

$$\rho \geqslant \frac{\log(np-n+1)}{\log p} - 1$$

If $\rho = \infty$, the theorem is obvious; let us assume therefore that $\rho < \infty$, which implies that the graph possesses a centre x_0. The number of vertices which are a distance 1 from x_0 is less than or equal to p; the number of vertices distant 2 units from x_0 is less than or equal to p^2, etc.... Therefore we have

$$n \leqslant 1+p+p^2+\ldots+p^\rho = \frac{p^{\rho+1}-1}{p-1}$$

which gives

$$n(p-1)+1 \leqslant p^{\rho+1}$$

or:

$$\log(np-n+1) \leqslant (\rho+1)\log p$$

which is the given formula. Clearly, this is the best result possible.

Theorem 4. *If a finite graph (X, Γ) is complete (that is, every pair of vertices is connected by an arc in at least one direction), its radius ρ cannot exceed 2; in addition, any vertex x_0 such that*

$$|\Gamma x_0 - \{x_0\}| = \max_{x \in X} |\Gamma x - \{x\}|$$

is a centre.

If $\rho = 1$, the theorem is obvious.

If $\rho > 1$, let us consider a vertex x_0 for which $|\Gamma x_0 - \{x_0\}|$ is a maximum (since the graph is Γ-bounded, such a vertex must exist). If we can prove that $e(x_0) = 2$, then since $e(x) \geqslant 2$ for all x, we shall have proved both that x_0 is a centre of the graph and that $\rho = 2$.

Let us assume $e(x_0) > 2$, and show that this leads to an absurdity: then a vertex $y \neq x_0$ exists which cannot be reached from x_0 either by a path of length 1 or 2. Since $y \notin \Gamma x_0$, we must have

$$x_0 \in \Gamma y$$

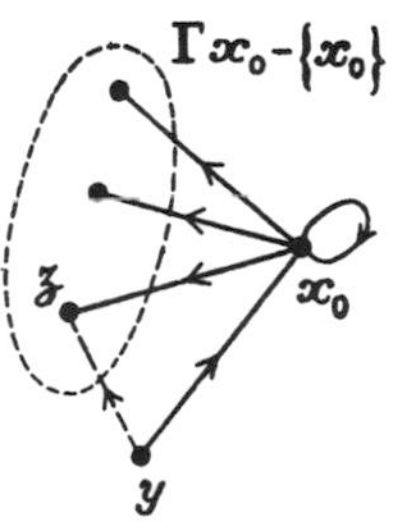

Fig. 12.2

On the other hand, if z is a vertex of $\Gamma x_0 - \{x_0\}$, then $y \notin \Gamma z$ (for otherwise y could be reached from x_0 by a path of length 2), and hence $z \in \Gamma y$, which gives

$$z \in \Gamma y - \{y\}$$

Finally, this gives

$$\Gamma x_0 - \{x_0\} \subset \Gamma y - \{y\}$$

Since $x_0 \in \Gamma y - \{y\}$, this is a relationship of strict inclusion:

$$\Gamma x_0 - \{x_0\} \subset \subset \Gamma y - \{y\}$$

Therefore

$$|\Gamma y - \{y\}| > |\Gamma x_0 - \{x_0\}| = \max_{x \in X} |\Gamma x - \{x\}|$$

which is an obvious contradiction.

EXAMPLE. Consider a tournament in which each contestant must play the same number of matches; how should the games be arranged in order that the player who is declared the winner shall be the one with the best right to that title?

Denote the set of players by X, and put $(x, y) \in U$ if player x beats player y when these two are matched against each other; this yields a complete graph (X, Γ) which we draw (Fig. 12.3). The function Γ is not transitive: x_1 is superior to x_3, and x_3 is superior to x_2, but x_1 is not superior to x_2; because of this, so-called 'elimination' tournaments can give very misleading results. An intuitive notion of fair play suggests that if the graph (X, Γ) has one or more centres, the corresponding players should be the winners; since the graph here is complete, a centre must exist (Theorem 3), and, in fact, we find two winners: x_1 and x_3. This is the simplest method of decision, and it still has its faults: Does not the winner x_1, who beat the other winner x_3, deserve to occupy the first place alone?—and does not player x_2 who beat the champion x_1, deserve something better than last place? Later on, we shall show how these factors may be taken into account (Chapter 14).

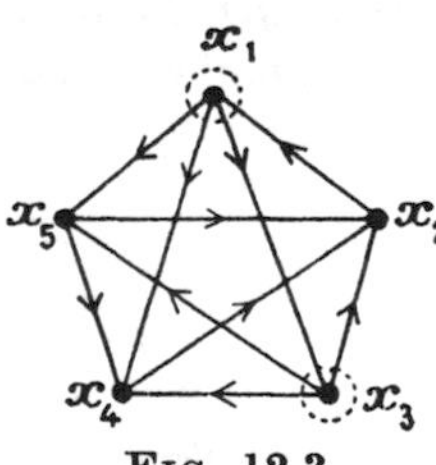

FIG. 12.3

A tournament consisting of m matches determines a partial graph of (X, Γ) with exactly m edges; if this partial graph contains any centres, we may take the corresponding players to be the winners of the competition.

13. The Diameter of a Strongly Connected Graph

General Remarks on Strongly Connected Graphs without Loops

Let $G = (X, U)$ be a strongly connected graph without loops, with $|X| > 1$. There is at least one path to and from every vertex $x \in X$, and therefore at least two arcs are incident to x, one leading into x, and the other going out from x. A vertex x to which more than two arcs are incident is called a *node*, the other vertices being *antinodes*. A path in which only the first and last vertices are nodes is called a *branch*. A strongly connected graph with no loops which has only one node is called a *rosette*: its structure is very simple, since the different branches all start and end at the single node.

If the outcome of deleting any arc whatsoever from a strongly connected graph G is that the graph loses this property, G is said to be *minimally connected*; clearly a rosette is a minimally connected graph. If $A \subset X$, we *shrink* the set A by deleting the arcs connecting any two points of A, and by identifying all the vertices of A with a single one of these vertices. This leads to an s-graph, and since in general $s \neq 1$, we no longer have a graph. Most of the following results are due to CHRISTIE, LUCE and MACY†.

THEOREM 1. *In a minimally connected graph G, let A be a set of vertices determining a strongly connected subgraph; the shrinkage of A leads to a minimally connected graph.*

1. We show first that the shrinkage of A yields a 1-graph. If this were not the case, then there would exist a vertex $x \notin A$, and two vertices $a, a' \in A$ such that (x,a), $(x,a') \in U$ [or else, which comes to the same thing for the purposes of the proof, (a,x), $(a',x) \in U$]. By deleting one of these two arcs, we would again get a strongly connected graph, and G therefore cannot have been minimally connected.

† Cf. CHRISTIE, LUCE and MACY [2]; these results should be supplemented by the following theorem of B. ROY, which came to our notice only after going to press: *A graph (X, Γ) is strongly connected if and only if there is no subset $A \subset \subset X$ (in the strict sense) which contains its image ΓA.* The proof is related to the theory of transport networks (cf. ROY, *C. R. Acad. Sciences*, **247**, 399, 28 July 1958).

2. Next we show that the shrinkage of A leads to a graph which is minimally connected. This graph is, in fact, strongly connected; if one arc u were to be deleted the resulting graph would not be strongly connected, since the graph $(X, U-\{u\})$ cannot be strongly connected.

THEOREM 2. *Let G be a minimally connected graph, and G' a graph derived from G by shrinking an elementary circuit; the cyclomatic numbers of these graphs satisfy*

$$\nu(G) = \nu(G')+1$$

A circuit is elementary if it does not intersect itself; G being minimally connected, the shrinkage of an elementary circuit μ_0 deletes only one circuit and therefore the maximum number of linearly independent circuits is diminished by 1. Since both G and G' are strongly connected (Theorem 1), they have $\nu(G)$ and $\nu(G')$ linearly independent circuits respectively (Theorem 3, Chapter 4), which proves the theorem.

THEOREM 3. *If G is a minimally connected graph, at least two of its vertices are antinodes.*

By reason of the hypothesis $|X| > 1$, the graph contains at least one circuit, and hence $\nu(G) \geqslant 1$; if $\nu(G) = 1$, the theorem is obvious, since G then consists of a single elementary circuit.

We shall now show that if the theorem is true for graphs with cyclomatic number $k-1$, then it is also true for graphs with cyclomatic number k.

Consider an elementary circuit μ of G; this circuit is adjacent to one, two, or more than two vertices of G. In the first case it must contain at least one antinode z_0, and the shrinkage of μ will yield an antinode. In the second case, if the shrinkage of μ yields an antinode, we may assume $l(\mu) \geqslant 3$, for any such circuit of length 2 can easily be replaced by an elementary circuit of length $\geqslant 3$ (passing through the vertices of μ and the two vertices adjacent to μ). In the third case the shrinkage of μ yields a node.

Now shrink the circuit μ: by hypothesis, the minimally connected graph G' with $\nu(G') = k-1$ so obtained contains at least two antinodes x_0 and y_0; one of these vertices, x_0 say, is the image of μ (for if this is not so, both x_0 and y_0 belong to G, which proves the theorem). Therefore, in G, μ contains at least one antinode z_0, since either $l(\mu) = 2$ which can only occur in the first case, or else $l(\mu) \geqslant 3$, and any elementary circuit of length $\geqslant 3$ contains at least one antinode. The graph G must therefore contain at least two antinodes, y_0 and z_0.

COROLLARY [1]. *If G is a strongly connected graph without loops possessing at least one node, a branch exists in which the arcs and antinodes can be deleted leaving a graph which is still strongly connected.*

If the graph has one node only, it is a rosette, and the graph which remains after the deletion of any branch whatsoever (which is itself a rosette) is still strongly connected.

If the graph has several nodes, consider a graph $\bar{G}$ whose vertices are the nodes of G and whose arcs are the branches of G.

$\bar{G}$ is strongly connected, but it is not minimally connected since it contains no antinodes (Theorem 3); therefore the graph will remain strongly connected even after the deletion of at least one of its arcs.

Q.E.D.

If $x, y \in X$, $x \neq y$, a *track from x to y* is defined to be a path of minimum length going from x to y; since the graph is strongly connected there is always a path between two given points; if u is an arc whose initial vertex is x, we denote by $S(u)$ the set of vertices z such that a track from x to z exists which begins with the arc u. The following theorems are due to BRATTON [1].

EXAMPLE. Consider a graph showing lines of communication by post: X is a set of towns, and $(x,y) \in U$ if there is a direct postal link between x and y.

For a sorting office at x, the set $S(x,y)$ represents the list of (final) destinations of letters which must be sent through y.

THEOREM 4. *A track is an elementary path.* By definition, a path from x to y which is not elementary passes at least twice through one of its vertices: it can therefore be replaced by a shorter path.

THEOREM 5. *If $\mu = [x_1, x_2, \ldots, x_i, \ldots, x_j, \ldots, x_h]$ is a track, then so also is* $\mu[x_i, x_j] = [x_i, \ldots, x_j]$. (Self-evident.)

THEOREM 6. *A path μ is a track if and only if*

$$\bigcap_{u \in \mu} S(u) \neq \emptyset$$

If μ is a track from x to y, the intersection of the $S(u)$ contains y, and therefore it is not empty.

Conversely, suppose that a vertex y exists which belongs to all the $S(u)$ with $u \in \mu = [x_1, x_2, \ldots, x_h]$.

If $d(x_1, y) = p$, we have $d(x_2, y) = p - 1$, since $y \in S(x_1, x_2)$; similarly, $d(x_3, y) = p - 2$, etc.... Therefore μ is a part of a track going from x_1 to y, therefore it is a track (Theorem 5).

COROLLARY. *If μ is a circuit, we have*

$$\bigcap_{u \in \mu} S(u) = \emptyset$$

This follows immediately.

Diameter

Let G be a finite, strongly connected graph without loops; the *diameter* of G is the length δ of the longest track of G; in other words, we have:

$$\delta = \max_{x, y} d(x, y)$$

EXAMPLE. The number of arcs m, the radius ρ, and the diameter δ are shown for each of the several different strongly connected graphs of Fig. 13.1, where n, the number of vertices, is 5 throughout; the centres of the graph have been ringed, and a square is drawn round the peripheral points.

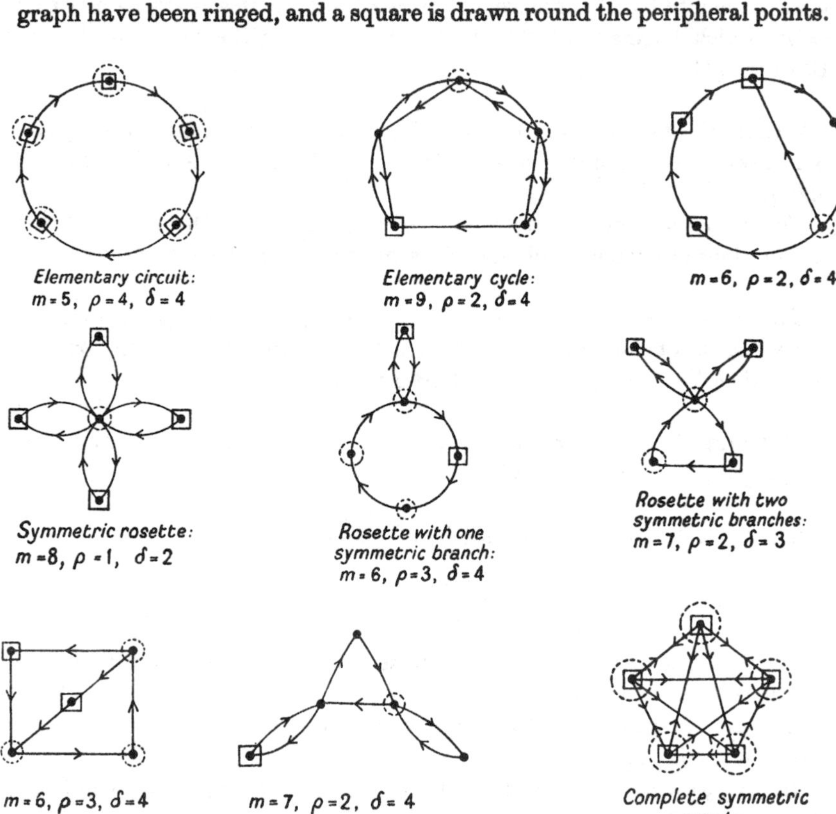

FIG. 13.1

Note that for $n = 5$, we always have $\delta + m \geqslant 9$. The complete symmetric graph is the only one for which the diameter is equal to 1.

The graph being finite, $\delta < \infty$; also, clearly, the diameter must be at least as large as the radius, that is: $\delta \geqslant \rho$. In a graph showing the lines of communication between different members of an organization, the diameter plays an important role, since certain messages must be relayed δ times before they reach their destination; if there is a possibility that the message be distorted at each relay, it is obviously desirable to choose a graph with as small a diameter as possible. On the other hand, one is also interested, for reasons of economy, in keeping the number of arcs as small as possible.

THEOREM 7. *In a strongly connected graph without loops, we have:*

$$n \leqslant m \tag{1}$$

$$m \leqslant n(n-1) \tag{2}$$

$$\delta \leqslant n-1 \tag{3}$$

Inequality (1) follows since we can associate with every vertex x at least one arc which is incident out from x.

Inequality (2) follows, since the largest possible value of m equals the number of arrangements of n vertices taken two at a time. Finally, (3) follows, since the longest track contains $\delta + 1$ distinct vertices, and therefore $n \geqslant \delta + 1$.

Remark. Strict equality holds in (1) and (3) whenever we have an elementary circuit; strict equality holds in (2) for a complete symmetric graph: consequently, none of the inequalities can be improved if it is considered in isolation from the other two.

On the other hand, an improvement can be made in the *system* of inequalities: there are numbers n, m, δ, which satisfy (1), (2) and (3) for which there is no strongly connected graph $G = (X, U)$ with $|X| = n$, $|U| = m$ and $\delta(G) = \delta$. This impossibility arises whenever too small a number is chosen for δ†.

It would be desirable therefore to find the lower bound $f(m,n)$ of the

† For example, LUCE [2] has shown that if numbers n, m, $\delta > 1$ are chosen which satisfy (1), (2) and (3), and if they also obey

$$\delta \geqslant 2n - m \tag{4}$$

then a strongly connected graph G without loops exists in which $|X| = n$, $|U| = m$ and $\delta(G) = \delta$. Unfortunately, there are many graphs for which (4) is not satisfied (in particular those of Figs. 13.2 and 13.3).

diameters of a strongly connected graph without loops with n vertices and m arcs, and to complete our system by the inequality:

$$(4) \qquad \delta \geqslant f(m,n)$$

Although it has not yet been possible to determine this number $f(m,n)$, a close approximation to it can be made; it is with this aim in mind that we shall establish the following theorem:

THEOREM 8. *Let q be the quotient when $n-1$ is divided by $m-n+1$, and let r be the remainder. Put:*

$$\delta(m,n) \begin{cases} = 2q & \text{if } r = 0 \\ = 2q+1 & \text{if } r = 1 \\ = 2q+2 & \text{if } r \geqslant 2 \end{cases}$$

If G is a rosette, n, m and δ satisfy the inequalities (1) *and* (2) *of the preceding theorem, and, in addition:*

$$(3') \qquad 1 < \delta(m,n) \leqslant \delta \leqslant 2n-m$$

Conversely, if m, n and δ are numbers satisfying (1), (2) *and* (3'), *a rosette $G=(X,U)$ exists with $|X|=n$, $|U|=m$, $\delta(G)=\delta$.*

It is sufficient to note that a rosette with n vertices and m arcs has exactly $\nu = m-n+1$ branches, and that an antinode may be transferred from one branch to another without affecting the number of arcs. By putting one antinode on each branch, except the last where we put $n-\nu$ antinodes, we get the maximum diameter:

$$\delta_0 = n-\nu+1 = n-(m-n+1)+1 = 2n-m$$

The minimum diameter δ_1 is found by dividing the $(n-1)$ antinodes as evenly as possible between the ν branches, giving $\delta_1 = \delta(m,n)$.

CONSEQUENCE: BRATTON'S CONJECTURE, concerning the construction of a strongly connected graph without loops with m arcs and n vertices, m and n being fixed, with as small a diameter as possible. BRATTON [1] has been able to show that graphs with a diameter smaller than $\delta(m,n)$ are infrequent. Such graphs however certainly exist; although the rosette of Fig. 13.2 and the

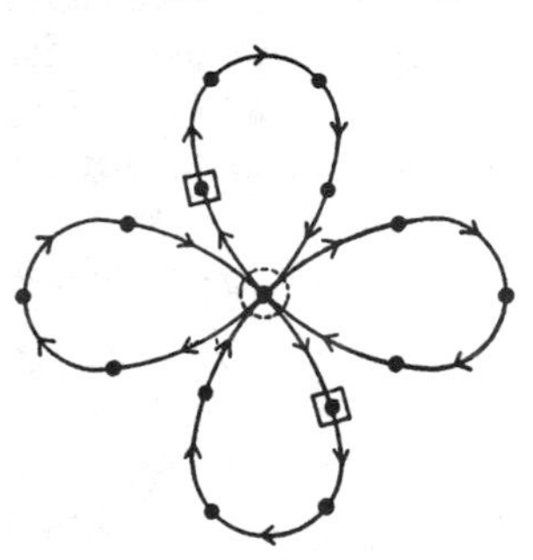

FIG. 13.2

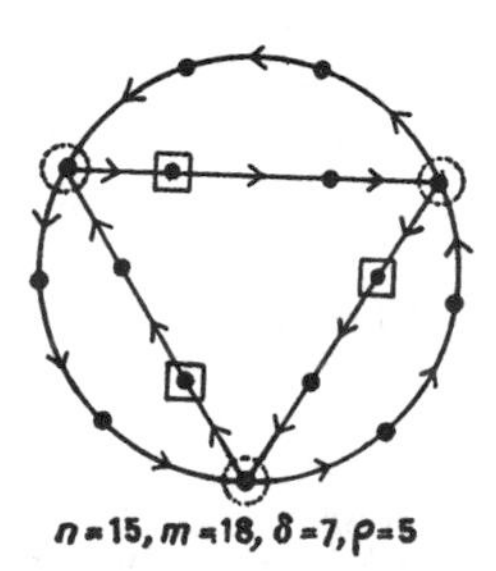

FIG. 13.3

graph of Fig. 13.3 each have the same number of arcs and vertices, the diameter of the latter is one less than that of the rosette.

Graphs like that of Fig. 13.3 are such that every branch is a track between two nodes, all branches being of the same length, and this length is an uneven number $\geqslant 3$. It can be proved [1] that the diameter of such graphs is always $\delta(m,n)-1$, and that they are the only strongly connected graphs with the following properties:

1. Their diameter is always $< \delta(m,n)$.
2. Each branch is a track.
3. Every pair of nodes is connected by exactly one branch in each direction.

These results appear to confirm the following conjecture, as yet unproved: *every strongly connected graph without loops with m arcs and n vertices has a diameter which is greater than or equal to* $\delta(m,n)-1$.

14. The Matrix Associated with a Graph

The Use of Standard Matrix Operations

For greater generality, let us consider an s-graph $G = (X, U)$, and let x_1, $x_2, \ldots, x_n$ be its vertices. Let a^i_j be the number of arcs of U going from x_i to x_j. The square matrix (a^i_j) with n rows and n columns is called the matrix *associated* with the s-graph G; according to standard practice, the coefficient a^i_j is the element located at the intersection of the i-th row and the j-th column. The i-th row vector will be denoted by $\boldsymbol{a}^i = (a^i_1, a^i_2, \ldots, a^i_n)$, and the j-th column vector by $\boldsymbol{a}_j = (a^1_j, a^2_j, \ldots, a^n_j)$.

EXAMPLE. If G is a graph, all the elements of the associated matrix $A = (a^i_j)$ are equal to 0 or 1, as shown in the matrix associated with the graph of Fig. 14.1.

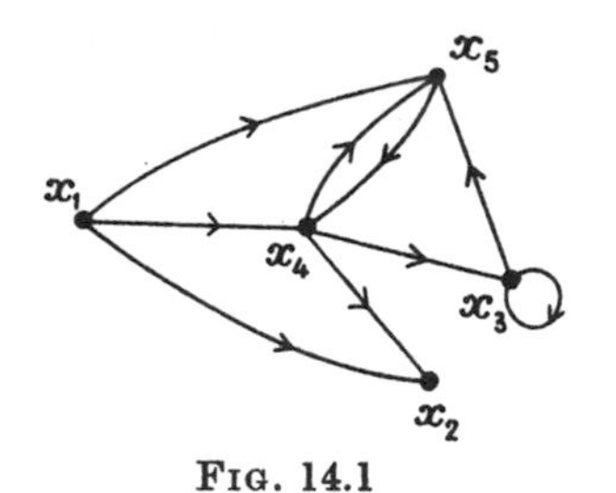

Fig. 14.1

$$
\begin{array}{c c}
j = \begin{matrix} 1 & 2 & 3 & 4 & 5 \end{matrix} & \\
A = \begin{bmatrix} 0 & 1 & 0 & 1 & 1 \\ 0 & 0 & 0 & 0 & 0 \\ 0 & 0 & 1 & 0 & 1 \\ 0 & 1 & 1 & 0 & 1 \\ 0 & 0 & 0 & 1 & 0 \end{bmatrix} & \begin{matrix} i = 1 \\ 2 \\ 3 \\ 4 \\ 5 \end{matrix}
\end{array}
$$

Note that in the matrix A, the element a^3_3 on the principal diagonal is 1, thus indicating that there is a loop at the vertex x_3.

Let $A = (a^i_j)$ be a matrix associated with an s-graph. The *transpose* $A^* = (a^{*i}_j)$ of A is obtained by reflecting A about its main diagonal (interchanging rows and columns); therefore it is the matrix associated with the s-graph obtained by reversing the orientation of all the arcs.

The *complementary* matrix $A' = (a'^i_j)$ is defined by $a'^i_j = s - a^i_j$; for a graph (X, Γ) it is obtained from A by replacing every zero element with a 1, and every unit element with a 0: therefore it is the matrix associated with the graph (X, Γ'), where Γ' is defined by

$$\Gamma' x = X - \Gamma x \qquad (x \in X)$$

THEOREM 1. *Let G be a graph, and A its associated matrix. The matrix A is symmetric ($A = A^*$, or: $a^i_j = a^j_i$) if and only if the graph G is symmetric. The matrix A is anti-symmetric ($A \leqslant (A')^*$, or: $a^i_j + a^j_i \leqslant 1$) if and only if the graph G is anti-symmetric. The matrix A is complete ($A \geqslant (A')^*$, or: $a^i_j + a^j_i \geqslant 1$) if and only if the graph G is complete.*

The proof is immediate.

THEOREM 2. *Consider two s-graphs $G = (X, U)$ and $H = (X, V)$ which have the same set X of vertices and let $A = (a^i_j)$ and $B = (b^i_j)$ be the associated matrices of G and H respectively; the matrix $A + B$ will correspond to the 2s-graph formed by the union of the arcs of U and V; the matrix AB will correspond to an s-graph defined in the following manner: the number of arcs going from a vertex x to a vertex y is put equal to the number of distinct paths going from x to y, which consist of an arc of U followed by an arc of V.*

1. In the graph $(X, U \cup V)$, the number of arcs going from x_i to x_j is

$$a^i_j + b^i_j$$

and this is simply the general element of the matrix $A + B$.

2. The number of distinct paths of the form $[x_i, x_k, x_j]$, with

$$(x_i, x_k) \in U, \qquad (x_k, x_j) \in V$$

is equal to $a^i_k b^k_j$; therefore in order to go from x_i to x_j, the total number of paths which are made up of an arc of U followed by an arc of V is

$$\sum_{k=1}^{n} a^i_k b^k_j = \langle \boldsymbol{a}^i, \boldsymbol{b}_j \rangle$$

where $\langle \boldsymbol{a}^i, \boldsymbol{b}_j \rangle$ stands for the scalar product of the row vector $\boldsymbol{a}^i$ and the column vector $\boldsymbol{b}_j$: this, however, is the general element of the matrix AB.

COROLLARY 1. *If G is an s-graph and A its associated matrix, the element p^i_j of the matrix $P = A^\lambda$ (obtained by taking the product of A with itself λ times) is equal to the number of distinct paths of length λ which go from x_i to x_j.*

The theorem is trivial for $\lambda = 1$; if we assume that it has been proved for a value $\lambda - 1$, then it must also be true for the value λ since $A^\lambda = A(A^{\lambda-1})$ shows, from the preceding theorem, the number of paths of length $1 + (\lambda - 1) = \lambda$ which go from a vertex x_i to a vertex x_j.

COROLLARY 2. *A graph G contains a path of length λ, if and only if $A^\lambda \neq 0$; it contains no circuits if and only if, for all λ sufficiently large, $A^\lambda = 0$.*

This can be proved immediately, using Corollary 1.

These results enable us to identify certain problems concerning *s*-graphs with problems in the theory of matrices. We shall now examine some of these in detail.

Problems of Enumeration

The practical value of describing a graph by means of its associated matrix becomes evident whenever we wish to enumerate the paths of a graph which satisfy certain given conditions. First, let us take the following problem:

PROBLEM 1. *Find the number of circuits of length 3 in a complete anti-symmetric graph (X, U).*

EXAMPLE [2]. The following scheme was used in testing the response of a dog to six different types of dog food: each morning the dog was given two plates of different food and the plate which it finished first was noted. After having tried the $\binom{6}{2} = 15$ possible pairs, the graph (X, U) was drawn, where $X = \{x_1, x_2, \ldots, x_6\}$ represents the six different foods, and $(x_i, x_j) \in U$ if food x_i is preferred to food x_j. This graph, which is complete, could be that of Fig. 14.2, and we shall denote its associated matrix by A.

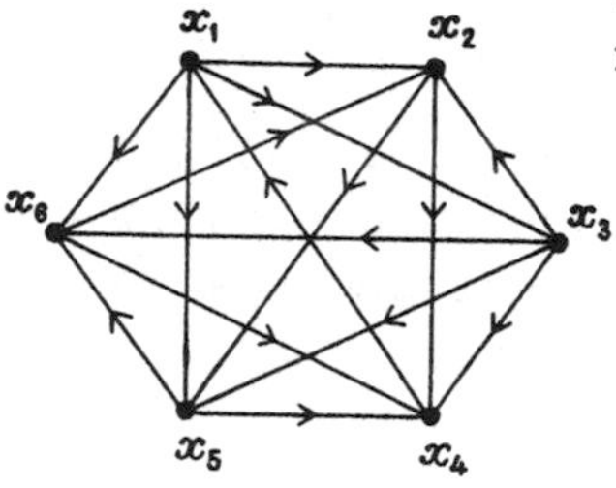

FIG. 14.2

$$A = \begin{bmatrix} 0 & 1 & 1 & 0 & 1 & 1 \\ 0 & 0 & 0 & 1 & 1 & 0 \\ 0 & 1 & 0 & 1 & 1 & 1 \\ 1 & 0 & 0 & 0 & 0 & 0 \\ 0 & 0 & 0 & 1 & 0 & 1 \\ 0 & 1 & 0 & 1 & 0 & 0 \end{bmatrix}$$

Now let us continue the test by offering the dog three different foods instead of two; if the foods are x_2, x_3 and x_4, we can predict that the dog's preference will be for x_3, but if they are x_1, x_2 and x_4, which form a circuit, we are unable to make a prediction because of the inconsistency of the dog's choice in the preliminary trials. If ξ is the number of circuits of length 3 in the graph (X, U), and if $g(n)$ is the maximum number of circuits of length 3 in a complete antisymmetric graph of order n, we can take as a *coefficient of inconsistency* for the dog the ratio $\dfrac{\xi}{g(n)}$.

It can be shown that

$$g(n) = \begin{cases} \dfrac{n^3-n}{24} & \text{if } n \text{ is odd} \\ \dfrac{n^3-4n}{24} & \text{if } n \text{ is even} \end{cases}$$

In this case, we find

$$g(6) = \tfrac{1}{24}6(6^2-4) = 8$$

Now we must find the number ξ, that is to say, we must solve Problem 1 for the graph (X, U).

THEOREM 3. *Let G be a complete antisymmetric graph and let $A = (a_j^i)$ be its associated matrix; if $r_i = \sum_j a_j^i$ denotes the sum of the elements of the i-th row, the number of circuits of length 3 is:*

$$\xi = \tfrac{1}{12}n(n-1)(2n-1) - \tfrac{1}{2}\sum_{i=1}^{n}(r_i)^2$$

In fact, the number of *cycles* of length 3 is

$$\binom{n}{3} = \tfrac{1}{6}n(n-1)(n-2)$$

A cycle of length 3 will not be a circuit if and only if two of its arcs are incident out from the same vertex x_i; therefore the total number of cycles of length 3 which are not circuits is exactly:

$$\sum_{i=1}^{n}\binom{r_i}{2} = \sum_i \tfrac{1}{2}r_i(r_i-1) = \tfrac{1}{2}\sum_i (r_i)^2 - \tfrac{1}{2}\sum_i r_i$$

We note that

$$\sum_{i=1}^{n} r_i = |U| = \binom{n}{2} = \tfrac{1}{2}n(n-1)$$

Combining these results

$$\xi = \tfrac{1}{6}n(n-1)(n-2) - \tfrac{1}{2}\sum_i (r_i)^2 + \tfrac{1}{4}n(n-1)$$

$$= \tfrac{1}{2}n(n-1)\left[\frac{n-2}{3} + \tfrac{1}{2}\right] - \tfrac{1}{2}\sum_i (r_i)^2$$

which, when simplified, yields the formula given in the statement of the theorem.

EXAMPLE. Return to the preceding example, in which $r_1 = 4$, $r_2 = 2$, $r_3 = 4$, $r_4 = 1$, $r_5 = 2$, $r_6 = 2$.

Therefore

$$\xi = \tfrac{1}{12}.6.5.11 - \tfrac{1}{2}(16+4+16+1+4+4) = \tfrac{1}{2}(55-45) = 5$$

The circuits are effectively:

$$[x_1, x_2, x_4, x_1], \quad [x_1, x_3, x_4, x_1], \quad [x_1, x_5, x_4, x_1]$$
$$[x_1, x_6, x_4, x_1], \quad [x_2, x_5, x_6, x_2]$$

PROBLEM 2. *Consider a graph G and three integers α, β and γ; a path*

$$\mu = [y_0, y_1, \ldots, y_m]$$

is said to satisfy the 3-tuple $\alpha\beta\gamma$ if there are two equal vertices y_i and y_j $(i < j)$ on this path, with $l(\mu[y_0, y_i]) = \alpha$, $l(\mu[y_i, y_j]) = \beta$, $l(\mu[y_j, y_m]) = \gamma$. Find the number of paths satisfying the 3-tuple $\alpha\beta\gamma$.

It may be seen that the preceding Problem 1 is a special case of this problem with $\alpha = 0$, $\beta = 3$, $\gamma = 0$. If M is a matrix, let us denote by $d(M)$ a matrix which has the same leading diagonal as M, all its other elements being zero. Then we have:

THEOREM 4. *For a graph G with associated matrix A, let s^i_j be the number of paths satisfying a 3-tuple $\alpha\beta\gamma$ and going from x_i to x_j; the matrix $S = (s^i_j)$ is given by the product*

$$S = A^\alpha d(A^\beta) A^\gamma$$

From Theorem 2, the number of circuits of length β is given by the diagonal matrix $d(A^\beta)$, and therefore the number of paths satisfying $0\beta\gamma$ is $d(A^\beta).A^\gamma$, and finally therefore the number of paths satisfying $\alpha\beta\gamma$ is given by the matrix $A^\alpha.d(A^\beta).A^\gamma$.

If k is an integer which is not too large, it is possible to make use of this result to enumerate all the elementary paths of length k in a given graph: it is sufficient to consider separately all possible 3-tuples $\alpha\beta\gamma$ (cf. [1], [4], [7]).

EXAMPLE. Let us now enumerate the different non-elementary paths of length 4 in a graph with associated matrix A. The different 3-tuples which must be considered are: 031, 130, 022, 220, 121; the corresponding matrices are: $d(A^3)A$, $Ad(A^3)$, $d(A^2)A^2$, $A^2d(A^2)$, $Ad(A^2)A$.

It will be noted that there are some paths of length 4 which can satisfy two 3-tuples simultaneously; to enumerate these, we shall use HADAMARD's notation for the *element by element product* of two matrices $A = (a_j^i)$ and $B = (b_j^i)$ and write $A \times B = (a_j^i \times b_j^i)$; we then get the following table:

PATHS	3-TUPLES	NUMBER OF PATHS FROM x_i TO x_j	MATRIX
$[c, x, b, c, b]$	031, 220	$\sum_k a_k^i a_j^k (a_i^j a_j^i)$	$A^2 \times (A^* \times A)$
$[c, b, x, c, b]$	130, 031	$\sum_k (a_j^i) a_k^j a_i^k$	$A \times (A^2)^*$†
$[c, b, c, x, b]$	130, 022	$\sum_k (a_j^i a_i^j) a_k^i a_j^k$	$(A \times A^*) \times A^2$
$[x, c, b, c, b]$	121, 220	$\sum_k a_k^i (a_j^k a_k^j)$	$A(A \times A^*)$
$[c, b, c, b, x]$	022, 121	$\sum_k (a_k^i a_i^k) a_j^k$	$(A \times A^*) A$

In a graph without loops the number of non-elementary paths of length 4 can therefore be found by forming the matrix sum

$$d(A^3)A + Ad(A^3) + \ldots + Ad(A^2)A$$

and subtracting from this the matrices given in the table.

The Problem of the Leader

Another case in which the associated matrix plays an important role occurs when X represents a set of people, and $(x, y) \in U$ if the individual x *dominates* the individual y.

When we say that x *dominates* y, we shall be using the word in its widest sense: x could be a competitor in a tournament who defeats player y; he may be a councillor who can count on the support of councillor y in a municipal election; in a moral society he may be an individual who is respected by individual y. The problem which we shall set here is to find a leader x_0, that is to say someone who may be said to have won the tournament, or to have been elected mayor, or to be the most representative member of the moral society under consideration.

EXAMPLE. Consider the graph of Fig. 14.3, where the vertices stand for the individuals in a social group, and an arc goes from x to y if the influence

† The complete formula here is $A \times (A^2)^* \times A$, but it can be contracted as shown since, for a graph, $A \times A = A$. Note that this does not hold for s-graphs.

of x on y is sufficiently great; such graphs have been studied in sociology under the name of 'sociogrammes' (cf. MORENO [6]).

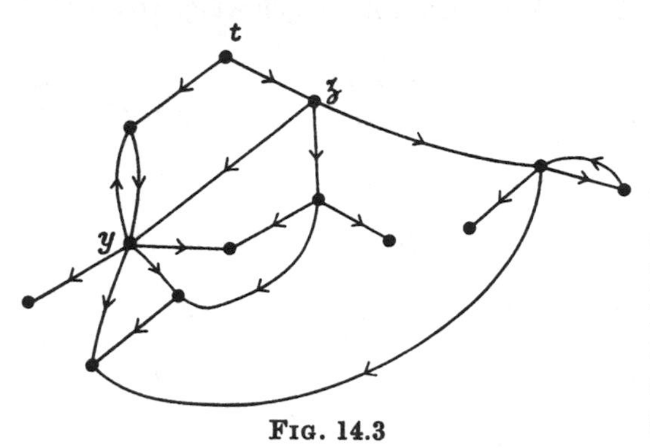

FIG. 14.3

One is often tempted to select the most dominant individual as the leader; in the graph of Fig. 14.3, y is the person who has the most direct influence, since he dominates five people; however these five people are not themselves influential. The individual z is the most *powerful*: he is chosen directly by only three people, but these people are themselves very influential. We can now perceive the complexity of the problem with which we are dealing.

In Chapter 12, it was suggested that one should choose the centre of the graph as the leader, and the centre is certainly *powerful*; nevertheless, in many cases, this suggestion proves to be inadequate: what is required is that the power $\pi(x)$ of every individual x should be measured, even for graphs which do not have a centre. As soon as a clear definition has been made of this function $\pi(x)$, we shall be able to choose as the leader the person x_0 for whom the power $\pi(x_0)$ is a maximum.

To take a concrete example, let us consider a chess tournament with five competitors: x_1, x_2, x_3, x_4 and x_5; if x_i defeats x_j, we join x_i and x_j by two arcs, both oriented in the direction x_i to x_j, and if the match results in a draw, we again join x_i and x_j by two arcs, but this time one is oriented in one direction and one in the other. Finally, a loop is drawn at each vertex. In a competition where every player meets every other player, we thus get a 2-graph G with associated matrix A (Fig. 14.4).

$$A = \begin{bmatrix} 1 & 0 & 2 & 2 & 1 \\ 2 & 1 & 1 & 0 & 0 \\ 0 & 1 & 1 & 2 & 2 \\ 0 & 2 & 0 & 1 & 1 \\ 1 & 2 & 0 & 1 & 1 \end{bmatrix}$$

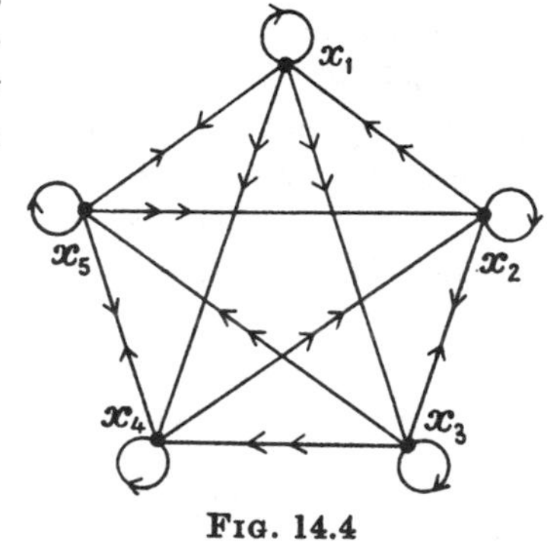

FIG. 14.4

The elements of the matrix A satisfy $a^i_j + a^j_i = 2$; in addition, A can be expressed as the sum of a skew symmetric matrix B and a matrix E the elements of which are all equal to one.

Let us denote by $p^i_j(k)$ the general element of the matrix A^k, *that is to say, the number of paths of length k going from x_i to x_j*, and put

$$p^i(k) = p^i_1(k) + p^i_2(k) + \ldots + p^i_5(k)$$

The number $p^i(k)$ will be called the *iterated power of order k of the player x_i*; now let us justify this terminology.

The iterated power of order 1 is formed by summing the elements of a row of A, giving

$$\begin{aligned}
p^1(1) &= 1+0+2+2+1 = 6\\
p^2(1) &= 2+1+1+0+0 = 4\\
p^3(1) &= 0+1+1+2+2 = 6\\
p^4(1) &= 0+2+0+1+1 = 4\\
p^5(1) &= 1+2+0+1+1 = 5
\end{aligned}$$

With this power $p^i(1)$, we get two winners x_1 and x_3, who are the most dominant individuals [in a tournament where draws are not permitted, this rule would yield *all* the centres of the graph, by reason of Theorem 3 (Chapter 12)].

The power of order 2 of each player is found by summing the scores of each player with whom the given player has had a drawn match and the scores *multiplied by* 2 of each player whom he defeated; here we get:

$$\begin{aligned}
p^1(2) &= 6+0+(2\times 6)+(2\times 4)+5 &&= 31\\
p^2(2) &= (2\times 6)+4+6+0+0 &&= 22\\
p^3(2) &= 0+4+6+(2\times 4)+(2\times 5) &&= 28\\
p^4(2) &= 0+(2\times 4)+0+4+5 &&= 17\\
p^5(2) &= 6+(2\times 4)+0+4+5 &&= 23
\end{aligned}$$

This time player x_1 is the sole winner, an event due to the fact that the players whom he defeated were stronger players than those defeated by x_3.

Let us carry out another iteration:

$$\begin{aligned}
p^1(3) &= 31+0+(2\times 28)+(2\times 17)+23 &&= 144\\
p^2(3) &= (2\times 31)+22+28+0+0 &&= 112\\
p^3(3) &= 0+22+28+(2\times 17)+(2\times 23) &&= 130\\
p^4(3) &= 0+(2\times 22)+0+17+23 &&= 84\\
p^5(3) &= 31+(2\times 22)+0+17+23 &&= 115
\end{aligned}$$

The classification is the same as the preceding one: x_1, x_3, x_5, x_2, x_4; it appears to be fixed in this pattern, and it can be seen that it will not be altered no matter how many more iterations are carried out.

This suggests that we should define the *power* of player x_i as the limit when $k \to \infty$ of

$$\pi_k^i = \frac{p^i(k)}{p^1(k)+p^2(k)+\ldots+p^n(k)}$$

From the theorem of PERRON and FROBENIUS, this limit always exists; further, it is very easy to find it, since the vector $\pi_k = (\pi_k^1, \pi_k^2, \ldots, \pi_k^n)$ tends towards a latent vector of the matrix A†.

The Use of Boolean Operations

Let us define two operations in the set of real non-negative numbers, which we shall call the *generalized sum* and the *generalized product*.

Given two numbers λ_1 and λ_2, their generalized sum defines a number $\lambda \geqslant 0$ which is written $\lambda_1 \dot{+} \lambda_2 = \lambda$; the generalized product of λ_1 and λ_2 defines a number $\lambda' \geqslant 0$ which is written $\lambda' = \lambda_1 \dot{\times} \lambda_2$.

Now let us consider two matrices $A = (a_j^i)$ and $B = (b_j^i)$, with non-negative integral elements; their *generalized sum* is the matrix $S = A \dot{+} B = (s_j^i)$, where $s_j^i = a_j^i \dot{+} b_j^i$; their *generalized product* is a matrix $P = A \,.\, B = (p_j^i)$, where

$$p_j^i = (a_1^i \dot{\times} b_j^1) \dot{+} (a_2^i \dot{\times} b_j^2) \dot{+} \ldots \dot{+} (a_n^i \dot{\times} b_j^n)$$

'Generalized' operations on matrices may be defined in exactly the same way as ordinary operations. We shall assume that, like ordinary addition and multiplication, the operations $\dot{+}$ and $\dot{\times}$ satisfy:

$$(1) \qquad \begin{cases} \lambda_1 \dot{+} \lambda_2 = \lambda_2 \dot{+} \lambda_1 \\ \lambda_1 \dot{+} (\lambda_2 \dot{+} \lambda_3) = (\lambda_1 \dot{+} \lambda_2) \dot{+} \lambda_3 \\ \lambda \dot{+} 0 = \lambda \\ \lambda \dot{+} (-\lambda) = 0 \\ \lambda_1 \dot{\times} \lambda_2 = \lambda_2 \dot{\times} \lambda_1 \\ \lambda_1 \dot{\times} (\lambda_2 \dot{\times} \lambda_3) = (\lambda_1 \dot{\times} \lambda_2) \dot{\times} \lambda_3 \\ \alpha \dot{\times} (\lambda_1 \dot{+} \lambda_2) = (\alpha \dot{\times} \lambda_1) \dot{+} (\alpha \dot{\times} \lambda_2) \end{cases}$$

(in other words, we are dealing here with a 'commutative ring').

† This remark, which is due to WEI, can be stated in the following way:
Consider the associated matrix A of a p-graph (with loops at each vertex), and form the determinant

$$\operatorname{Det}(A - \lambda I) = f(\lambda)$$

1. *The root of the characteristic equation $f(\lambda) = 0$ with the greatest absolute value, is a simple root; let λ_1 be this simple root, and let $t_1 = (t_1^1, t_1^2, \ldots, t_1^n)$ be the vector such that*

$$At_1 = \lambda_1 t_1, \qquad \sum_{i=1}^{n} t_1^i = 1$$

2. *The vector π_k tends to t_1 when k increases without limit.*

1. This first part of the statement is simply the theorem of PERRON and FROBENIUS, which is well known in the theory of stochastic processes.

2. The second part can be verified immediately if A is reduced to diagonal form by a suitable transformation of the co-ordinate system.

There is, in fact, an enormous literature on these generalized matrix operations; we shall only point out those properties which have direct application to questions in which we are interested.

EXAMPLE. In a network G, where a *length* $l(u)$ is associated with each arc u, consider a matrix $A = (a_j^i)$, where

$$a_j^i \begin{cases} = l(x_i, x_j) & \text{if } (x_i, x_j) \in U \\ = \infty & \text{if } (x_i, x_j) \notin U, i \neq j \\ = 0 & \text{if } i = j \end{cases}$$

Put:

$$\lambda_1 \dot{+} \lambda_2 = \min \{\lambda_1, \lambda_2\}$$
$$\lambda_1 \dot{\times} \lambda_2 = \lambda_1 + \lambda_2$$

These operations certainly satisfy the system (1); in addition the general element of $A.A = A^2$ is:

$$p_j^i = \min_k (a_k^i + a_j^k)$$

This number represents the length of the shortest path going from x_i to x_j using two arcs or less (since $a_j^i + a_j^j = a_j^i$).

More generally, the general element of A^α is the *length* of the shortest path containing α or fewer arcs which goes from vertex x_i to vertex x_j. A value α_0 exists such that:

$$A^{\alpha_0} = A^{\alpha_0+1} = A^{\alpha_0+2} = \dots$$

The matrix A^{α_0} consists of elements representing the shortest distance from any one vertex to any other.

THEOREM 5. *Using the generalized operations* $\dot{+}$ *and* $\dot{\times}$, *matrices obey the following rules:*

$$(1) \quad A \dot{+} B = B \dot{+} A$$
$$(2) \quad A \dot{+} (B \dot{+} C) = (A \dot{+} B) \dot{+} C$$
$$(3) \quad A.(B \dot{+} C) = (A.B) \dot{+} (A.C)$$
$$(4) \quad (A \dot{+} B)^* = B^* \dot{+} A^*$$
$$(5) \quad (A.B)^* = B^*.A^*$$

(1), (2) and (4) follow immediately. To prove (3), we simply write down the general element of the matrix $A.(B \dot{+} C)$ which is:

$$\langle \boldsymbol{a}^i, \boldsymbol{b}_j \dot{+} \boldsymbol{c}_j \rangle = [a_1^i \dot{\times} (b_j^1 \dot{+} c_j^1)] \dot{+} \dots$$
$$= [(a_1^i \dot{\times} b_j^1) \dot{+} (a_1^i \dot{\times} c_j^1)] \dot{+} \dots$$
$$= \langle \boldsymbol{a}^i, \boldsymbol{b}_j \rangle \dot{+} \langle \boldsymbol{a}^i, \boldsymbol{c}_j \rangle$$

To prove (5), it is sufficient to observe that the element p^{*i}_j of $(A.B)^* = P^*$ satisfies:

$$p^{*i}_j = p^j_i = \langle a^j, b_i \rangle = \langle a^*_j, b^{*i} \rangle = \langle b^{*i}, a^*_j \rangle$$

Note that in general $A.B \neq B.A$.

When the matrices under consideration are the matrices associated with a graph, all the elements are either 0 or 1, and we shall define the operations $\dot{+}$ and $\dot{\times}$ by:

$$\begin{array}{ll} 1 \dot{+} 1 = 1 & 1 \dot{\times} 1 = 1 \\ 1 \dot{+} 0 = 1 & 1 \dot{\times} 0 = 0 \\ 0 \dot{+} 0 = 0 & 0 \dot{\times} 0 = 0 \end{array}$$

In other words, we shall write:

$$\lambda_1 \dot{+} \lambda_2 = \max\{\lambda_1, \lambda_2\}$$
$$\lambda_1 \dot{\times} \lambda_2 = \min\{\lambda_1, \lambda_2\}$$

The operations $\dot{+}$ and $\dot{\times}$ are called *Boolean addition* and *Boolean multiplication*, respectively.

THEOREM 6. *If two graphs $G = (X, \Gamma)$ and $H = (X, \Delta)$ have associated matrices A and B, the Boolean sum $A \dot{+} B$ corresponds to the graph $(X, \Gamma \cup \Delta)$ and the Boolean product to the graph $(X, \Delta.\Gamma)$, where, as usual, we put:*

$$(\Gamma \cup \Delta)x = \Gamma x \cup \Delta x$$
$$(\Delta.\Gamma)x = \Delta(\Gamma x)$$

This theorem is the analogue of Theorem 2; the element p^i_j of the product $P = A.B$ is

$$p^i_j = (a^i_1 \dot{\times} b^1_j) \dot{+} (a^i_2 \dot{\times} b^2_j) \dot{+} \ldots$$

Therefore if a path of the type $[x_i, x_k, x_j]$ with $x_k \in \Gamma x_i$, $x_j \in \Delta x_k$ exists, we have $a^i_k \dot{\times} b^k_j = 1$, from which $p^i_j = 1$; in all other cases, $p^i_j = 0$. Therefore $p^i_j = 1$ if and only if $x_j \in \Delta(\Gamma x_i)$, that is to say if a path exists going from x_i to x_j which consists of an arc of G followed by an arc of H.

COROLLARY. *Let G be a graph with associated matrix A. The Boolean product $A^\alpha = A.A.\ldots.A$, corresponds to a graph which contains the arc (x, y) if and only if G contains a path of length α going from x to y* ('graph of the α-domination').

The proof is immediate.

We observe that by means of the algebra of Boolean matrices, which is developed parallel to that of ordinary matrices, we can solve certain problems in the theory of graphs, for example: *Knowing the graph of the α-domination, find the initial graph from which this graph is derived.*

Note the similarity of this problem with that of finding the α-th root of a matrix.

15. Incidence Matrices

Matrices with the Unimodular Property

Let us denote the *arcs* of a graph G by $u_1, u_2, \ldots, u_m$, and its vertices by $x_1, x_2, \ldots, x_n$, and put:

$$s_j^i = \begin{cases} +1 \text{ if } u_j \text{ starts from } x_i, \text{ and is not a loop} \\ -1 \text{ if } u_j \text{ terminates at } x_i, \text{ and is not a loop} \\ 0 \text{ if } x_i \text{ is not a vertex of } u_j, \text{ or if } u_j \text{ is a loop.} \end{cases}$$

The matrix $S = (s_j^i)$ is called the *incidence matrix of the arcs* of the graph. Now let $u_1, u_2, \ldots, u_m$, stand for the *edges* of the graph, and let us put:

$$r_j^i = \begin{cases} +1 \text{ if } x_i \text{ is a vertex of } u_j, \text{ and } u_j \text{ is not a loop} \\ 0 \text{ in all other cases.} \end{cases}$$

The matrix $R = (r_j^i)$ is, by definition, the *incidence matrix of the edges* of the graph.

A matrix $A = (a_j^i)$ is said to have the *unimodular property* (or: *to be totally unimodular*) if the determinant of any square submatrix of A, is either 0, + 1 or − 1.

Any element a_j^i of a matrix with the unimodular property must be either 0, +1 or −1, since it is the determinant of a minor of order 1 of the matrix; what we need therefore are criteria whereby we can recognize if a matrix of which all the elements are either 0, +1 or −1, has the unimodular property.

The graph of Fig. 15.1 has S as the incidence matrix of its arcs; the incidence matrix of its edges may be deduced from S by replacing all the elements equal to −1 by elements equal to +1.

It can be shown, either by direct proof or by using the following theorems, that all these matrices are totally unimodular.

$$S = \begin{array}{c} \begin{array}{cccccc} u_1 & u_2 & u_3 & u_4 & u_5 & u_6 \end{array} \\ \left[\begin{array}{cccccc} 0 & +1 & 0 & +1 & 0 & 0 \\ +1 & 0 & +1 & 0 & 0 & 0 \\ -1 & -1 & 0 & 0 & 0 & 0 \\ 0 & 0 & -1 & -1 & +1 & +1 \\ 0 & 0 & 0 & 0 & -1 & 0 \\ 0 & 0 & 0 & 0 & 0 & -1 \end{array}\right] \end{array} \begin{array}{c} \\ x_1 \\ x_2 \\ x_3 \\ x_4 \\ x_5 \\ x_6 \end{array}$$

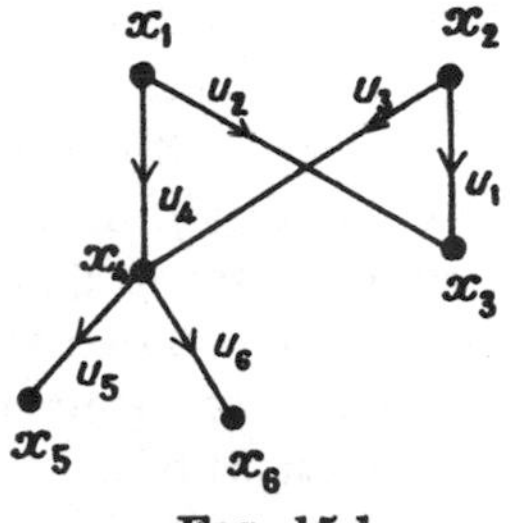

Fig. 15.1

THEOREM 1. (HELLER–TOMPKINS–GALE). *Let A be a matrix the elements of which are 0, $+1$ or -1, such that every column contains at most two non-zero elements; then A has the unimodular property if and only if its rows can be divided into two disjoint sets I_1 and I_2, which obey the following two conditions:*

(1) *if the two non-zero elements of a column are both of the same sign, one of them is in I_1 and the other is in I_2;*

(2) *if the two non-zero elements of a column are of opposite sign, both of them are in I_1, or both of them are in I_2.*

1. Let A be a matrix whose rows may be divided into two parts as described in the statement of the theorem; then any square submatrix B of A will also have this property. To show that A has the unimodular property, we have to show that the determinant $\det(B) = 0, +1$ or -1.

This proposition is certainly true for matrices B of order 1; let us assume that it has been proved for matrices of order $q-1$, and thence deduce that it is true for a square matrix B of order q, for which the two disjoint sets are I_1 and I_2.

If every column vector $\boldsymbol{b}_j$ has exactly two non-zero elements, we have

$$\sum_{i \in I_1} \boldsymbol{b}^i = \sum_{i \in I_2} \boldsymbol{b}^i$$

Therefore the rows of B are not linearly independent, and we must have $\det(B) = 0$.

If a column vector of B contains no non-zero elements, we again have $\det(B) = 0$.

Finally, if there is a column vector $\boldsymbol{b}_j$ with exactly one non-zero element, $b_j^i = +1$, say, let us denote by C the square matrix obtained by deleting row i and column j from B; since we have assumed that the theorem is true for matrices of order $q-1$, we have

$$\det(B) = \pm \det(C) = +1, -1 \text{ or } 0$$

The proposition is therefore proved for all cases.

2. Let A be a matrix with the unimodular property in which each column contains at most two non-zero terms. We now show that a partition I_1, I_2 exists which satisfies the conditions of the theorem. We can assume that every column of A contains exactly two non-zero elements, since a column containing fewer than two non-zero elements may be deleted without affecting the statement of the theorem. Now construct a (non-oriented) graph G to correspond with the matrix A in the following manner: associate a vertex x_i with the row i, and an edge $\boldsymbol{u}_j$ with the column j; this edge connects the vertices x_h and x_k where $a_j^h \neq 0$ and $a_j^k \neq 0$. Finally, we shall say

that an edge u_j is *special* if the two non-zero elements of the column vector a_j are of the same sign; we shall show that *every elementary cycle of the graph contains an even number of special edges.*

Suppose in fact that the cycle $\mu = [x_1, x_2, \ldots, x_k, x_1] = (u_1, u_2, \ldots, u_k)$ is an elementary cycle which does not possess this property, and let us find the determinant of the square submatrix which corresponds to μ.

$$B = \begin{array}{c} \begin{matrix} u_1 & u_2 & \ldots & u_k \end{matrix} \\ \begin{bmatrix} \alpha_1 & 0 & \ldots & \beta_k \\ \beta_1 & \alpha_2 & \ldots & 0 \\ 0 & \beta_2 & \ldots & 0 \\ \cdot & \cdot & \cdot & \cdot \\ 0 & 0 & \ldots & \alpha_k \end{bmatrix} \end{array} \begin{matrix} x_1 \\ x_2 \\ \vdots \\ \vdots \\ x_k \end{matrix}$$

We have $\alpha_j \neq 0$, $\beta_j \neq 0$; further $\alpha_j = -\beta_j$ for the indices j corresponding to edges other than special edges, of which there are in all $k-(2p+1)$.

If the number of inversions present in a permutation $(i_1, i_2, \ldots, i_k)$ is denoted by $\epsilon(i_1, i_2, \ldots, i_k)$, we can write:

$$\begin{aligned} \det(B) &= (-1)^{\epsilon(1,2,\ldots,k)} \alpha_1 \alpha_2 \ldots \alpha_k + (-1)^{\epsilon(2,\ldots,k,1)} \beta_2 \ldots \beta_k \beta_1 \\ &= (+1) \alpha_1 \alpha_2 \ldots \alpha_k + (-1)^{k-1}[(-1)^{k-(2p+1)} \alpha_1 \alpha_2 \ldots \alpha_k] \\ &= 2\alpha_1 \alpha_2 \ldots \alpha_k \\ &= \pm 2 \end{aligned}$$

But this contradicts our assumption that the matrix has the unimodular property.

Therefore, every elementary cycle must contain an even number of special edges, and so also must every cycle of the graph. If we shrink each non-special edge by regarding its pair of terminal vertices as one vertex, we get a new graph containing no cycles of uneven length, and from König's theorem (Chapter 4), this graph will be bi-chromatic; if we denote by I_1 the set of the indices of the vertices x_i which are coloured blue, and by I_2 the set of the indices of the vertices coloured red, the disjoint sets I_1 and I_2 must satisfy the requirements of the theorem.

Corollary 1. *The incidence matrix $S = (s_j^i)$ of the arcs of a graph has the unimodular property.*

This follows from the fact that the column vector s_j, if it is not the null vector, contains one element which equals $+1$, one element which equals -1, and all its other elements are zero. Therefore we can take:

$$I_1 = \{1, 2, \ldots, n\}; \qquad I_2 = \emptyset$$

Corollary 2. *The incidence matrix $R = (r_j^i)$ of the edges of a graph has the unimodular property if and only if the graph is bi-chromatic.*

Since all the non-zero elements of R are equal to $+1$, R is totally unimodular if and only if the vertices of the graph can be divided into two disjoint internally stable sets, defined by:

$$\{x_i \mid i \in I_1\} \quad \text{and} \quad \{x_i \mid i \in I_2\}$$

Theorem 2 (Heller). *Let $A = (a_j^i)$ be a matrix the elements of which are either 0, $+1$ or -1; for two row vectors $\boldsymbol{a}^i$ and $\boldsymbol{a}^j$, we put $\boldsymbol{a}^i \succ \boldsymbol{a}^j$ if each non-zero element a_k^j of $\boldsymbol{a}^j$ is equal to the element a_k^i of $\boldsymbol{a}^i$ in the same column; the matrix A has the unimodular property if we have:*

$$(1) \qquad a_k^i = a_k^j \neq 0 \quad \Rightarrow \boldsymbol{a}^i \succ \boldsymbol{a}^j \quad \text{or} \quad \boldsymbol{a}^j \succ \boldsymbol{a}^i$$

$$(2) \qquad a_k^i = -a_k^j \neq 0 \Rightarrow \boldsymbol{a}^i \succ -\boldsymbol{a}^j \quad \text{or} \quad \boldsymbol{a}^j \succ -\boldsymbol{a}^i$$

The relation $\succ$ is a partial order which can easily be visualized; for example:

$$(0, -1, -1, 0, +1, +1, +1) \succ (0, -1, 0, 0, 0, 0, +1)$$

Any square submatrix B of A will also have the properties (1) and (2); we shall now show that the determinant of B, $\det(B) = 0, -1$ or $+1$.

This proposition is true for matrices B of order 1; let us assume that it is satisfied by matrices of order $q-1$, and show that it must then also be true for a matrix B of order q.

If B contains a column vector all of whose elements are zero, then certainly $\det(B) = 0$; if B contains a column vector with only one non-zero element, $\det(B)$ equals (apart from the sign) the minor corresponding to this element, and therefore $\det(B) = 0, -1$ or $+1$. The only case which remains to be investigated is that in which each column vector has at least two non-zero elements.

Let b_1^1 and b_1^2 be two non-zero elements of the column vector $\boldsymbol{b}_1$; to be definite, let us assume $b_1^1 = +1$, $b_1^2 = +1$. From (1), either $\boldsymbol{b}^1 \succ \boldsymbol{b}^2$ or $\boldsymbol{b}^2 \succ \boldsymbol{b}^1$. Suppose that $\boldsymbol{b}^1 \succ \boldsymbol{b}^2$, changing the order of the rows if necessary; in the matrix B, change each row vector $\boldsymbol{b}^i$, such that $\boldsymbol{b}^i \succ \pm \boldsymbol{b}^2$, to $\boldsymbol{b}^i \mp \boldsymbol{b}^2 = \bar{\boldsymbol{b}}^i$; all the elements of the new matrix $\bar{B}$ will be 0, -1 or $+1$ (note that it will contain more zero elements than B), and $\det(\bar{B}) = \det(B)$. We shall now show that $\bar{B}$ satisfies (1) and (2).

A row vector $\bar{\boldsymbol{b}}^i$ of $\bar{B}$ may either be an unaltered row of B, in which case we shall write: $\boldsymbol{b}^i$, or else it may be a modified row, which we shall denote by: $\boldsymbol{c}^i$.

Let us suppose, for example, $c_k^1 = +1$, $\bar{b}_k^i = +1$, and show that

$$\boldsymbol{c}^1 \succ \bar{\boldsymbol{b}}^i \quad \text{or} \quad \bar{\boldsymbol{b}}^i \succ \boldsymbol{c}^1$$

Since $c_k^1 = b_k^1 - b_k^2 = +1$, we must have $b_k^2 = 0$, and therefore

$$b_k^1 = c_k^1 = +1 = \bar{b}_k^i = b_k^i$$

from which $\boldsymbol{b}^i \succ \boldsymbol{b}^1$ or $\boldsymbol{b}^1 \succ \boldsymbol{b}^i$. Now:

$$\boldsymbol{b}^i \succ \boldsymbol{b}^1 \Rightarrow \boldsymbol{b}^i \succ \boldsymbol{b}^1 \succ \boldsymbol{b}^2 \Rightarrow \bar{\boldsymbol{b}}^i = \boldsymbol{b}^i - \boldsymbol{b}^2 \succ \boldsymbol{b}^1 - \boldsymbol{b}^2 = \boldsymbol{c}^1$$

On the other hand, if $\boldsymbol{b}^1 \succ \boldsymbol{b}^i$, we cannot have $\boldsymbol{b}^i \succ -\boldsymbol{b}^2$ (for then $\boldsymbol{b}^1 \succ -\boldsymbol{b}^2$, which is absurd), nor can we have $\boldsymbol{b}^2 \succ \pm \boldsymbol{b}^i$ (because $b_k^i = 1$ and $b_k^2 = 0$); this leaves two cases to consider: either $\boldsymbol{b}^i \succ \boldsymbol{b}^2$, or else the non-zero elements of $\boldsymbol{b}^i$ and $\boldsymbol{b}^2$ belong respectively to two disjoint sets of columns. Therefore:

$$\boldsymbol{b}^1 \succ \boldsymbol{b}^i \Rightarrow \begin{cases} \boldsymbol{b}^1 \succ \boldsymbol{b}^i \succ \boldsymbol{b}^2 \Rightarrow \boldsymbol{c}^1 = \boldsymbol{b}^1 - \boldsymbol{b}^2 \succ \boldsymbol{b}^i - \boldsymbol{b}^2 = \bar{\boldsymbol{b}}^i \\ \quad \text{(first case)} \\ \text{or:} \\ \boldsymbol{b}^1 \succ \boldsymbol{b}^i = \bar{\boldsymbol{b}}^i \Rightarrow \boldsymbol{c}^1 = \boldsymbol{b}^1 - \boldsymbol{b}^2 \succ \boldsymbol{b}^i = \bar{\boldsymbol{b}}^i \\ \quad \text{(second case)} \end{cases}$$

Therefore (1) has been shown to hold in all cases; the proof that (2) holds may be made in exactly the same way.

By repeating the transformation $B \to \bar{B}$ using other row vectors, we increase the number of zeros in the matrix $\bar{B}$, and we continue this until there is at least one column vector of $\bar{B}$ which has at most one non-zero element. Since we have assumed the theorem to hold for matrices of order $q-1$, we now have

$$\det(B) = \det(\bar{B}) = 0, +1 \text{ or } -1$$

Q.E.D.

THEOREM 3. *Let A and B be two matrices the elements of which are either* 0 *or* $+1$, *and which satisfy condition* (1) *of the above theorem; the matrix C which is formed by the union of the row vectors of A and B has the unimodular property.*

Consider a square submatrix $\bar{C}$ of C, which consists of a submatrix $\bar{A}$ of A and a submatrix $\bar{B}$ of B; we shall show that $\det(\bar{C}) = 0$, $+1$ or -1. Following the method of Theorem 2, we diminish the number of non-zero elements of the matrix $\bar{A}$ (and of the matrix $\bar{B}$) until every column vector of $\bar{A}$ (and of $\bar{B}$) has at most one non-zero element; the matrix $\bar{C}$ therefore contains at most two non-zero elements in every column (both of which are equal to $+1$); if one of these is in $\bar{A}$, the other must be in $\bar{B}$; therefore, from Theorem 1, $\bar{C}$ has the unimodular property, and we must have $\det(\bar{C}) = 0$, $+1$ or -1.

EXAMPLES. By direct application of the results of this paragraph, it may be shown that the following matrices have the unimodular property:

$$A = \begin{bmatrix} -1 & +1 & 0 & 0 \\ 0 & +1 & -1 & 0 \\ 0 & 0 & +1 & -1 \\ +1 & 0 & 0 & -1 \end{bmatrix} \qquad B = \begin{bmatrix} -1 & +1 & 0 & +1 \\ -1 & 0 & 0 & +1 \\ +1 & -1 & +1 & -1 \end{bmatrix}$$

$$C = \begin{bmatrix} +1 & 0 & +1 \\ +1 & +1 & +1 \\ 0 & +1 & +1 \end{bmatrix}$$

Systems with the Unimodular Property

Consider the system of homogeneous linear equations:

$$(1) \qquad \begin{cases} a_1^1 x^1 + a_2^1 x^2 + \ldots + a_m^1 x^m = 0 \\ a_1^2 x^1 + a_2^2 x^2 + \ldots + a_m^2 x^m = 0 \\ \cdots\cdots\cdots\cdots\cdots \\ a_1^n x^1 + a_2^n x^2 + \ldots + a_m^n x^m = 0 \end{cases}$$

If we put $\mathbf{x} = (x^1, x^2, \ldots, x^m)$, $\mathbf{0} = (0,0,\ldots,0)$, and $A = (a_j^i)$, this system may also be written: $A\mathbf{x} = \mathbf{0}$. We shall study here the vectors $\mathbf{x}$ which satisfy this equation when the matrix A has the unimodular property.

Let us follow the well-known method of solution, and consider the largest non-zero determinant which can be selected from A; let this be:

$$\Delta = \begin{vmatrix} a_1^1 & a_2^1 & \ldots & a_{n-p}^1 \\ a_1^2 & a_2^2 & \ldots & \\ \vdots & & & \\ a_1^{n-p} & a_2^{n-p} & \ldots & a_{n-p}^{n-p} \end{vmatrix} \neq 0$$

We first solve the *principal* system:

$$(1') \qquad \begin{cases} a_1^1 x^1 + a_2^1 x^2 + \ldots + a_{n-p}^1 x^{n-p} = -(a_{n-p+1}^1 x^{n-p+1} + \ldots + a_m^1 x^m) \\ a_1^2 x^1 + a_2^2 x^2 + \ldots + a_{n-p}^2 x^{n-p} = -(a_{n-p+1}^2 x^{n-p+1} + \ldots + a_m^2 x^m) \\ \cdots\cdots\cdots\cdots\cdots \\ a_1^{n-p} x^1 + a_2^{n-p} x^2 + \ldots + a_{n-p}^{n-p} x^{n-p} = -(a_{n-p+1}^{n-p} x^{n-p+1} + \ldots \\ \qquad\qquad + a_m^{n-p} x^m) \end{cases}$$

First let us give the non-principal variables $x^{n-p+1}, x^{n-p+2}, \ldots, x^m$ the values 1, 0, 0, ..., 0; then the vector $\boldsymbol{c}_1 = (c_1^1, c_1^2, \ldots, c_1^m)$ defined by (1′) is well-determined, its i-th element ($i \leqslant n-p$) being given by the formula:

$$c_1^i = \frac{1}{\Delta}\begin{vmatrix} a_1^1 & \cdots & a_{i-1}^1 & -a_{n-p+1}^1 & a_{i+1}^1 & \cdots & a_{n-p}^1 \\ a_1^2 & \cdots & & -a_{n-p+1}^2 & & \cdots & a_{n-p}^2 \\ \vdots & & & \vdots & & & \vdots \\ a_1^{n-p} & \cdots & & -a_{n-p+1}^{n-p} & & \cdots & a_{n-p}^{n-p} \end{vmatrix} = 0, +1 \text{ or } -1$$

It may be seen that all the elements of the vector $\boldsymbol{c}_1$ are 0, $+1$ or -1; and, since the system is homogeneous, $\boldsymbol{c}_1$, which is a solution of (1′), is also a solution of (1). This leads us to the following result:

THEOREM 4†. *If $A = (a_j^i)$ has the unimodular property, the equation $A\boldsymbol{x} = \boldsymbol{0}$ has as its solutions the vectors of a linear manifold of $m-n+p$ dimensions determined by the vectors $\boldsymbol{c}_1, \boldsymbol{c}_2, \ldots, \boldsymbol{c}_{m-n+p}$, the elements of which are all either* 0 *or* $+1$ *or* -1.

By giving $(x^{n-p+1}, x^{n-p+2}, \ldots, x^m)$ the values $\boldsymbol{e}_1 = (1, 0, 0, \ldots, 0)$, $\boldsymbol{e}_2 = (0, 1, 0, \ldots, 0), \ldots, \boldsymbol{e}_{m-n+p} = (0, 0, 0, \ldots, 1)$ in turn, we may use the above method to define vectors $\boldsymbol{c}_1, \boldsymbol{c}_2, \ldots, \boldsymbol{c}_{m-n+p}$, all of whose elements are either 0, $+1$ or -1. These vectors are linearly independent, since

$$\lambda_1 \boldsymbol{e}_1 + \lambda_2 \boldsymbol{e}_2 + \ldots + \lambda_{m-n+p} \boldsymbol{e}_{m-n+p} = \boldsymbol{0} \quad \Rightarrow \quad \lambda_1 = \ldots = \lambda_{m-n+p} = 0$$

Since the number of these vectors is equal to the dimensions of the solution space of the equations $A\boldsymbol{x} = \boldsymbol{0}$, the vectors $\boldsymbol{c}_k$ certainly form a fundamental basis of the solutions.

In Chapter 4 it was shown that we could construct a vector $(c^1, c^2, \ldots, c^m)$ which corresponds to a cycle μ of an antisymmetric graph by using the method: if in following μ, the arc u_j is traversed r times in the sense of its orientation, and s times in the opposite direction, we put $c^j = r - s$. We have also seen that the maximum number of linearly independent cycles is $\nu(G) = m - n + p$, where m is the number of arcs, n is the number of vertices and p is the number of components (Theorem 2, Chapter 4).

† HOFFMAN and KRUSKAL have proved a variant of this theorem which can be used in the theory of linear programming: let A be a matrix with integral elements, and let $\boldsymbol{b}, \boldsymbol{b}', \boldsymbol{a}$ and $\boldsymbol{a}'$ be vectors of integers; then if the matrix A has the unimodular property each face of the convex polyhedron

$$\{\boldsymbol{x} | \boldsymbol{b} \leqslant A\boldsymbol{x} \leqslant \boldsymbol{b}'; \boldsymbol{a} \leqslant \boldsymbol{x} \leqslant \boldsymbol{a}'\}$$

contains a point all of whose co-ordinates are integers. (In particular, the vertices of the polyhedron, which are faces of dimension 0, are points with integer co-ordinates.)

Theorem 5 (Poincaré–Veblen–Alexander). *Let G be an antisymmetric graph and $S = (s_j^i)$ the incidence matrix of its arcs; the necessary and sufficient condition that a vector $\mathbf{c} = (c^1, c^2, \ldots, c^m)$ represent a cycle or a sum of cycles is that $S\mathbf{c} = \mathbf{0}$.*

1. *Necessity of the condition.* We simply need to show that a vector $\mathbf{c}$ which corresponds to an elementary cycle satisfies $S\mathbf{c} = \mathbf{0}$. Therefore let

$$\mu = [x_1, x_2, \ldots, x_k, x_1]$$

be an elementary cycle, $\mathbf{c} = (c^1, c^2, \ldots, c^m)$ the corresponding vector, and let us show

$$(1) \qquad s_1^i c^1 + s_2^i c^2 + \ldots + s_m^i c^m = 0 \qquad (i = 1, 2, \ldots, n)$$

If the cycle μ does not go through x_i, the i-th vertex of the graph, then it does not use any arc u_j which is incident to x_i, and therefore $c^j = 0$ for every index j such that $s_j^i \neq 0$, which proves (1) for all vertices which are not on the cycle.

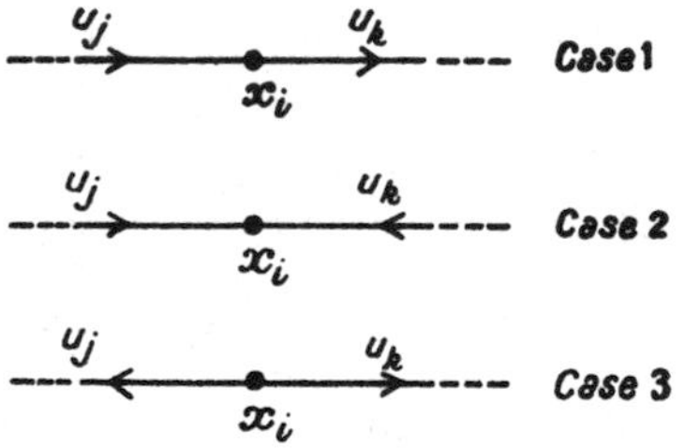

Fig. 15.2

If the cycle μ goes through the vertex x_i, it uses exactly two arcs u_j and u_k which are incident to x_i; these arcs may be oriented in three different ways as shown in Fig. 15.2. In case 1, we have

$$c^j = c^k \neq 0; \qquad s_j^i = -s_k^i \neq 0$$

and from this $s_j^i c^j + s_k^i c^k = 0$, which satisfies (1).

In case 2 or in case 3, we have

$$c^j = -c^k \neq 0; \qquad s_j^i = s_k^i \neq 0$$

and from this $s_j^i c^j + s_k^i c^k = 0$, which again satisfies (1).

Therefore we have shown that (1) holds for all vertices.

2. *Sufficiency of the condition.* Since any vector $\mathbf{c}$ such that $S\mathbf{c} = \mathbf{0}$ is a linear combination of the vectors $\mathbf{c}_1, \mathbf{c}_2, \ldots, \mathbf{c}_{m-n+p}$ as defined in Theorem 4, we simply need to show that these vectors represent cycles of the graph.

Since all its elements are 0, +1 or −1, the vector $\boldsymbol{c}_1$ corresponds to a simple chain μ, which never uses the same arc more than once (or to the sum of several simple disjoint chains). Since $s_j^i c_1^j$ can only equal 0 or +1 or −1, and as $\sum_{j=1} s_j^i c_k^j = 0$ (for all k), we must have $|s_j^i c_1^j| = 1$ for an even number of indices j; in other words, the chain μ uses an even number of the arcs u_j incident to any one vertex x_i of the graph. Therefore if we travel along the chain μ, starting from some arbitrary vertex, we may be sure that at some stage we shall return to this vertex: μ is therefore a cycle (or a sum of disjoint cycles).

Q.E.D.

This theorem provides an analytic method for finding the independent cycles of a graph; however, in the following chapter a geometric method will be given which is a great deal simpler.

Cyclomatic Matrices

Let $\boldsymbol{c}_1 = (c_1^1, c_1^2, \ldots, c_1^m)$, ..., $\boldsymbol{c}_\nu = (c_\nu^1, c_\nu^2, \ldots, c_\nu^m)$, be the vectors of a fundamental basis of the cycles of a graph G, and let us consider the matrix

$$C = \begin{bmatrix} c_1^1 & c_2^1 & \ldots & c_\nu^1 \\ c_1^2 & c_2^2 & \ldots & c_\nu^2 \\ \vdots & & & \\ c_1^m & c_2^m & \ldots & c_\nu^m \end{bmatrix}$$

A matrix of this sort is said to be *cyclomatic* for the graph G. If the column vectors $\boldsymbol{c}_k$ are chosen according to Theorem 4, all the elements of the matrix C are either 0, +1 or −1, and, moreover, C is of the form

$$C = \begin{bmatrix} c_1^1 & \ldots & \ldots & c_\nu^1 \\ \vdots & & & \\ c_1^{m-\nu} & \ldots & \ldots & c_\nu^{m-\nu} \\ 1 & 0 & 0 & 0 \\ 0 & 1 & 0 & 0 \\ \cdots & \cdots & \cdots & \cdots \\ 0 & 0 & 0 \ldots & 1 \end{bmatrix}$$

The result of pre-multiplying C by S is the matrix 0, since

$$SC = \begin{bmatrix} \langle \boldsymbol{s}^1, \boldsymbol{c}_1 \rangle & \langle \boldsymbol{s}^1, \boldsymbol{c}_2 \rangle & \ldots & \langle \boldsymbol{s}^1, \boldsymbol{c}_\nu \rangle \\ \langle \boldsymbol{s}^2, \boldsymbol{c}_1 \rangle & \langle \boldsymbol{s}^2, \boldsymbol{c}_2 \rangle & \ldots & \\ \cdots & \cdots & \cdots & \cdots \\ \langle \boldsymbol{s}^n, \boldsymbol{c}_1 \rangle & \langle \boldsymbol{s}^n, \boldsymbol{c}_2 \rangle & \ldots & \langle \boldsymbol{s}^n, \boldsymbol{c}_\nu \rangle \end{bmatrix} = \begin{bmatrix} 0 & 0 & \ldots & 0 \\ 0 & & & \\ \vdots & & & \\ 0 & & \ldots & 0 \end{bmatrix}$$

Therefore if the cyclomatic matrix C is known, the equation $SC = 0$ may be used to find the incidence matrix of the arcs; we have to solve the linear equation $C^*\mathbf{s} = \mathbf{0}$ to obtain the different row vectors of S. This remark has many applications in the theory of relays.

EXAMPLE (S. OKADA). An unknown electrical relay consists of 8 switches $u_1, u_2, \ldots, u_8$ (which we shall denote more simply by 1, 2, ..., 8). These switches are connected in some way to the terminals $x_1, x_2, x_3, x_4, x_5, x_6$. We want to find the geometric configuration of the relay; to do this, we observe that by closing a set of switches simultaneously, the relay allows the current to pass and a lamp lights up.

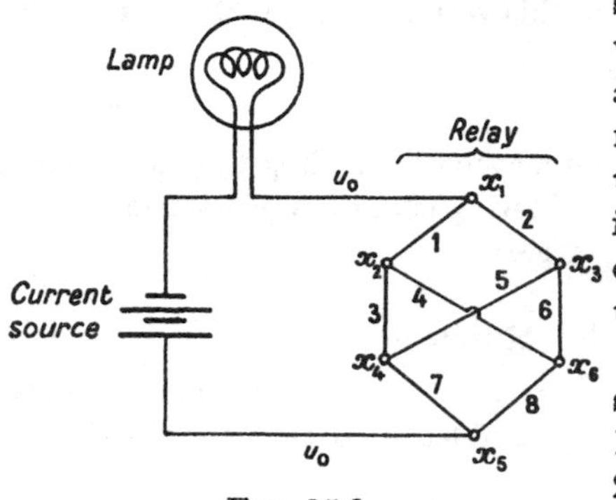

FIG. 15.3

The following combinations of switches allow the lamp to light: 13568, 137, 14567, 148, 23458, 23467, 257, 268.

Therefore we can form the matrix C, the elements of which are only known modulo 2:

$$C = \begin{array}{c} \begin{matrix} c_1 & c_2 & c_3 & c_4 & c_5 & c_6 & c_7 & c_8 \end{matrix} \\ \begin{bmatrix} 1 & 1 & 1 & 1 & 1 & 1 & 1 & 1 \\ 1 & 1 & 1 & 1 & 0 & 0 & 0 & 0 \\ 0 & 0 & 0 & 0 & 1 & 1 & 1 & 1 \\ 1 & 1 & 0 & 0 & 1 & 1 & 0 & 0 \\ 0 & 0 & 1 & 1 & 1 & 1 & 0 & 0 \\ 1 & 0 & 1 & 0 & 1 & 0 & 1 & 0 \\ 1 & 0 & 1 & 0 & 0 & 1 & 0 & 1 \\ 0 & 1 & 1 & 0 & 0 & 1 & 1 & 0 \\ 1 & 0 & 0 & 1 & 1 & 0 & 0 & 1 \end{bmatrix} \end{array} \begin{matrix} u_0 \\ u_1 \\ u_2 \\ u_3 \\ u_4 \\ u_5 \\ u_6 \\ u_7 \\ u_8 \end{matrix}$$

By eliminating the columns which are linear combinations of the other columns, transposing the result and sweeping out the new columns, we get:

$$C^* = \begin{bmatrix} 1 & 1 & 0 & 1 & 0 & 1 & 1 & 0 & 1 \\ 0 & 1 & 1 & 0 & 1 & 1 & 0 & 1 & 1 \\ 0 & 0 & 0 & 1 & 1 & 0 & 0 & 1 & 1 \\ 0 & 0 & 0 & 0 & 0 & 1 & 1 & 1 & 1 \end{bmatrix}$$

We now have to solve the equation $C^*\mathbf{t} \equiv \mathbf{0}$ (mod. 2), that is to say

$$\begin{aligned}
t_1+t_2+t_4+t_6+t_7+t_9 &\equiv 0 \pmod{2}\\
t_2+t_3+t_5+t_6+t_8+t_9 &\equiv 0 \pmod{2}\\
t_4+t_5+t_8+t_9 &\equiv 0 \pmod{2}\\
t_6+t_7+t_8+t_9 &\equiv 0 \pmod{2}
\end{aligned}$$

This gives the fundamental system of independent solutions which are the rows of the matrix:

$$\begin{bmatrix}
1 & 1 & 1 & 0 & 0 & 0 & 0 & 0 & 0\\
0 & 1 & 0 & 1 & 1 & 0 & 0 & 0 & 0\\
1 & 1 & 0 & 0 & 0 & 1 & 1 & 0 & 0\\
0 & 0 & 0 & 1 & 0 & 1 & 0 & 1 & 0\\
1 & 0 & 0 & 1 & 0 & 1 & 0 & 0 & 1
\end{bmatrix}$$

By making linear combinations of the rows, we can so arrange it that each column contains at most two elements equal to 1; we then have the incidence matrix R of the edges of the graph G of Fig. 15.3:

$$R = \begin{array}{c} u_0 \;\; u_1 \;\; u_2 \;\; u_3 \;\; u_4 \;\; u_5 \;\; u_6 \;\; u_7 \;\; u_8 \\
\begin{bmatrix}
1 & 1 & 1 & 0 & 0 & 0 & 0 & 0 & 0\\
0 & 1 & 0 & 1 & 1 & 0 & 0 & 0 & 0\\
0 & 0 & 1 & 0 & 0 & 1 & 1 & 0 & 0\\
0 & 0 & 0 & 1 & 0 & 1 & 0 & 1 & 0\\
1 & 0 & 0 & 0 & 0 & 0 & 0 & 1 & 1\\
0 & 0 & 0 & 0 & 1 & 0 & 1 & 0 & 1
\end{bmatrix}\end{array}
\begin{array}{c} \\ x_1\\ x_2\\ x_3\\ x_4\\ x_5\\ x_6 \end{array}$$

16. Trees and Arborescences

Trees

From now on, we shall be principally interested in the study of finite non-oriented graphs. $U = \{u_1, u_2, \ldots, u_m\}$ denotes the set of the *edges*, and U_x denotes the set of *edges* incident to a vertex x. A vertex x of a graph is said to be *pendant* if there is only one edge incident to x, that is: $|U_x| = 1$; a vertex x is said to be *isolated* if no edge is incident to x, that is: $|U_x| = 0$. Likewise we represent the set of vertices adjacent to x by Γx.

A *tree* is defined to be a finite connected graph with no cycles, and possessing at least two vertices. This important concept, which was first introduced by CAYLEY, can also be defined in other ways, and we shall in future make use of the six equivalent definitions given in the following theorem:

THEOREM 1. *Let H be a graph of order $|X| = n > 1$; any one of the following equivalent properties characterizes a tree:*

(1) *H is connected and does not possess any cycles;*
(2) *H contains no cycles and has $n-1$ edges;*
(3) *H is connected and has $n-1$ edges;*
(4) *H contains no cycles, and if an edge is added which joins two non-adjacent vertices, one (and only one) cycle is thereby formed;*
(5) *H is connected but loses this property if any edge is deleted;*
(6) *every pair of vertices is connected by one and only one chain.*

(1) $\Rightarrow$ (2), for if p is the number of components, and m the number of edges, we have $p = 1$, $\nu = m - n + p = 0$, and therefore $m = n - p = n - 1$.

(2) $\Rightarrow$ (3), for $\nu = 0$, $m = n - 1$, and therefore $p = \nu - m + n = 1$, and H is connected.

(3) $\Rightarrow$ (4), for $p = 1$, $m = n - 1$, and therefore $\nu = m - n + p = 0$: H contains no cycles; further, if we add an edge which connects any two non-adjacent vertices, ν becomes equal to 1, and therefore the graph now contains exactly one cycle.

(4) $\Rightarrow$ (5), for if H is not connected, the vertices x and y not being connected, then adding the edge $[x, y]$ does not form a cycle; therefore $p = 1$,

$\nu = 0$, and hence $m = n-1$. On the other hand, by deleting an edge, we have $m = n-2$, $\nu = 0$, and hence $p = \nu - m + n = 2$, and H is no longer connected.

(5) $\Rightarrow$ (6), because, since H is connected, a chain exists joining any two vertices, say x and y; there cannot be two chains joining x and y since the deletion of any edge which belongs only to the second chain does not disconnect the graph.

(6) $\Rightarrow$ (1), for if H contains a cycle, at least one pair of vertices must be connected by two distinct chains.

THEOREM 2. *A tree contains at least two pendant vertices.*

Let H be a tree with less than two pendant vertices. Consider a man who sets out to traverse the graph and who starts from any arbitrary point (if there are no pendant vertices) or from the pendant vertex (if there is exactly one): if he refrains from using an edge more than once, he will not be able to return twice to any vertex (since H contains no cycles); on the other hand, if he has reached a point x, he will always be able to leave by a new edge (since $|\boldsymbol{U}_x| \geqslant 2$ if x is not a pendant vertex). Therefore he can go on indefinitely, but this is impossible since H is a finite graph.

THEOREM 3. *A graph $G = (X, \boldsymbol{U})$ possesses a partial graph which is a tree, if and only if it is connected and $|X| \geqslant 2$.*

If G is not connected, none of its partial graphs is connected, and therefore G cannot possess any partial graph which is a tree.

If G is connected, let us see if there is any edge which may be deleted without disconnecting G. If no such edge exists, G is a tree, by virtue of the property (5); if such an edge does exist, it is deleted, and we search for a new edge to delete, etc....

When no further edges can be deleted, we shall have a tree (property 5), which has X as its set of vertices.

Q.E.D.

This theorem thus gives us an algorithm whereby we can construct a partial tree of a graph. We can generalize the foregoing by means of the following problem:

PROBLEM 1. *Consider a connected graph $G = (X, \boldsymbol{U})$, and let us associate a number $l(\boldsymbol{u}) \geqslant 0$ with every edge $\boldsymbol{u}$, which we shall call its length; find a partial tree $H = (X, V)$ of the graph which has a minimum total length $\sum_{\boldsymbol{u} \in V} l(\boldsymbol{u})$.*

EXAMPLE. This problem is often encountered in telecommunications and on divers other occasions. For example, let us consider the following question: what is the shortest length of cable needed to link together n given towns? The towns are the vertices of the graph, and $l(x,y)$ is the distance in miles separating the towns x and y. The required network of cables must be connected, and, since it is of minimum length, it will not contain any cycles: therefore it must be a tree. What we want here is the 'shortest' tree possible which is a partial graph of the complete graph with n vertices.

We shall first establish a lemma.

LEMMA. *If $G = (X, U)$ is a complete graph, and if the lengths $l(\boldsymbol{u})$ associated with the edges are all different, then Problem 1 possesses a unique solution (X, V); the set $V = \{\boldsymbol{v}_1, \boldsymbol{v}_2, \ldots, \boldsymbol{v}_{n-1}\}$ is found in the following manner: $\boldsymbol{v}_1$ is the shortest edge; $\boldsymbol{v}_2$ is the shortest edge such that $\boldsymbol{v}_1 \neq \boldsymbol{v}_2$ and $V_2 = \{\boldsymbol{v}_1, \boldsymbol{v}_2\}$ contains no cycles; $\boldsymbol{v}_3$ is the shortest edge such that $\boldsymbol{v}_3 \neq \boldsymbol{v}_1$, $\boldsymbol{v}_3 \neq \boldsymbol{v}_2$, and $V_3 = \{\boldsymbol{v}_1, \boldsymbol{v}_2, \boldsymbol{v}_3\}$ contains no cycles, etc....*

Using the method described in the lemma, find the sets V_1 $(= \{\boldsymbol{v}_1\})$, V_2, V_3, ..., V_k, and let us assume that when V_k has been found no further edge can be added without making at least one cycle. From the characteristic property (4), the graph (X, V_k) is a tree, and therefore from property (2), $k = n-1$.

Let (X, V) be a tree for which $\sum_{\boldsymbol{u} \in V} l(\boldsymbol{u})$ is a minimum; we shall show that $V = V_{n-1}$.

If $V \neq V_{n-1} = \{\boldsymbol{v}_1, \boldsymbol{v}_2, \ldots, \boldsymbol{v}_{n-1}\}$, let $\boldsymbol{v}_k$ be the first of the edges $\boldsymbol{v}_1$, $\boldsymbol{v}_2$, ... which is not in V.

Using the characteristic property (4), there is one and only one cycle in $V \cup \{\boldsymbol{v}_k\}$; and this cycle contains an edge $\boldsymbol{u}_0$ such that $\boldsymbol{u}_0 \notin V_{n-1}$. If we put $W = (V \cup \{\boldsymbol{v}_k\}) - \{\boldsymbol{u}_0\}$, the graph (X, W) contains no cycles and has $n-1$ edges and therefore is a tree [property (2)].

$V_{k-1} \cup \{\boldsymbol{u}_0\}$ contains no cycles (because it is contained in V), and therefore $l(\boldsymbol{u}_0) > l(\boldsymbol{v}_k)$. Therefore the total length of the tree (X, W) which was deduced from the tree (X, V) simply by changing the edge $\boldsymbol{u}_0$ to $\boldsymbol{v}_k$, must be less than that of (X, V); but as (X, V) is the 'shortest' tree, this leads to a contradiction.

ALGORITHM FOR PROBLEM 1 (KRUSKAL [6]). We proceed by stages, choosing at each step the shortest edge which does not complete any cycles with the edges already chosen.

In this way, a set $V_{n-1} = \{\boldsymbol{v}_1, \ldots, \boldsymbol{v}_{n-1}\}$ of $n-1$ edges is chosen, and the graph (X, V_{n-1}) is a tree of minimum total length.

We can, in fact, arrange matters so that all the edges are of different length; if, for example, $l(u_1) = l(u_2) = l(u_3)$, we make the following alterations:

$$l(u_1) \to l'(u_1) = l(u_1) + \epsilon$$
$$l(u_2) \to l'(u_2) = l(u_2) + 2\epsilon$$
$$l(u_3) \to l'(u_3) = l(u_3) + 3\epsilon$$

and by taking ϵ sufficiently small, the positions of these edges relative to the remaining edges will not be affected.

Likewise, we can ensure that the graph is complete by adding edges w_k as required, whose lengths are great enough to satisfy

$$l(w_k) > \sum_{u \in U} l(u)$$

This new graph contains one and only one tree $H = (X, V_{n-1})$ of minimum total length; from Theorem 3, V_{n-1} does not contain any of the edges $w_k \notin U$, since

$$l(w_k) > \sum_{u \in U} l(u)$$

Therefore H must be a partial tree of G of minimum total length.

ALGORITHM FOR FINDING A FUNDAMENTAL SYSTEM OF INDEPENDENT CYCLES. We can assume that the graph G is connected (if this is not so, then we imagine that each component is treated separately). Using the above method, we first construct a tree (X, V_{n-1}). Let $u_1, u_2, \ldots, u_{m-n+1}$ be the edges of U which are not in V_{n-1}; by adding any one of these edges, u_k say, to the tree (X, V_{n-1}), we create exactly one cycle μ_k [property (4)]. The cycles $\mu_1, \mu_2, \ldots, \mu_{m-n+1}$ are certainly linearly independent, since each of them contains an edge which does not appear in any of the others; on the other hand, there are $m-n+1 = \nu(G)$ of them, and hence we must have found a fundamental system of linearly independent cycles.

It should be noted that, when the graph is planar, there is a much simpler method for finding a fundamental system of independent cycles, as will be shown later (Theorem 1, Chapter 21).

EXAMPLE. Consider a symmetric transport network (X, U), with a source x_0 and a sink z; let us join the source and the sink by an edge $[x_0, z]$, and give each edge an arbitrary orientation. The traffic $\phi(u)$ along the edge u is positive if the flow is in the same direction as its orientation, and negative otherwise; further, for each vertex, the inflow must equal the outflow, that is:

$$\sum_{u \in U_x^+} \phi(u) = \sum_{u \in U_x^-} \phi(u), \ (x \in X)$$

We now wish to solve the following problem:

Find a flow ϕ which has given values $\phi(\boldsymbol{u}_1) = \lambda_1, \phi(\boldsymbol{u}_2) = \lambda_2, \ldots, \phi(\boldsymbol{u}_k) = \lambda_k$, for the edges $\boldsymbol{u}_1, \boldsymbol{u}_2, \ldots, \boldsymbol{u}_k$, respectively.

Suppose, for example, that in the graph G of Fig. 16.1 we wish to get:

$$\phi(\boldsymbol{u}_1) = 6; \qquad \phi(\boldsymbol{u}_2) = \phi(\boldsymbol{u}_3) = 1; \qquad \phi(\boldsymbol{u}_4) = \phi(\boldsymbol{u}_5) = \phi(\boldsymbol{u}_6) = 2$$

In this graph, we get a tree (shown by the heavy lines) if we cut out the selected edges $\boldsymbol{u}_1, \boldsymbol{u}_2, \ldots, \boldsymbol{u}_6$, and the edge $\boldsymbol{u}_7$; if each edge $\boldsymbol{u}_i$ (with $i \leqslant 6$) is considered with this tree, exactly one cycle $\boldsymbol{\mu}_i$ is determined, and we send along this cycle an amount of traffic equal to $\phi(\boldsymbol{u}_i) = \lambda_i$ [the cycle containing $\boldsymbol{\mu}_7$ is traversed by a flow of traffic of arbitrary value $\phi(\boldsymbol{u}_7) = \lambda$].

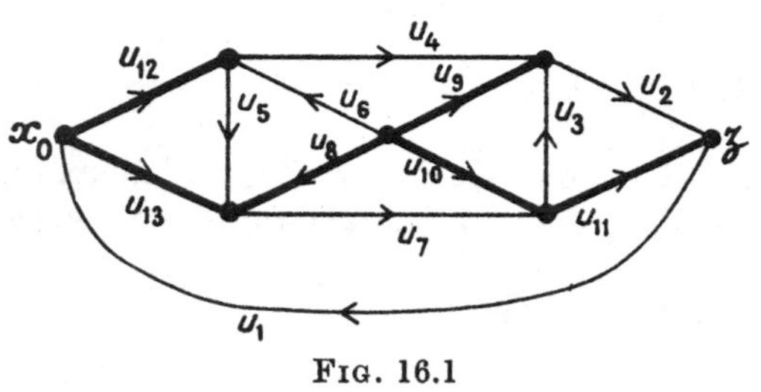

FIG. 16.1

By summing the individual traffic flows edge by edge, we shall get the required overall flow (which in this case will depend on the parameter λ). The traffic $\phi(\boldsymbol{u})$ along each edge is therefore given by the following table:

Traffic round	u_1	u_2	u_3	u_4	u_5	u_6	u_7	u_8	u_9	u_{10}	u_{11}	u_{12}	u_{13}
μ_1:	+6							−6		+6	+6		+6
μ_2:		+1							+1	−1	−1		
μ_3:			+1						−1	+1			
μ_4:				+2				+2	−2			+2	−2
μ_5:					+2							+2	−2
μ_6:						+2		−2				−2	+2
μ_7:							$+\lambda$	$+\lambda$		$-\lambda$			
$\phi =$	6	1	1	2	2	2	λ	$\lambda-6$	(-2)	$6-\lambda$	5	2	4

In general there is no solution to the problem if the graph remaining after the selected edges have been deleted is disconnected.

Analytic Treatment

Consider a matrix $A = (a_j^i)$ with m rows and n columns, and let us put $N = \{1,2,\ldots,n\}$, $M = \{1,2,\ldots,m\}$. It is often convenient to denote the matrix A by a_N^M; with this notation, if $I \subset M, J \subset N$, a_J^I stands for the submatrix obtained by deleting the columns whose indices are not elements of J, and the rows whose indices are not elements of I. First let us quote two formulae from the theory of determinants, proofs of which may be found in books on that topic:

LAPLACE'S FORMULA. *Let a_N^N be a matrix of order n, and let $k \leqslant n$; then we have*

$$\det(a_N^N) = \sum_{1 \leqslant i_1 < i_2 < \ldots < i_k \leqslant n} (-1)^{i_1 + \ldots + i_k + 1 + \ldots + k} \det(a_{\{1,2,\ldots,k\}}^{\{i_1, i_2, \ldots, i_k\}}) \times \times \det(a_{N-\{1,2,\ldots,k\}}^{N-\{i_1,i_2,\ldots,i_k\}})$$

(on putting $k = 1$, this reduces to the formula for developing a determinant by means of the elements in its first row).

FORMULA OF BINET AND CAUCHY. *Let a_N^N and b_N^N be two matrices of order n, and let $k \leqslant n$; then we have*

$$\det(a_N^K b_K^N) = \sum_{\substack{|I|=k \\ I \subset N}} \det(a_I^K) \det(b_K^I)$$

In this formula, $K = \{1,2,\ldots,k\}$. If $k = n$, we get the formula for the determinant of the product of two square matrices.

Consider a graph G with n vertices and m edges ($m \leqslant n$); by giving each edge an arbitrary orientation, we can construct an incidence matrix of the arcs: $S = s_M^N$. The square matrix s_M^M corresponds to the vertices $x_1, x_2, \ldots, x_m$, and to the edges $u_1, u_2, \ldots, u_m$; the pair formed by such a set of vertices and such a set of arcs is defined to be a *linear configuration* H. In general, this will not be a graph, since an edge may appear in H with neither of its terminal vertices ('floating' edge) or else with only one of them ('pendant' edge). We now have:

THEOREM 4. *Let H be a linear configuration consisting of m vertices x_1, $x_2, \ldots, x_m$, and m arcs $u_1, u_2, \ldots, u_m$; if H contains the arcs of a cycle, $\det(s_M^M) = 0$.*

Let us assume, in fact, that $u_1, u_2, \ldots, u_k$ form a cycle, and that the vertices of H on this cycle are numbered $x_1, x_2, \ldots, x_i$ (with $i \leqslant k$); then we can always assume that the arcs of the cycle are all oriented in the same direction. If this is not the case, the orientation of some of the arcs must be reversed. The only effect of such a change is that the sign of $\det(s_M^M)$ may be altered. We may now write:

$$s_M^M = \begin{bmatrix} s_K^K & s_{M-K}^K \\ 0 & s_{M-K}^{M-K} \end{bmatrix}$$

From Laplace's formula, we have

$$\det(s_M^M) = \pm\det(s_K^K)\det(s_{M-K}^{M-K})+0$$

If $i = k$, we have

$$\det(s_K^K) = \begin{vmatrix} 1 & 0 & 0 & \dots & & -1 \\ -1 & 1 & 0 & \dots & & 0 \\ 0 & -1 & 1 & \dots & & \\ \dots & \dots & \dots & \dots & \dots & \dots \\ & & & & 1 & 0 \\ 0 & 0 & 0 & \dots & -1 & 1 \end{vmatrix} = 0$$

(since the sum of the rows is zero). On the other hand, if $i < k$, we again have $\det(s_K^K) = 0$ since in this case the k-th row is empty. Therefore $\det(s_M^M) = 0$ for all cases.

THEOREM 5. *A graph G which has $m+1$ vertices $x_1, x_2, \dots, x_{m+1}$ and m edges $u_1, u_2, \dots, u_m$ (here given an arbitrary orientation) is a tree if and only if the determinant of the square matrix s_M^M, which can be formed from the incidence matrix s_M^N by omitting the $(m+1)$st row, is equal to $+1$ or to -1; in every other case, this determinant is zero.*

1. Let G be a graph of order $n = m+1$, with $\det(s_M^M) \neq 0$; from Theorem 4, G contains no cycles; but since G contains $n-1$ edges, it must be a tree [Theorem 1, property 2].

2. Let G be a tree with vertices $x_1, x_2, \dots, x_{m+1}$; we shall now alter the numbering of all the arcs and of all the vertices except x_{m+1} (this can only affect the sign of $\det(s_M^M)$, since altering the numbering is equivalent to altering the order of the rows and of the columns of the determinant). As in Theorem 2, the tree G must contain at least one pendant vertex which is distinct from x_{m+1}; we shall call this vertex x_1, and its incident edge u_1. Now let us delete x_1 and u_1, and again look for a pendant vertex which we shall label x_2, and give its incident edge the label u_2; delete this edge and vertex, and begin again, etc. With this new system of numbering, we have

$$\det(s_M^M) = \begin{vmatrix} \pm 1 & 0 & 0 & \dots & 0 \\ s_1^2 & \pm 1 & 0 & \dots & 0 \\ s_1^3 & s_2^3 & \pm 1 & \dots & 0 \\ \vdots & & & & \vdots \\ s_1^m & & & \dots & \pm 1 \end{vmatrix} = \pm 1$$

Q.E.D.

THEOREM 6 (Proof as given by TRENT). *In a graph G without loops, the number of distinct trees which are partial graphs of G equals the minor (of any term on the principal diagonal) of a square matrix of order n, the elements of which are*

$$b_j^i = \begin{cases} |\Gamma x_i| & \text{if } i = j \\ -1 & \text{if } i \neq j, \quad [x_i, x_j] \in U \\ 0 & \text{if } i \neq j, \quad [x_i, x_j] \notin U \end{cases}$$

Give an arbitrary orientation to the edges of G, and form the incidence matrix of the arcs $S = s_M^N$. The general element of the matrix SS^* is

$$\langle s^i, s_j^* \rangle = \langle s^i, s^j \rangle = s_1^i s_1^j + s_2^i s_2^j + \ldots + s_m^i s_m^j$$

$$= \begin{cases} |\Gamma x_i| & \text{if } i = j \\ -1 & \text{if } i \neq j, \quad [x_i, x_j] \in U \\ 0 & \text{if } i \neq j, \quad [x_i, x_j] \notin U \end{cases}$$

Therefore, we have $\langle s^i, s_j^* \rangle = b_j^i$; if we put $K = \{1, 2, \ldots, k = n-1\}$, and use the formula of BINET and CAUCHY, the minor relative to the element b_n^n is equal to

$$\Delta_n = \det(b_K^K) = \det(s_M^K . s^{*M}_K) = \sum_{\substack{I \subset M \\ |I| = n-1}} \det(s_I^K) \det(s^{*I}_K)$$

$$= \sum_{\substack{I \subset M \\ |I| = n-1}} |\det(s_I^K)|^2$$

From Theorem 5, the last expression is equal to the number of trees with n vertices which can be extracted from G.

EXAMPLE. The graph of Fig. 16.2 has 16 partial trees; this corresponds exactly to the value of the minor:

$$\Delta_1 = \begin{vmatrix} 3 & -1 & -1 \\ -1 & 3 & -1 \\ -1 & -1 & 3 \end{vmatrix} = 16$$

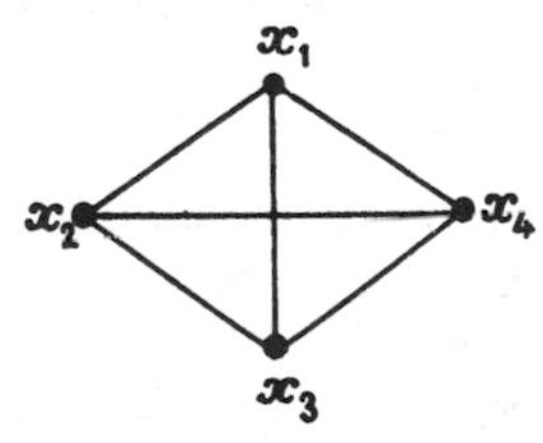

FIG. 16.2

As the graph is complete, this result could have been obtained more directly by noting that the total number of distinct trees with $n = 4$ vertices is $n^{n-2} = 4^2 = 16$.

The 16 partial trees are shown in Fig. 16.2 *bis* on the following page.

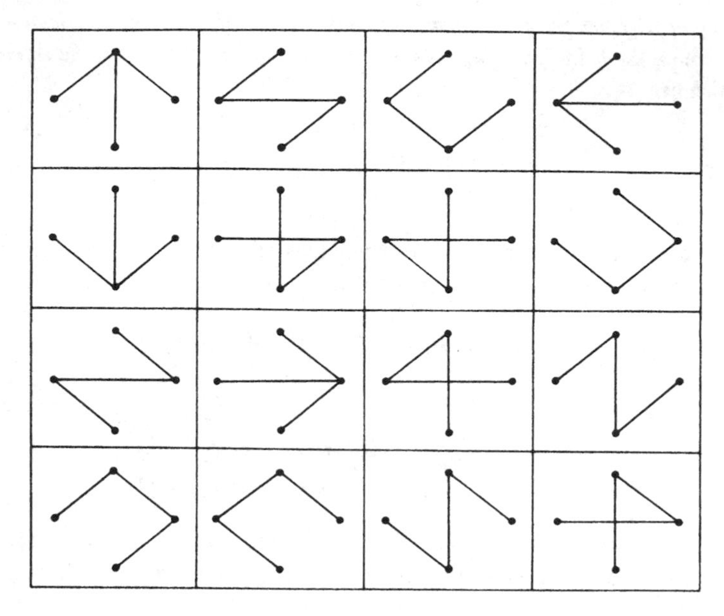

FIG. 16.2 *bis*

Arborescences

A finite graph (X, U) is an *arborescence with root* $x_1 \in X$ if $n > 1$ and if:

1. every vertex $\neq x_1$ is the terminal vertex of a single arc;
2. x_1 is not the terminal vertex of any arc;
3. (X, U) contains no circuits.

THEOREM 7. *Every arborescence is a tree.*

Let H be an arborescence with root x_1; from (3), there are as many arcs as there are edges. If n is the number of vertices, m the number of edges, we have therefore:

$$m = n-1$$

On the other hand, H contains no cycles, since there are no circuits, and in any cycle which is not a circuit, there is a vertex which is the terminal vertex of at least two arcs. Therefore H must be a tree.

It is easy to visualize an arborescence; it is a tree, in which every edge has a unique orientation; further, there is a path going from x_1 to every other vertex x (because if a traveller, placed at x, wanted to traverse the arcs of the graph in the opposite direction to their orientation, he could only be held up at x_1).

EXAMPLE 1 (Plane geometry). Suppose that $n-1$ disjoint rings are drawn on a plane. Every possible position of these $n-1$ rings can be represented by an arborescence with n vertices (having as its root x_1 a ring of infinite radius). With $n = 4$, there are four possible arrangements, corresponding to the four arborescences of Fig. 16.3.

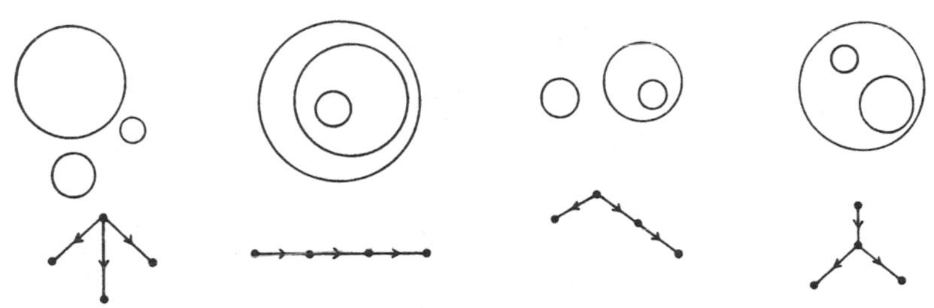

FIG. 16.3

EXAMPLE 2 [Theory of the parentheses (LUKASIEWICZ)]. Consider a set $A = \{a, b, c, \ldots\}$, and an operation . defined on A, which is commutative but not associative: for example, $a.(b.c) \neq (a.b).c$. In algebra, we call the result of a certain number of operations a *monoid*, for example:

$$u = [(a.b).c].([d.(e.f)].g)$$

With this monoid, we can associate the 'bifurcating' arborescence shown in Fig. 16.4: for every vertex x, either $|\Gamma x| = 0$ or $|\Gamma x| = 2$.

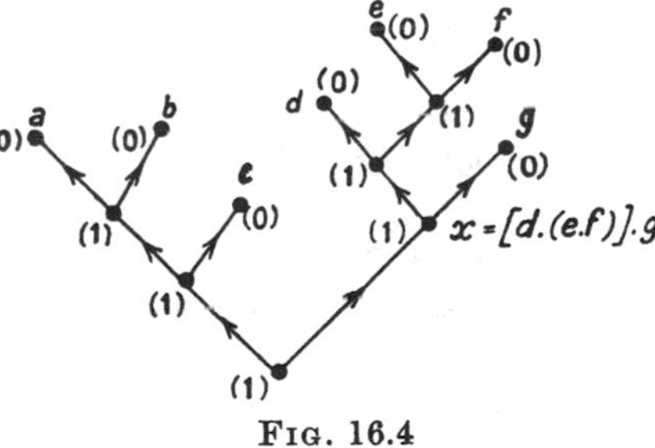

FIG. 16.4

If the operation . is not commutative, we have to use a topological graph, that is to say a graph which can only be effectively represented in an oriented plane. LUKASIEWICZ has suggested that the monoid u can be represented by the sequence *abcdefg*, followed by the sequence 1110001101000 (obtained by labelling the pendant vertices, with a 0, and all other vertices with a 1, and then reading the tree from the bottom (root) to the top, and from left to right)†.

† From this remark, it is easy to deduce that the number of distinct bifurcating arborescences with $n + 1$ pendant vertices is

$$\frac{1}{2n+1}\binom{2n+1}{n}$$

We shall now investigate for arborescences the problems which we have already considered for trees; in particular: how many partial graphs of the (oriented) graph G are arborescences? Let $A = (a_j^i)$ be the matrix associated with G, and let us form the diagonal matrix $D = (d_j^i)$, where

$$d_j^i \begin{cases} = 0 & \text{if } i \neq j \\ = |\Gamma^{-1}x_i| & \text{if } i = j \end{cases}$$

The matrix $D - A = (d_j^i - a_j^i)$ can therefore be written:

$$D-A = \begin{bmatrix} \sum\limits_{i\neq 1} a_1^i & -a_2^1 & \dots & -a_n^1 \\ -a_1^2 & \sum\limits_{i\neq 2} a_2^i & \dots & -a_n^2 \\ -a_1^3 & -a_2^3 & \dots & -a_n^3 \\ \vdots & \vdots & & \vdots \\ -a_1^n & -a_2^n & \dots & \sum\limits_{i\neq n} a_n^i \end{bmatrix}$$

The determinant of this matrix is equal to 0 (since the rows sum to zero).

We shall denote the minor of this determinant obtained by deleting the k-th row and the k-th column by $\Delta_k = \Delta_k(\boldsymbol{a}_1, \boldsymbol{a}_2, \dots, \boldsymbol{a}_n)$; it does not depend on the vector $\boldsymbol{a}_k$.

Remark. $\Delta_k(\boldsymbol{a}_1, \boldsymbol{a}_2, \dots, \boldsymbol{a}_n)$ *is a linear function of the column vectors* $\boldsymbol{a}_j$*; that is to say:*

$$\Delta_k(\boldsymbol{a}_1 + \boldsymbol{a}_1', \boldsymbol{a}_2, \dots, \boldsymbol{a}_n) = \Delta_k(\boldsymbol{a}_1, \boldsymbol{a}_2, \dots, \boldsymbol{a}_n) + \Delta_k(\boldsymbol{a}_1', \boldsymbol{a}_2, \dots, \boldsymbol{a}_n)$$
$$\Delta_k(\lambda \boldsymbol{a}_1, \boldsymbol{a}_2, \dots, \boldsymbol{a}_n) = \lambda \Delta_k(\boldsymbol{a}_1, \boldsymbol{a}_2, \dots, \boldsymbol{a}_n).$$

The proof is immediate.

THEOREM 8. *Let* G *be a graph without loops, with associated matrix* A*, which has* $m = n-1$ *arcs;* G *is an arborescence with root* x_1 *if and only if* $\Delta_1 = 1$*; in every other case* $\Delta_1 = 0$.

1. If G is an arborescence with root x_1, the elements d_i^i are equal to $|\Gamma^{-1}x_i| = 1$ (for $i = 2, 3, \dots, n$); let us label the vertices in such a way that their indices are increasing whenever we are following a path (this is always possible if G is an arborescence); we then have:

$$\Delta_1 = \begin{vmatrix} 1 & -a_3^2 & -a_4^2 & \dots & -a_n^2 \\ 0 & 1 & -a_4^3 & \dots & -a_n^3 \\ 0 & 0 & 1 & \dots & -a_n^4 \\ \dots & \dots & \dots & \dots & \dots \\ 0 & 0 & 0 & \dots & 1 \end{vmatrix} = 1$$

2. Let us assume that $\Delta_1 \neq 0$, and thence show that G is an arborescence with root x_1. Every vertex x_k (with $k \neq 1$) is the terminal vertex of at least one arc of G (for otherwise the k-th column vector of Δ_1 is zero, and $\Delta_1 = 0$).

As in Theorem 5, we see that G is connected and antisymmetric (for otherwise Δ_1 will be zero). Since the graph is connected and has $n-1$ edges, it is a tree [property (3)]. Any vertex other than x_1 is the terminal vertex of exactly one arc (since there are only $n-1$ arcs altogether); therefore G must be an arborescence with root x_1.

THEOREM 9 (TUTTE, BOTT). *Let G be a graph without loops with associated matrix A; the number of arborescences with root x_1 which are partial graphs of G is exactly equal to Δ_1.*

Let us denote by $\boldsymbol{e}_k = (0,0,\ldots,1,0,\ldots,0)$ the n-dimensional vector whose k-th element is 1, all the rest being zero. Put:

$$K_i = \{k \mid (x_k, x_i) \in U\}$$

Then we have

$$\Delta_1(\boldsymbol{a}_2, \boldsymbol{a}_3, \ldots, \boldsymbol{a}_n) = \Delta_1\Big(\sum_{k_2 \in K_2} \boldsymbol{e}_{k_2}, \sum_{k_3 \in K_3} \boldsymbol{e}_{k_3}, \ldots\Big) = \sum_{\substack{k_2 \in K_2 \\ k_3 \in K_3 \\ \cdots}} \Delta_1(\boldsymbol{e}_{k_2}, \boldsymbol{e}_{k_3}, \ldots, \boldsymbol{e}_{k_n})$$

Now, from Theorem 8, $\Delta_1(\boldsymbol{e}_{k_2}, \boldsymbol{e}_{k_3}, \ldots, \boldsymbol{e}_{k_n})$ is equal to 1 if the $n-1$ arcs (x_{k_2}, x_2), $(x_{k_3}, x_3), \ldots$, form an arborescence with root x_1, and to 0 in all other cases. The required result follows immediately.

Remark 1. The argument above actually proves a much stronger result, which can be stated as follows: consider a network formed by a graph G without loops and with capacities $c(u)$, and let $C = (c_j^i)$ be the associated matrix with $c_j^i = c(x_i, x_j)$. Form the diagonal matrix $D = (d_j^i)$ with

$$d_i^i = \sum_{k \neq i} c(x_k, x_i)$$

and the determinant $\Delta = \det(D - C)$; if $H = (X, \Gamma)$ is an arborescence with root x_1, we put

$$c(H) = \prod_{\substack{x \in X \\ y \in \Gamma x}} c(x, y)$$

Then: *The minor Δ_1 of the determinant Δ is equal to the sum of the $c(H)$, when we take as H all arborescences with root x_1 which are partial graphs of G.*

This result was established by BOTT and MAYBERRY [2] in order to calculate the determinants of certain matrices met with in economics.

EXAMPLE. Consider the determinant

$$\Delta_1 = \begin{vmatrix} 4 & -1 & -2 \\ -1 & 3 & -1 \\ -2 & -2 & 5 \end{vmatrix}$$

This corresponds to a matrix $D-C$ in which only the first column is arbitrary, and leads to the network of Fig. 16.5.

x_1
x_2
x_3
x_4
1
2
2
2
1
1
1
2

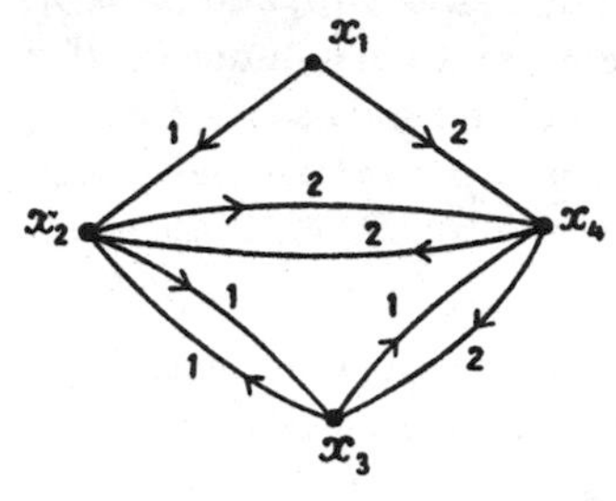

FIG. 16.5

$$D-C = \begin{bmatrix} d_1^1 & -1 & 0 & -2 \\ -c_1^2 & 4 & -1 & -2 \\ -c_1^3 & -1 & 3 & -1 \\ -c_1^4 & -2 & -2 & 5 \end{bmatrix}$$

By summing all the values $c(H)$ shown in the table in Fig. 16.6, we get 29; the reader can confirm that this is precisely the value of the determinant Δ_1.

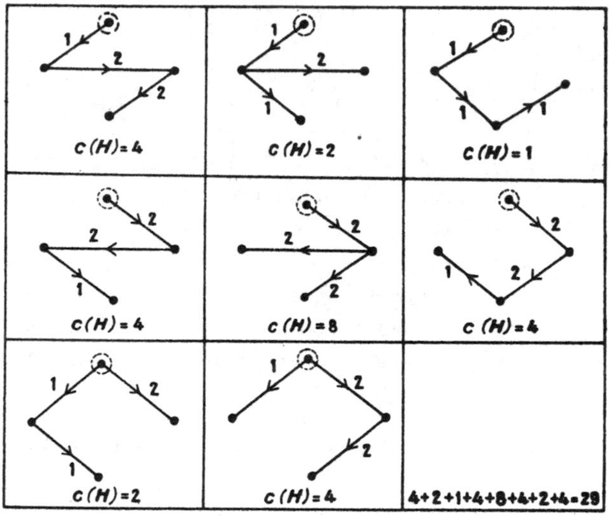

c(H)=4
c(H)=2
c(H)=1
c(H)=4
c(H)=8
c(H)=4
c(H)=2
c(H)=4
4+2+1+4+8+4+2+4=29

FIG. 16.6

Remark 2. Let G be a symmetric graph without loops and with associated matrix A; the number of symmetric trees which are partial graphs of G is equal to the number of arborescences with root x_k, or alternatively, by virtue of Theorem 8, to

$$\Delta_1 = \Delta_2 = \ldots = \Delta_n$$

This leads us back to the statement of Theorem 6.

17. Euler's Problem

Eulerian Cycles

One of the oldest problems in the geometry of situations, which was set by Euler, can be formulated as follows: a chain (cycle) is called an *Eulerian chain* (*Eulerian cycle*) if it uses every edge once and once only. *How can we recognize if a graph possesses an Eulerian chain or cycle?*

Fig. 17.1

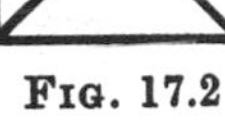

Fig. 17.2

EXAMPLE 1. Consider the graph of Fig. 17.1; is it possible to draw this without lifting one's pen from the paper or going twice over the same line? After having made several attempts, any reader who has not met the problem before will end up by admitting that it is impossible. On the other hand, the graph in Fig. 17.2 can be drawn by one continuous movement of the pen: why?

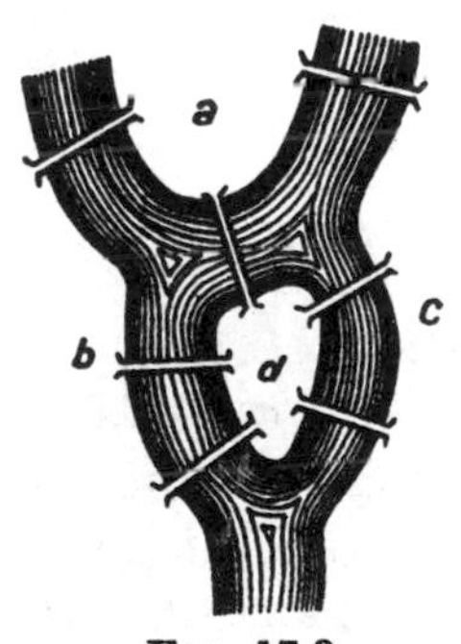

Fig. 17.3

EXAMPLE 2 (Euler). The river Pregel runs through the town of Königsberg (now called Kaliningrad) and flows on either side of the island of Kneiphof; there are seven bridges over the river as shown in Fig. 17.3. Is it possible to go for a walk and cross each bridge once and once only?

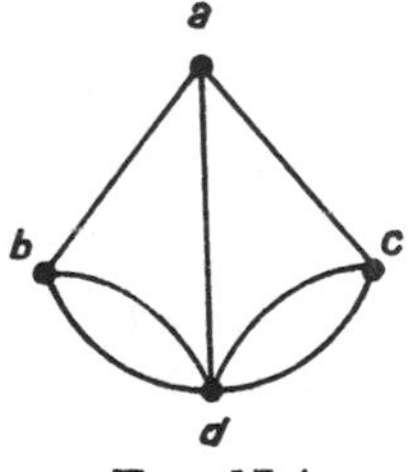

Fig. 17.4

This problem intrigued the townsfolk of Königsberg in 1736, until Euler proved that it was impossible. If we consider the graph of Fig. 17.4, where the vertices represent the regions a, b, c, d, and where each edge represents a bridge, we see that the desired answer to the problem is an Eulerian chain.

Theorem 1. *An s-graph G possesses an Eulerian chain if and only if it is* *connected (apart from isolated points), and if the number of vertices of uneven degree is* 0 *or* 2.

1. *Necessity of the condition;* if the graph has an Eulerian chain μ, then clearly it is connected. Further, only the two terminal vertices of μ (if they are distinct) can be of uneven degree: therefore there can only be 0 or 2 vertices of uneven degree.

2. *Sufficiency of the condition.* More precisely, we shall prove: *if there are two vertices a and b whose degree is uneven, then an Eulerian chain exists which starts at a and finishes at b: if there are no points whose degree is uneven, then the graph possesses an Eulerian cycle.*

We shall assume this statement to be true for graphs with fewer than m edges, and prove that it also holds for a graph G with m edges. To be more definite, let us assume that G has two vertices a and b of uneven degree.

The chain μ will be defined for us by a traveller who sets out from a in any arbitrary direction and moves in such a way that he never covers the same edge twice. If he reaches a vertex $x \neq b$, he will have been along an uneven number of the edges incident to x, and therefore he will be able to leave by an edge which has not yet been used; when this is no longer possible, he must necessarily have arrived at b. However, it is possible that this arbitrary route which goes from a to b has not made use of all the edges; a partial graph G' remains all of whose vertices are of even degree.

Let $C_1, C_2, C_3, \ldots, C_k$ be the components of G' containing at least one edge; by hypothesis, they contain Eulerian cycles $\mu_1, \mu_2, \ldots$; since G is connected, the chain μ meets each of the C_i in turn, at, say, the vertices $x_1 \in C_1, x_2 \in C_2, \ldots, x_k \in C_k$ (in this order).

Now consider the chain:

$$\mu[a, x_1] + \mu_1 + \mu[x_1, x_2] + \mu_2 + \ldots + \mu[x_k, b]$$

This is certainly an Eulerian chain which goes from a to b.

Q.E.D.

If there are no points whose degree is uneven, choose two points a and b connected by an edge, remove edge ab and use the previous argument.

The reader can now satisfy himself immediately that no solution exists to the Königsberg problem (Example 2).

Remark. In the same way, it can be seen that a connected s-graph with four vertices of uneven degree can be drawn completely by two continuous lines, and that a connected s-graph with $2q$ vertices of uneven degree can be drawn by q continuous movements of the pen.

ALGORITHM FOR TRACING AN EULERIAN CYCLE DIRECTLY (FLEURY). Consider a connected s-graph G, all of whose vertices are of even degree, and let us set out to follow every edge in one movement of the pen, *without ever having to correct our route*. The rules we have to obey are:

1. *Start from any vertex a; each time an edge has been followed, erase it.*

2. *Never use an edge if, at that particular moment, it is an isthmus* (that is to say: the deletion of this edge would divide the graph into two connected components, each of which contains at least one edge).

If these rules are obeyed, we are always in a good position; for if we are at a vertex $x \neq a$, the graph has two vertices of uneven degree, x and a; if the isolated points are deleted, a connected graph is left, and from Theorem 1 this graph must contain an Eulerian chain starting from x.

It remains to be shown that the rules can always be obeyed, that is to say: having arrived at a junction x, there is always an unused edge leading from the junction which is not an isthmus. In fact, if all the edges incident to x were isthmuses, there must be at least two (for if $|U_x| = 1$, the edge incident to x is simply 'pendant'); the two isthmuses lead to two disjoint components, each of which contains a vertex of uneven degree. Since a is the only vertex of uneven degree (apart from x), this must lead to a contradiction.

Eulerian Circuits

It is of interest to extend these notions to the study of oriented graphs; in an oriented graph (X, U), an *Eulerian circuit* is a circuit which uses every arc once and once only. Which graphs possess such circuits?

A graph is said to be *pseudo-symmetric* if for every vertex the number of arcs which leave the vertex is the same as the number which arrive, or, in other words, if

$$|U_x^+| = |U_x^-| \qquad (x \in X)$$

This terminology is justified by the fact that every symmetric graph is also pseudo-symmetric; we can now state:

THEOREM 2. *A graph possesses an Eulerian circuit if and only if it is connected and pseudo-symmetric.*

(The proof follows exactly the same lines as in Theorem 1.)

EXAMPLE (IR. K. POSTHUMUS). What is the longest circular sequence which can be formed from the digits 0 and 1 such that the same group of k consecutive figures does not appear more than once in the sequence? Since there are 2^k distinct k-tuples formed from 0 and 1, the sequence cannot have more than 2^k figures; we shall now use Theorem 2 to show *that a circular sequence can be formed which has this property and contains 2^k figures.*

Consider the graph G whose vertices are the different $(k-1)$-tuples formed from 0 and 1, and join every vertex $(\alpha_1, \alpha_2, \ldots, \alpha_{k-1})$ by an arc to each of the vertices $(\alpha_2, \alpha_3, \ldots, \alpha_{k-1}, 0)$ and $(\alpha_2, \alpha_3, \ldots, \alpha_{k-1}, 1)$. Since the graph G is pseudo-symmetric, it has an Eulerian circuit. If the first vertex on this circuit is $(\alpha_1, \alpha_2, \ldots, \alpha_{k-1})$, $(\alpha_2, \alpha_3, \ldots, \alpha_k)$ the second, $(\alpha_3, \alpha_4, \ldots, \alpha_{k+1})$ the third, etc., the required sequence is $\alpha_1 \alpha_2 \alpha_3 \alpha_4 \ldots$.

With $k = 4$, the graph of Fig. 17.5 provides several circular sequences of 16 figures, such as:

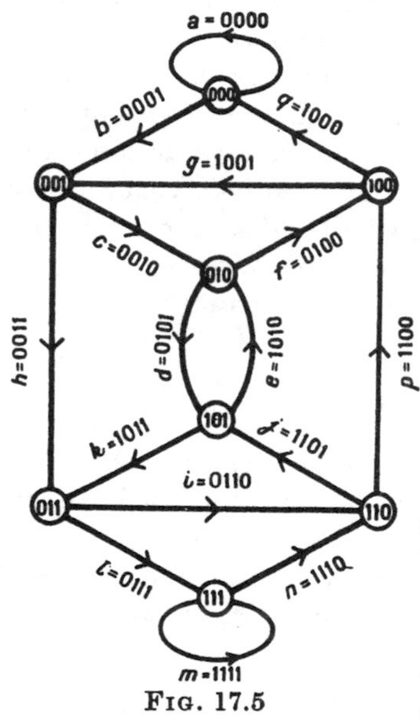

FIG. 17.5

abcdefgh*i*jk*lmn*pq = 0000101001101111

abcdk*i*jefgh*lmn*pq = 0000101101001111

abcdk*i*pgh*lmn*jefq = 0000101100111101

abcfgh*i*jedk*lmn*pq = 0000100110101111

abh*i*jk*lmn*pgcdefq = 0000110111100101

abh*i*jedk*lmn*pgcfq = 0000110101111001

abh*i*jefgcdk*lmn*pq = 0000110100101111

abh*i*pgcdk*lmn*jefq = 0000110010111101

(We have omitted the other circular sequences which are found by permuting i and lmn.) There are in all 16 solutions†.

A problem of interest to some authors is that of enumerating the different Eulerian circuits of a pseudo-symmetric graph; this problem is closely connected with the enumeration of the arborescences of a graph, which was solved earlier (Chapter 16). There are very close connections between Eulerian circuits and arborescences, and we shall first derive the following result:

THEOREM 3. *Let G be a connected, pseudo-symmetric graph, and let x_1 be one of its vertices; G has as a partial graph an arborescence H with root x_1, which may be found as follows: we follow all the arcs of G which start from x_1, and build up H from those arcs which have led us to a vertex for the first time.*

† This problem has been encountered in telecommunications in locating the position of a drum carrying two different sorts of marks by means of a reading head capable of reading k marks at a time. The same problem is also met in cryptography in the following form: to find a word in which each arrangement of k letters of the alphabet appears once and once only.

H is certainly an arborescence with root x_1, since:

1. every vertex $\neq x_1$ is the terminal vertex of exactly one arc;
2. x_1 is not the terminal vertex of any arc;
3. H does not contain any circuits (for if a path which is formed from the arcs of H goes from x to y, then the vertex x must have been visited (during the formation of H) before the vertex y, and therefore no paths exist in H going from y to x).

THEOREM 4 (AARDENNE–EHRENFEST, DE BRUIJN). *In a graph G which is pseudo-symmetric and connected, let Δ_1 be the number of arborescences with root x_1 which are partial graphs, and let r_k be the external (or internal) demi-degree of the vertex x_k; there are exactly*

$$\Delta_1 \prod_{k=1}^{n} (r_k - 1)!$$

distinct Eulerian circuits.

Note that two Eulerian circuits are not considered as distinct if one is obtained by a circular permutation of the arcs of the other. Note also that Δ_1 is known from Theorem 8 of Chapter 16.

Let us consider an arborescence H with root x_1 which is a partial graph of G, and show that there are exactly

$$\prod_{k=1}^{n} (r_k - 1)!$$

Eulerian circuits in which the arcs which lead us to a vertex for the first time are the arcs of H. Let us label the arcs leading to x_k from 1 to r_k, in such a way that we can write:

$$U_{x_k}^{-} = \{u_k(1), u_k(2), \ldots, u_k(r_k)\}$$

Let this numbering be such that $u_k(r_k)$ is the arc of the arborescence H which leads to x_k (if $k \neq 1$). If we assume that the arc $u_1(1)$ is fixed once and for all, there are exactly

$$\prod_{k=1}^{n} (r_k - 1)!$$

possible numberings. Each numbering corresponds to one and only one circuit in the following manner: starting from $u_1(1)$, we follow, in the opposite direction to their orientation, all the arcs of the graph; whenever we arrive at x_k, we choose the arc of $U_{x_k}^{-}$ which has the lowest number and has not yet been used.

This certainly defines a circuit, since the route can only be terminated by arrival at the point of departure x_1 (and as the graph is pseudo-symmetric there must always be an arc whereby we can leave any other vertex which we encounter). Now let us show that this circuit is Eulerian.

In fact, if this were not so, there would be an arc $u_k(j)$ which has not been used; as $r_k \geqslant j$, the arc $u_k(r_k)$ has also not been used; consequently an arc $u_p(r_p)$ which arrives at the initial vertex x_p of the arc $u_k(r_k)$ has also not been used, etc. All these unused arcs belong to the arborescence H and hence we shall eventually arrive back at the root x_1; and this is absurd, since we were stopped at x_1 only because all arcs incident to x_1 had been used in our route.

To sum up, with each numbering there is associated a well-defined Eulerian circuit which terminates with the fixed arc $u_1(1)$, and which is such that the arc leading to a given vertex for the first time is that of the arborescence: therefore the theorem has been proved†.

COROLLARY. *In a pseudo-symmetric graph, the number of arborescences with root x_k which are partial graphs is independent of the chosen vertex x_k.*

We could, in fact, have stated Theorem 3 just as well by using x_k in place of x_1; as the number of Eulerian circuits cannot alter, we must have $\Delta_1 = \Delta_k$.

(In this way, the reader may show that the number of Eulerian circuits of the graph of Fig. 17.5 actually is 16.)

† If the number of arcs of the graph $m \geqslant 2n+1$, then by listing the Eulerian circuits, we find a curious combinatorial property which was recently formulated by M. P. SCHUTZENBERGER. Let us regard an Eulerian circuit μ as a permutation of the arcs of the graph, and say that μ is *even* or *odd* according to whether the number of inversions in the permutation is even or odd; then we have: *In a graph G of order n in which the number of arcs $m \geqslant 2n+1$, the number of odd Eulerian circuits which start with a given arc of the graph is equal to the number of even Eulerian circuits starting with the same arc.*

Note that this is not true if $m = 2n$, as can be seen from the following example: let us construct a graph with the arcs $u_1, u_2, \ldots, u_n$ of an elementary circuit going through n vertices, and with loops $u_{n+1}, u_{n+2}, \ldots, u_{2n}$ at each of these vertices; there is only one Eulerian circuit beginning with any given arc, so that the above statement certainly breaks down.

It may be shown that this property of graphs of order n is equivalent to a property of families of matrices of order n which is of some importance in algebra (see also: the theorem of A. S. AMITSUR and J. LEVITZKI, *Proc. Amer. Math. Soc.*, **1**, 1950, p. 449).

18. Matching in the General Case

The Theory of Alternating Chains

In this section, we shall consider a non-oriented graph $\bar{G}$ with vertices $\bar{a}, x_1, x_2, \ldots, x_n$. The edges of $\bar{G}$ fall into two categories: those which we shall call *heavy* and those which we shall call *light*. The former will be represented by a heavy line and the latter by a light one. The point $\bar{a}$ is called the *origin* of the graph $\bar{G}$, and it is assumed that *all edges incident to it are heavy*.

An *alternating chain* is a chain $\mu = (v_1, v_2, \ldots, v_k)$ of $\bar{G}$ such that:

1. if two edges v_j and v_{j+1} are consecutive, one of them is heavy and the other is light;
2. the chain is simple; that is to say: a given edge appears at most once in the sequence μ.

If there exists an alternating chain μ going from $\bar{a}$ to x, we give its last edge an orientation by directing it towards x; an edge of $\bar{G}$ may thus be oriented in both directions, in one only, or not at all. A vertex x is said to be *heavy* if it is the terminal vertex of a heavy edge oriented towards x and is not the terminal vertex of a light edge oriented in the same direction; the set of all heavy vertices will be denoted by H.

Similarly, a vertex x is said to be *light* if it is the terminal vertex of a light edge oriented towards x and not of a heavy edge oriented towards x; the set of such vertices will be denoted by L.

A vertex x which is the terminal vertex of a light and of a heavy edge, both of which are oriented towards x, is said to be *mixed*; the set of such vertices will be denoted by M.

Finally, a vertex x which is not the terminal vertex of any edge oriented in the direction of x, is said to be *inaccessible*; the set of these vertices will be denoted by I.

The vertex $\bar{a}$, which we singled out as being the *origin* of the graph $\bar{G}$, is so far unclassified: if there are no alternating chains which start at $\bar{a}$ and return to $\bar{a}$ it is *light*; otherwise it is *mixed*.

Since every vertex of $\bar{G}$ must be either heavy, light, mixed or inaccessible, we have $H \cup L \cup M \cup I = X$; there follows immediately:

THEOREM 1. *Two heavy vertices can only be connected by a heavy edge; two light vertices can only be connected by a light edge; there can be no edge connecting a mixed vertex and an inaccessible vertex; a heavy vertex and an inaccessible vertex can only be connected by a heavy edge; a light vertex and an inaccessible vertex can only be connected by a light edge.*

This may be summarized by the table:

	H	L	M	I
H	heavy			heavy
L		light		light
M				X
I	heavy	light	X	

Let the connected components of the subgraph of $\bar{G}$ determined by the set M be written $M_1, M_2, \ldots$; we shall say that a chain μ enters M_1 by the edge $[a,b]$ if μ is of the form:

$$\mu = [\bar{a}, a_1, a_2, \ldots, a_k = a, b_1 = b, b_2, \ldots, b_l]$$

with

$$\bar{a}, a_1, a_2, \ldots, a_k \notin M_1$$
$$b_1, b_2, b_3, \ldots, b_l \in M_1$$

LEMMA 1. *Let $\bar{a} \notin M_1$, $x \in M_1$; if an alternating chain $\mu[\bar{a},x]$ enters M_1 by an edge $[a,b]$ and terminates with a heavy (light) edge, then another alternating chain $\mu'[\bar{a},x]$ exists which also enters M_1 by means of the edge $[a,b]$ and terminates with a light (heavy) edge.*

Since x is a mixed vertex, an alternating chain $\nu[\bar{a},x]$ which terminates with a light edge certainly exists. Further, as $\bar{a} \notin M_1$, $x \in M_1$, $\nu[\bar{a},x]$ must contain edges of which one vertex is in $X - M_1$, and the other is in M_1. Let $[c,d]$ be the last such edge.

Finally, let y be the first point of μ which is on $\nu[d,x]$ (such a point always exists). If $\mu[b,y]$ and $\nu[d,y]$ terminate with edges belonging to the same category, the alternating chain:

$$\mu' = \mu[\bar{a},y] + \nu[y,x]$$

satisfies the requirements of the theorem.

But it can be assumed that this is not the case; it then follows that $[a,b] = [c,d]$, for if not, $\mu[\bar{a},y] + \nu[y,c]$ would be a simple alternating chain, the point c would be mixed, and we would have $c \in M_1$. From this, we may conclude that the chain

$$\mu' = \mu[\bar{a},b] + \nu[d,x]$$

is alternating, and therefore satisfies the theorem.

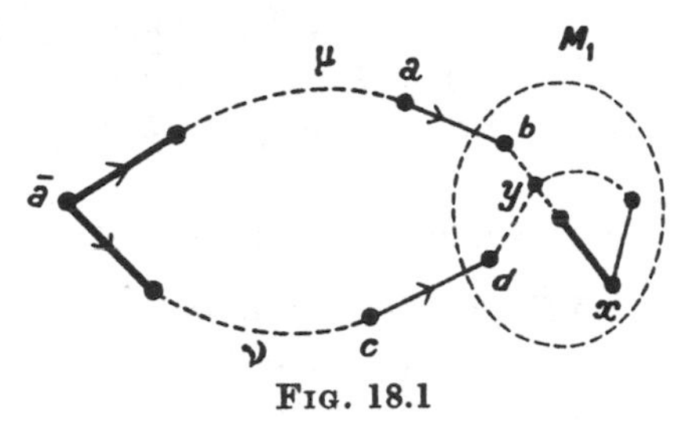

Fig. 18.1

Lemma 2. *Let $\bar{a} \notin M_1$, $x \in M_1$, and let $[a,b]$ be an edge which is incident into M_1: then there exists an alternating chain $\mu[\bar{a},x]$ from $\bar{a}$ to x which enters M_1 by the edge $[a,b]$.*

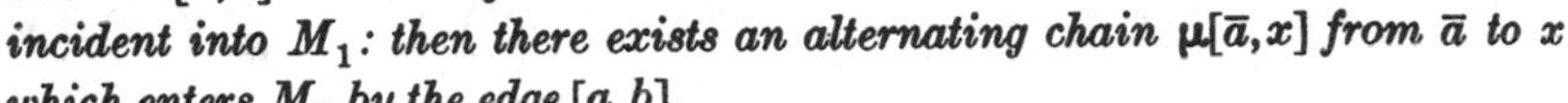

Let Y be the set of vertices of M_1 which are accessible by an alternating chain entering by $[a,b]$, and let Z be the set of vertices of M_1 which are not so accessible. $Y \neq \emptyset$ (since $b \in Y$), and if $Z \neq \emptyset$, then there must be an edge $[y,z]$ with $y \in Y$, $z \in Z$. We shall now show that this leads to a contradiction.

From Lemma 1, there are two alternating chains $\mu_1[\bar{a},y]$ and $\mu_2[\bar{a},y]$ leading to y which enter by the edge $[a,b]$, and terminate with a light and a heavy edge respectively; but z must then be accessible from $\bar{a}$ by an alternating chain which enters by $[a,b]$, which contradicts the assumption that $z \in Z$.

Theorem of Petersen and Gallai. *Let M_1 be a component of the subgraph determined by the mixed points; if $\bar{a} \notin M_1$, there is one and only one edge incident to M_1, which is incident into M_1. If $\bar{a} \in M_1$, no edge exists which is incident into M_1.*

1. If $\bar{a} \notin M_1$, let μ be an alternating chain going from $\bar{a}$ to M_1, and let b be the first vertex of M_1 on this chain. The alternating chain $\mu[\bar{a},b]$ enters M_1 by an edge $[a,b]$ which is incident into M_1.

Let $[c,d]$ be an edge other than $[a,b]$, which is also incident into M_1. From Lemma 2, $d \in M_1$ is accessible by an alternating chain which enters by $[a,b]$, and from Lemma 1, we can assume that this chain terminates with an edge in the opposite category from that of $[c,d]$: but c is then a mixed point, which contradicts the assumption that $c \notin M_1$.

Therefore $[a,b]$ is the only edge which is incident into M_1.

2. If $\bar{a} \in M_1$, we can get back to the preceding case by adding two vertices $\bar{a}_0$ and $\bar{b}_0$, a heavy edge $[\bar{a}_0, \bar{b}_0]$ and a light edge $[\bar{b}_0, \bar{a}]$. Using $\bar{a}_0$ as origin in place of $\bar{a}$, the only edge entering M_1 is $[\bar{b}_0, \bar{a}]$, and therefore the initial graph contains no edge which is incident into M_1.

Q.E.D.

In particular, this theorem shows that an alternating chain which has passed through M_1, will never be able to return to this component; some applications of this property will be given later.

The Determination of a Partial Graph with Given Degrees

Let $G = (X, U)$ be a non-oriented graph, and let us assign an integer $f(x)$ to every vertex x where $f(x)$ is less than or equal to the degree $d_x(U)$ of x; we intend here to construct a partial graph (X, W) in which the degree of x will be $d_x(W) = f(x)$. A set of edges $W \subset U$ is said to be *compatible* with the function $f(x)$ if

$$d_x(W) \leqslant f(x) \qquad (x \in X)$$

If W is a given set, we say that the edges of W are *heavy*, all other edges being *light*; the problem therefore becomes one of finding a set of heavy edges compatible with the function $f(x)$ and containing as many elements as possible.

This new problem is to a certain extent analogous with the following problem (which was considered in Chapter 11): *find a partial graph of a known oriented graph which has given inward and outward demi-degrees.* Now, however, the theory of maximum flow does not help us, and we shall use instead the tool provided by the theory of alternating chains.

Remark. Note that we have here a linear programming problem. Put $b_i = f(x_i)$, and let (r_j^i) be the incidence matrix of the edges. We require integers ξ_j which are equal to 1 if u_j is a heavy edge, and 0 otherwise, such that:

$$0 \leqslant \xi_j \leqslant 1 \qquad (j = 1, 2, \ldots, m)$$

$$\sum_{j=1}^{m} r_j^i \xi_j \leqslant b_i \qquad (i = 1, 2, \ldots, n)$$

$$\sum_{j=1}^{m} \xi_j \text{ is a maximum}$$

Unfortunately, the solution to this linear programme must be integral, and none of the methods known at present for solving such problems is very convenient to use. On the other hand, there is a 'graphical' algorithm by which solutions can easily be found.

We shall say that a point x is unsaturated if $d_x(\mathbf{W}) < f(x)$. Starting from G, we construct an s-graph $\bar{G}(\mathbf{W})$ in which every unsaturated point x is connected to an origin $\bar{a}$ by $f(x) - d_x(\mathbf{W})$ heavy edges. If this graph $\bar{G}(\mathbf{W})$ contains an alternating chain going from $\bar{a}$ to $\bar{a}$, then by making its heavy edges light, and its light edges heavy, we get a new compatible set which is larger than the preceding one. It will be shown below that if a compatible set cannot be enlarged in this way, it is a maximum.

Lemma. *Let $\mathscr{W}$ be the family of compatible sets $\mathbf{W}$ which do not contain alternating chains going from $\bar{a}$ to $\bar{a}$; then*

$$\mathbf{W}_1 \subset \mathbf{W}_2, \qquad \mathbf{W}_1, \mathbf{W}_2 \in \mathscr{W} \Rightarrow \mathbf{W}_1 = \mathbf{W}_2$$

If, in fact, $\mathbf{W}_2 \supset\supset \mathbf{W}_1$, then there is an edge $[x,y] \in \mathbf{W}_2 - \mathbf{W}_1$, and since the vertices x and y must therefore be unsaturated in the graph $\bar{G}(\mathbf{W}_1)$, $\bar{a}$ can be reached by the alternating chain $[\bar{a},x]+[x,y]+[y,\bar{a}]$, which contradicts the hypothesis $\mathbf{W}_1 \in \mathscr{W}$.

(Note that this reasoning is valid even if $x = y$.)

Theorem of Berge, Norman and Rabin. *A compatible set $\mathbf{W}_0$ is maximum if and only if it contains no alternating chains which start and finish at $\bar{a}$.*†

We simply have to show that a compatible set $\mathbf{W}_0$ which belongs to the family $\mathscr{W}$ defined above, is maximum. To do this, let us consider a maximum compatible set $\mathbf{W}_M$, and let this set be chosen so that $|\mathbf{W}_0 - \mathbf{W}_M|$ is as small as possible; if $|\mathbf{W}_0 - \mathbf{W}_M| = 0$, then $\mathbf{W}_0 \subset \mathbf{W}_M$, and therefore, from the lemma, $\mathbf{W}_0 = \mathbf{W}_M$, and our goal has been achieved. We shall assume therefore that $|\mathbf{W}_0 - \mathbf{W}_M| > 0$, and show that this leads to a contradiction.

1. *Among the edges of G which are incident to a given vertex x, there are at least as many edges belonging to $\mathbf{W}_M - \mathbf{W}_0$ as there are edges belonging to $\mathbf{W}_0 - \mathbf{W}_M$.* If this is not the case, we have

$$d_x(\mathbf{W}_M) < d_x(\mathbf{W}_0) \leqslant f(x)$$

If $\mathbf{u} = [x,y] \in \mathbf{W}_0 - \mathbf{W}_M$, then there is also an edge $\mathbf{v} = [y,z] \in \mathbf{W}_M - \mathbf{W}_0$ (because otherwise $\mathbf{W}_M \cup [x,y]$ would be a compatible set larger than $\mathbf{W}_M$); $\mathbf{W}' = (\mathbf{W}_M - \{\mathbf{v}\}) \cup \{\mathbf{u}\}$ is a compatible set, because

$$\begin{aligned} d_x(\mathbf{W}') &= d_x(\mathbf{W}_M)+1 \leqslant f(x) \\ d_y(\mathbf{W}') &= d_y(\mathbf{W}_M) \quad\;\; \leqslant f(y) \\ d_z(\mathbf{W}') &= d_z(\mathbf{W}_M)-1 < f(z) \end{aligned}$$

† The general statement given here contains both a theorem of Berge [1] and a theorem of Norman and Rabin [3]; the proof used is based on a suggestion by Rabin.

Therefore $|W'| = |W_M|$ and W' is a maximum compatible set; but this is absurd, since

$$|W_0 - W'| < |W_0 - W_M|$$

2. *No simple cycles of even length exist whose edges are alternately in* $W_0 - W_M$ *and* $W_M - W_0$. If μ were such a cycle, the set

$$W' = (W_M \cup \mu) - (W_M \cap \mu)$$

would be compatible; since $|W'| = |W_M|$, W' is a maximum compatible set, and this is absurd because

$$|W_0 - W'| < |W_0 - W_M|$$

3. Let $\mu = (u_1, u_2, \ldots, u_k)$ be the longest simple chain the edges of which are alternately in $W_0 - W_M$ and $W_M - W_0$, and let x_0 and x_k be the initial and final points of this chain; to begin with, we shall assume $x_0 \neq x_k$. From (1), the two terminal edges u_1 and u_k belong to $W_M - W_0$, and consequently

$$d_{x_0}(W_0) < d_{x_0}(W_M) \leqslant f(x_0)$$
$$d_{x_k}(W_0) < d_{x_k}(W_M) \leqslant f(x_k)$$

Therefore $\bar{G}(W_0)$ contains the alternating chain $[\bar{a}, x_0] + \mu[x_0, x_k] + [x_k, \bar{a}]$ which starts at $\bar{a}$ and returns to $\bar{a}$, which contradicts our original hypothesis.

4. If we now assume $x_0 = x_k$, then, from (2), the terminal edges must both be of the same sort, and therefore must both belong to $W_M - W_0$; then we have

$$d_{x_0}(W_0) \leqslant d_{x_0}(W_M) - 2 \leqslant f(x_0) - 2$$

Therefore there are at least two heavy edges connecting $\bar{a}$ to x_0, and we can construct an alternating chain which starts and finishes at $\bar{a}$, which contradicts our original hypothesis.

Q.E.D.

ALGORITHM FOR FINDING A MAXIMUM COMPATIBLE SET. We first form a compatible set W by putting as many edges as possible in the set of heavy edges; to increase the size of the set, we examine $\bar{G}(W)$ to see if it contains any alternating chains which originate and terminate at $\bar{a}$; this may be done as follows: starting from $\bar{a}$, we go as far as possible along an alternating chain, taking care to mark each edge used with an arrowhead. When it is no longer possible to go further, return to the nearest branching point (and delete the arrowheads on the edges no longer in the chain) and try a new direction.

EXAMPLE 1. In the prison of ..., the plan of which is given in Fig. 18.2, a warder placed in the corridor $[x, y]$ is able to keep watch over cells x and y

at the two ends of this corridor; if every cell is to be guarded in this way by at least one warder, what is the smallest number of warders needed?

Following R. NORMAN, we shall call a *cover* of a graph (X, U) a set $V \subset U$ such that every vertex x is a terminal of at least one edge of V. The problem then becomes one of finding a minimum cover. The set V is a cover if and only if the set $W = U - V$ satisfies

$$d_x(W) \leqslant d_x(U) - 1 \qquad (x \in X)$$

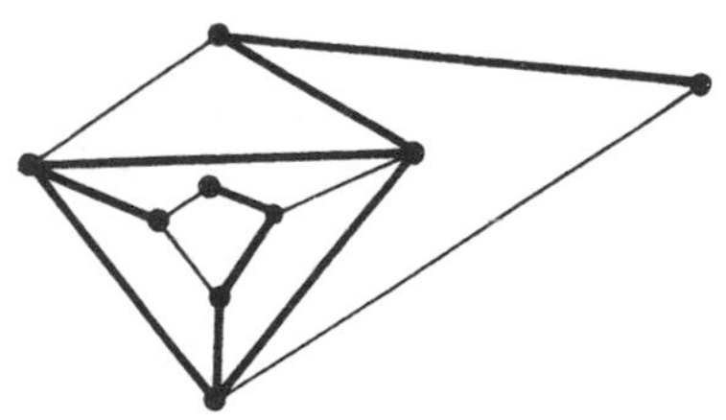

FIG. 18.2

If the preceding method is used to find a maximum set W which is compatible with these inequalities, then $V = U - W$ is a minimum cover; in the graph of Fig. 18.2, the minimum cover contains five edges and therefore five warders are needed.

EXAMPLE 2 (J. P. ROTH [7]). Suppose we have a relay system or some sort of control mechanism and we wish to draw the simplest possible plan which is equivalent to this mechanism. This problem is one of finding a minimum cover. More exactly, let us consider an electrical relay which is governed by three contacts A, B and C, the position of the relay being described by three variables ϵ_A, ϵ_B and ϵ_C: we put $\epsilon_A = 1$ if the contact A is depressed, and $\epsilon_A = 0$ otherwise, etc. To each possible vector $\epsilon = (\epsilon_A, \epsilon_B, \epsilon_C)$ with elements which are either 0 or 1 (that is to say, to each vertex of the unit cube in the 3-dimensional space R^3), we assign a number

$$\phi(\epsilon) = \phi(\epsilon_A, \epsilon_B, \epsilon_C)$$

which is equal to 1 if the current passes through the relay, and 0 otherwise. We need a diagram containing switches which are governed by the contacts A, B and C, and with branches in either series or parallel, such that:

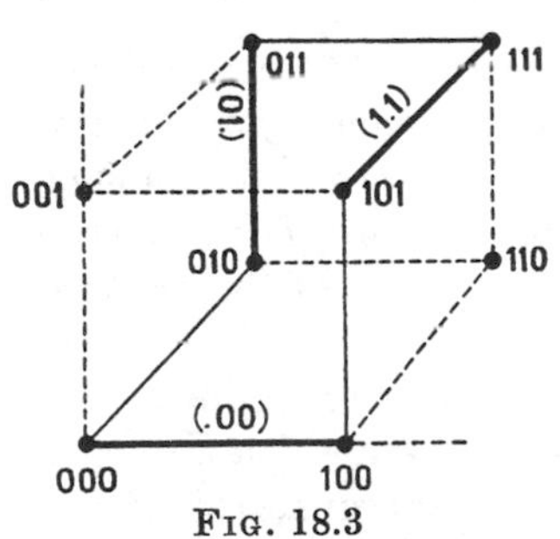

FIG. 18.3

1. the diagram is equivalent to the given relay, that is: it allows current to flow if $\phi(\epsilon_A, \epsilon_B, \epsilon_C) = 1$, and prevents it if

$$\phi(\epsilon_A, \epsilon_B, \epsilon_C) = 0$$

2. there are two switches on each branch;
3. there are as few branches as possible†.

The problem reduces to finding a minimum cover of the graph G which consists of the vertices ϵ such that $\phi(\epsilon) = 1$, two vertices

† This problem may be formulated in much more general terms by means of Boolean algebra; it is then known as 'Quine's problem' (cf. [7]).

being connected if they are adjacent corners of the unit cube (Fig. 18.3). Let us suppose that the current flows if $\epsilon = 000, 010, 011, 111, 101, 100$; a minimum cover of the graph G consists of the edges (.00), (1.1), (01.), which gives the required diagram (Fig. 18.4).

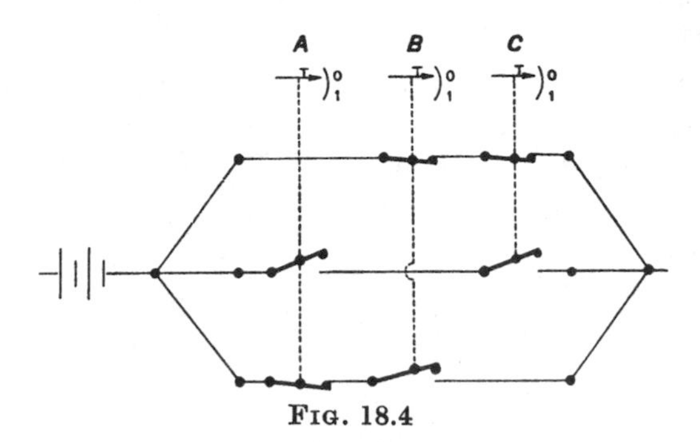

FIG. 18.4

(The reader can prove that the current flows only if the system is in one of the six given positions.)

Perfect Matching

Let $G = (X, U)$ be a graph in which, as before, the edges are classified as being either *light* or *heavy*. The set of heavy edges is denoted by $\boldsymbol{W}$. $\boldsymbol{W}$ is said to be a matching if no two heavy edges are adjacent.

If no heavy edge is adjacent to a given vertex x, the vertex is said to be *unsaturated*: we say that a matching is *perfect* if there are no unsaturated vertices. We shall first investigate the problem of finding in which graphs the maximum matching is perfect.

EXAMPLE 1. *The problem of the truncated chessboard.* A standard chessboard is truncated by removing two corner squares diagonally opposite one another, as in the left-hand diagram of Fig. 18.5. We also have 31 dominoes, each of which can cover two adjacent squares of the board. The problem is: can these 31 dominoes be used to cover all 62 squares of the truncated board?

This problem reduces to finding the maximum matching of a graph whose vertices are the squares of the chessboard, and in which we connect every pair of vertices which correspond to two adjacent squares (right-hand diagram of Fig. 18.5); it is easily seen that this matching is not perfect, and therefore the truncated board cannot be covered. There is another simple and elegant proof, which owes nothing to the theory of matching, which can be used to show that the problem has no solution: if we colour the squares of the board alternately black and white (as in the ordinary chessboard), we see that the two squares which have been removed are both of the same colour; since each domino necessarily covers one black and one white square, the truncated board would have to have 31 black squares and 31 white squares if it were to be covered by 31 dominoes.

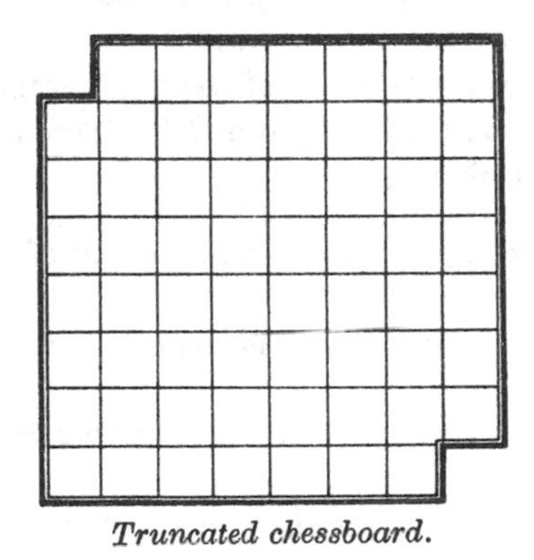

Truncated chessboard.

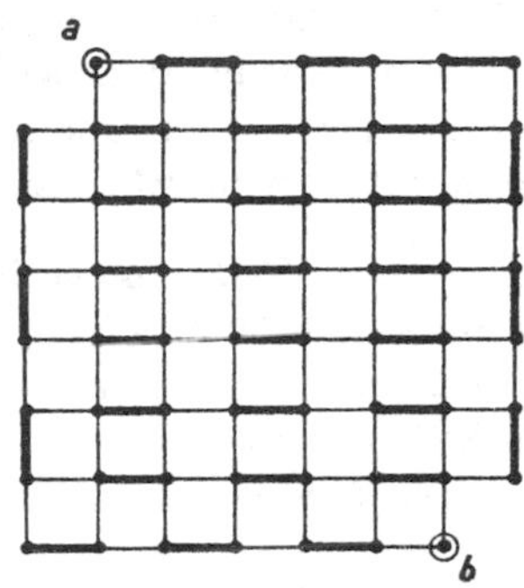

Maximum matching of the corresponding graph.

FIG. 18.5

EXAMPLE 2. A famous problem which LUCAS called 'the schoolgirls' walk' may be stated thus: if the mutual likes and dislikes of the pupils at a boarding school were known, could the whole school go for a walk in crocodile, each girl being placed with one whom she likes? Under what circumstances does this problem have a solution? If $X = \{x_1, x_2, \ldots, x_n\}$ denotes the students, and if we put $[x_i, x_j] \in \boldsymbol{U}$ if the students x_i and x_j like one another, the problem becomes one of discovering if the graph $(X, \boldsymbol{U})$ possesses a perfect matching.

In a given graph G, let $\boldsymbol{W}$ be a maximum matching, the elements of which are represented by heavy edges; as in the preceding paragraphs, let us define a graph $\bar{G}(\boldsymbol{W})$, which is built up from G by adding a point $\bar{a}$ which we connect to every unsaturated vertex by a heavy edge.

As before, if there is an alternating chain $\boldsymbol{\mu}$ which goes from $\bar{a}$ to a given vertex x, we place an arrow directed towards x on the last edge of $\boldsymbol{\mu}$. If x is the terminal vertex of a heavy oriented edge, but not of a light oriented edge, we say that x is a ***heavy*** vertex, and write $x \in H$; if x is the terminal vertex of a light oriented edge but not of a heavy oriented edge, we say that x is a ***light*** vertex, and put $x \in L$; if x is the terminal vertex of both a light oriented edge and a heavy oriented edge, we say that it is a ***mixed*** vertex, and write $x \in M$; finally, if x is not a terminal vertex of any oriented edge, we say that x is *inaccessible*, and write $x \in I$. It follows that every unsaturated vertex is either heavy or mixed. In the graph $\bar{G}(W)$, the point $\bar{a}$ will be considered to be a light vertex; if, in fact, W is a maximum matching, it follows from the theorem of BERGE, NORMAN and RABIN, that $\bar{a}$ cannot be a mixed vertex.

THEOREM 2. *A matching in G is maximum if and only if no two distinct unsaturated vertices can be connected by an alternating chain.*

This may be deduced immediately from the theorem of BERGE, NORMAN and RABIN; the above condition states that in the graph $\bar{G}(W)$ there are no alternating chains which start and finish at $\bar{a}$.

THEOREM 3. *If W is a maximum matching, let M_1 be a component of the subgraph defined by the mixed points.*

1. *In the graph $\bar{G}(W)$, there is exactly one heavy edge incident to M_1 directed towards the interior of M_1;*

2. *Every other edge which is incident to M_1 is light and is directed out from M_1.*

1. Since $\bar{a}$ is light, $\bar{a} \notin M_1$, and therefore from the theorem of PETERSEN and GALLAI, there is one and only one edge incident to M_1 which is directed towards the interior of M_1. Let this edge be $[c, x_0]$, where $c \notin M_1$, $x_0 \in M_1$.

Let us suppose that $[c, x_0]$ is a light edge: since x_0 is mixed, an alternating chain μ exists which goes from $\bar{a}$ to x_0, and ends with a heavy edge; this chain μ must use the edge $[c, x_0]$ (since this is the only way μ can enter M_1), followed by a heavy edge which is incident to x_0, and finally μ terminates with another heavy edge which is incident to x_0. But this is absurd, since there can only be one heavy edge incident to x_0. Therefore the edge $[c, x_0]$ must be heavy.

2. Let $[y, b]$ be another edge incident to M_1, with $y \in M_1$, $b \notin M_1$; from the theorem of PETERSEN and GALLAI, this edge must be directed out from M_1. On the other hand, this outgoing edge cannot be heavy, since y is mixed and must therefore be accessible by an alternating chain which terminates with a heavy edge.

COROLLARY 1. *If W is a maximum matching and if M_1 is a component of the subgraph determined by M, then $|M_1| \geqslant 3$, and $|M_1|$ is odd.*

Let $[c, x_0]$ be the heavy edge which is incident to M_1, with $x_0 \in M_1$. Then $|M_1| \geqslant 3$, for otherwise x_0 cannot be reached by an alternating chain which ends with a light edge. On the other hand, since M_1 consists of the point x_0 and a number of pairs of points connected by a heavy edge, $|M_1|$ must be odd.

COROLLARY 2. *If W is a maximum matching, any unmixed vertex which is adjacent to a mixed vertex, must be light.*

If $[c,x_0]$ is the heavy edge which is incident into a component M_1, with $c \notin M_1$, $x_0 \in M_1$, then since $c \notin I$, $c \notin H$, $c \notin M$, we must have $c \in L$.

Similarly, if $[x,y]$ is a light edge incident to M_1, with $x \notin M_1$, $y \in M_1$, then $x \notin I$, $x \notin H$, $x \notin M$, and therefore $x \in L$.

COROLLARY 3. *If W is a maximum matching, any vertex adjacent to a heavy vertex must be light.*

If $h \in H$, and if $[x,h]$ is a heavy edge, then $x \notin I$ (since it is a vertex of an oriented edge), $x \notin H$ (because if $x \in H$, h would be mixed), $x \notin M$ (from Corollary 2), and therefore $x \in L$.

On the other hand, if $[x,h]$ is a light edge, then $x \notin I$, $x \notin H$, and $x \notin M$ (from Corollary 2), and therefore $x \in L$.

THEOREM 4. *Let W be a maximum matching, and let I_1 be a component of the subgraph determined by the inaccessible points. All edges incident to I_1 are light and non-oriented. Every point $x \notin I_1$ which is adjacent to a point of I_1 is a light vertex. $|I_1|$ is even.*

1. If $[c,x]$ is an edge incident to I_1, with $x \in I_1$, then $c \notin M$, $c \notin I$, $c \notin H$ (from Corollary 3): therefore $c \in L$, and consequently $[c,x]$ must be a light edge.

2. I_1 contains neither the origin $\bar{a}$ nor any unsaturated points. It is therefore made up of pairs of points connected by a heavy edge. It follows that $|I_1|$ is even.

THEOREM 5 (TUTTE, generalized; BERGE [2]). *Given a connected graph G and a set of vertices S, let $p_i(S)$ be the number of components of uneven order in the subgraph determined by $X-S$; the number of unsaturated vertices in a maximum matching of G is*

$$\xi = \max_{S \subset X} [p_i(S) - |S|]$$

1. Consider a set $S \subset X$, and let $C_1, C_2, \ldots, C_p$ be the components of uneven order in the subgraph determined by $X-S$. If the component C_k contains no unsaturated points, then there must be at least one heavy edge leading out of C_k (since $|C_k|$ is odd) which terminates at a point $s_k \in S$; to two distinct components C_k there correspond two distinct vertices s_k; therefore, if n_0 is the number of unsaturated vertices, we have:

$p_i(S) - n_0 \leqslant$ (the number of C_k containing no unsaturated points) $\leqslant |S|$.
And hence

$$p_i(S) - |S| \leqslant n_0 \qquad (S \subset X)$$

2. We shall now show that if the set S is properly chosen, $p_i(S)-|S|=n_0$ (which completes the proof).

Let W be a maximum matching, and L the set of light vertices. From Theorem 3 and Corollaries 1, 2, 3, we have

$$p_i(L) = (\text{number of components } M_1)+|H|$$

In addition, the heavy edges of G establish a one–one correspondence between the set L on the one hand, and the components M_1 which contain no unsaturated points and the saturated heavy vertices, on the other; therefore we have

$$|L| = (\text{number of components } M_1)+|H|-n_0$$

Combining these results:

$$n_0 = p_i(L)-|L|$$

Q.E.D.

COROLLARY 1. *The number of edges in a maximum matching is*

$$\tfrac{1}{2}(n-\xi)$$

exactly, where:

$$\xi = \max_{S\subset X}[p_i(S)-|S|]$$

This follows immediately.

COROLLARY 2 (TUTTE's theorem). *The necessary and sufficient condition for a graph to possess a perfect matching is that*

$$p_i(S) \leqslant |S| \qquad (S \subset X)$$

This follows immediately.

If h is an integer, a graph is said to be *homogeneous of degree h* if every vertex is of degree h; if the graph G is homogeneous, some much simpler results can be proved.

THEOREM 6. *If a connected graph with an even number of vertices which is homogeneous of degree h, contains no set $S \subset\subset X$ with $|U_S| < h-1$, a perfect matching exists.*

Let S be a subset of X (in the strict sense), and let C_1 be a component of the subgraph determined by $X-S$, with an uneven number of elements. By hypothesis, $|U_{C_1}| \geqslant h-1$; we cannot have $|U_{C_1}| = h-1$, since the number of edges in C_1 would be

$$\tfrac{1}{2}[h|C_1|-(h-1)] = \tfrac{1}{2}[h(|C_1|-1)+1]$$

which is not an integer (since $|C_1|$ is odd). Therefore $|U_{C_1}| \geqslant h$, and hence

$$h|S| \geqslant |U_S| \geqslant \sum_k |U_{C_k}| \geqslant hp_i(S)$$

Therefore $p_i(S) \leqslant |S|$ for all $S \subset \subset X$; this inequality holds even if $S = X$, and therefore, from TUTTE's theorem, the graph possesses a perfect matching.

LEMMA. *If a homogeneous graph of degree h contains no set $S \subset \subset X$ with $|U_S| < h-1$, and if W is a maximum matching of heavy edges, every edge u_0 belongs to a simple cycle consisting of light and heavy edges alternately.*

From Theorem 6, the matching W is perfect; it is sufficient if we show that every *light* edge is on an alternating cycle (because then any heavy edge u_0 will be on the alternating cycle to which an adjacent light edge belongs).

Therefore let u_0 be a light edge, and let us assume that it does not belong to an alternating cycle; if, by an appropriate shrinkage, we attach u_0 to a vertex $\bar{a}$, then in the graph $\bar{G}$ so obtained, there are no alternating chains which start at $\bar{a}$ and finish at $\bar{a}$, and therefore the vertex $\bar{a}$ will be light.

On the other hand, from Theorem 3, a component M_1 of mixed points possesses a heavy entrant edge and several light outgoing edges. Since $|M_1|$ is odd, we can use the same argument as in Theorem 6 to show that $|U_{M_1}| \geqslant h$, and therefore there must be at least $h-1$ light edges leaving M_1.

If, by a shrinkage, we identify each component M_1 with a heavy vertex, we get a new graph which no longer contains any edges oriented in both directions; the number of light edges which are incident out from a heavy vertex will be $\geqslant h-1$, and the number of light edges which are incident into a light vertex will be $\leqslant h-1$. Now let us determine the number of heavy oriented edges in the graph $\bar{G}$ as a function of the number $|H|$ of heavy vertices, and also as a function of the number $|L|$ of light vertices; we get:

$$|L|+1 = \text{number of heavy oriented edges} = |H|$$

A similar determination for the light oriented edges gives:

$$(h-1)(|L|-1) \geqslant \text{number of light oriented edges} \geqslant (h-1)|H|$$

Thus we have:

$$\begin{cases} |L|+1 = |H| \\ |L|-1 \geqslant |H| \end{cases}$$

Since these two relations are incompatible, the lemma is proved.

THEOREM 7. *If a connected, homogeneous graph of degree h contains no set $S \subset \subset X$ with $|U_S| < h-1$, then given any arbitrary edge u_0, a perfect matching exists which uses this edge.*

From Theorem 6, a perfect matching exists; if the edge $\boldsymbol{u}_0$ is heavy, the theorem is proved. If, however, $\boldsymbol{u}_0$ is light, find the alternating cycle $\boldsymbol{\mu}$ which uses $\boldsymbol{u}_0$ as described in the preceding lemma.

Interchange the light and the heavy edges of $\boldsymbol{\mu}$; we now have a new perfect matching which contains $\boldsymbol{u}_0$.

Q.E.D.

Application to the Coefficient of Internal Stability

We can use the theory of matching to give us an algorithm for constructing an internally stable set with the maximum number of elements; for simplicity, we shall here describe a set as 'stable' if it is 'internally stable'.

LEMMA 1. *Let G be a graph, and $\boldsymbol{W}$ a maximum matching; if all vertices are either light or heavy, the set of heavy vertices is a maximum internally stable set of G.*

Consider the set H of heavy vertices (which contains the unsaturated points) and the set L of light vertices. A set $C \subset X$ is a *support* of the graph G if every edge has at least one vertex in C; the complement of a support is a stable set, and vice versa. If $\boldsymbol{W}$ is a matching and C is a support, then we always have

$$|\boldsymbol{W}| \leqslant |C|$$

Therefore if a given matching $\boldsymbol{W}$ and a given support C satisfy $|\boldsymbol{W}| = |C|$, $\boldsymbol{W}$ is a maximum matching and C is a minimum support. In G, two heavy vertices cannot be adjacent (Corollary 3), and therefore the set L is a support. Since $\boldsymbol{W}$ is a maximum matching, every point of L is the terminal of a heavy edge leading into H, and therefore $|L| = |\boldsymbol{W}|$, and L is a minimum support.

Therefore, the complement of L, $H = X - L$, is a maximum stable set.

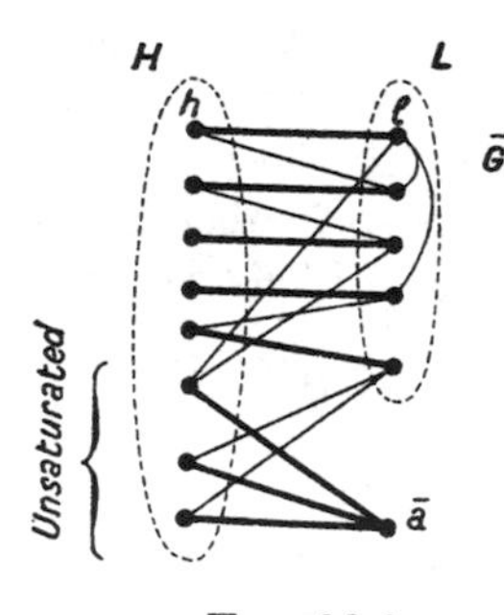

FIG. 18.6

LEMMA 2. *Let $\boldsymbol{W}$ be a maximum matching of a graph G which has only light or heavy vertices; a maximum stable set of the graph G' which is formed from G by deleting a heavy vertex h and all edges incident to h, is the set $H - \{h\}$.*

The set L is a support of the subgraph G' obtained by suppressing the heavy vertex h; we shall now show that it is a minimum support (in which case we shall also have proved that $H - \{h\} = X - L$ is a maximum stable set of G').

First let us suppose that the point h was saturated, and that $[h,l]$ was the heavy edge incident to h; then $W-[h,l]$ is a matching for G', but we no longer have a maximum matching, since an alternating chain exists which goes from the origin $\bar{a}$ to the unsaturated point l (Theorem 2). Therefore a maximum matching W' of G' satisfies $|W'| = |W|$.

(If the point h is unsaturated, W is still a maximum matching, and we again have $|W'| = |W|$.)

Since $|W'| = |L|$ for the matching W' and the support L, we have shown that L must be a minimum support of G'.

Q.E.D.

LEMMA 3. *Let G be a graph, and G' a new graph formed by adding some new edges to G; the coefficients of internal stability of these graphs satisfy $\alpha(G') \leqslant \alpha(G)$.*

Let $\mathscr{S}$ be the family of stable sets of G, and let $\mathscr{S}'$ be that of G'. Therefore we have $\mathscr{S}' \subset \mathscr{S}$, and therefore

$$\alpha(G') = \max_{S \in \mathscr{S}'} |S| \leqslant \max_{S \in \mathscr{S}} |S| = \alpha(G)$$

THEOREM 8 (BERGE [1]). *If $A \subset X$ is a set of vertices of a graph G, let $S(A)$ denote a maximum stable set of the subgraph determined by A; for a maximum matching W, let $M_1, M_2, \ldots,$ be the components of the subgraph determined by the mixed points, and $I_1, I_2, \ldots,$ those of the subgraph determined by the inaccessible points, and let us put:*

$$S = H \cup S(M_1) \cup S(M_2) \cup \ldots \cup S(I_1) \cup S(I_2) \cup \ldots$$

If the number of components consisting of mixed points is zero or one, S is a maximum stable set of G.

In fact, any vertex adjacent to a heavy vertex is light (Corollary 3), as is any vertex adjacent to the possible component M_1 (Corollary 2), or to a component I_k (Theorem 4); therefore S is a stable set.

On the other hand, if we delete all edges connecting a light vertex to a mixed or inaccessible point, we disconnect the graph into components $H \cup L, M_1, I_1, I_2, \ldots$.

H is a maximum stable set of the component $H \cup L$; for if we remove the inaccessible points from G and reduce the component M_1 (if it exists) to a point h_1 (by shrinkage), the new graph contains only heavy or light vertices, and we can apply Lemma 2. S is thus a maximum stable set of the disconnected graph. If we now replace the edges which were removed, thus regaining the graph G, the coefficient of internal stability cannot increase (Lemma 3); but as S is still a stable set in G, it must be a maximum stable set.

Remark 1. The statement of the theorem is valid even if there are several components consisting of mixed points, provided that no component M_j is accessible by an alternating chain which has already traversed another component M_k. In the case in which an alternating chain μ exists which leaves M_k and enters M_j, we must see if we can enlarge the stable set S by selecting the light vertices of μ instead of the heavy ones.

For example, in the graph of Fig. 18.7, there is a light vertex between the components M_1 and M_2 which can clearly be added to S to make a larger stable set (this set consists of the vertices of Fig. 18.7 which have been ringed).

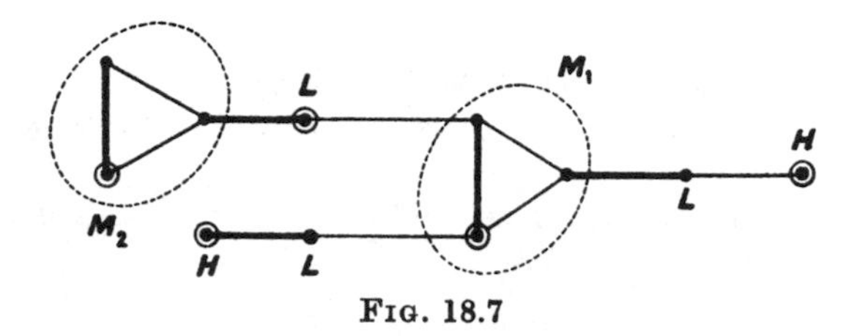

FIG. 18.7

Remark 2. To find a maximum stable set in a component M_j or I_k, we can again make good use of Theorem 8 and its lemmas. We simply have to add an unsaturated point judiciously connected to the rest of the graph, or break some carefully selected light edge. This is illustrated in the following example:

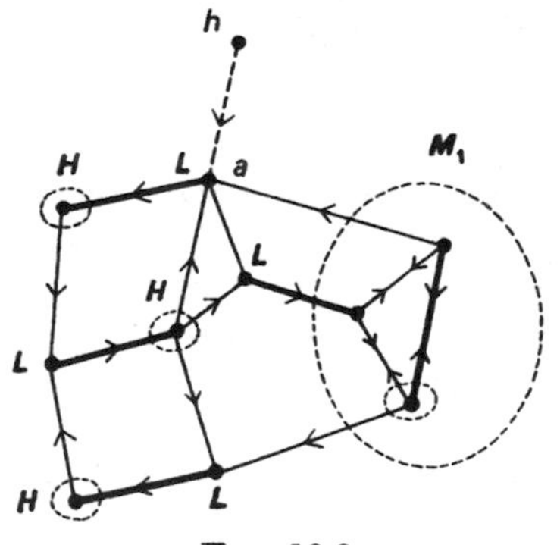

FIG. 18.8

EXAMPLE. Consider the graph G of Fig. 18.8, with a maximum matching W. There are no unsaturated points and therefore every vertex of the graph is inaccessible. Nevertheless, let us add a point h and an edge $[h,a]$ and put $Y = X \cup \{h\}$.

In the subgraph determined by $Y - M_1$, there are only light and heavy vertices: therefore it possesses a maximum stable set which consists of h and the three vertices labelled 'H'. From Lemma 2, the three vertices labelled 'H' constitute a maximum stable set of the subgraph determined by $X - M_1$; therefore the four vertices which have been ringed form a maximum stable set of a graph consisting of the disconnected components $X - M_1$ and M_1. From Lemma 3, these four vertices also form a maximum stable set of the graph G.

19. Semi-Factors

Hamiltonian Cycles and Semi-Factors

In a finite graph G, a *Hamiltonian chain* is a chain which meets every vertex exactly once; a *Hamiltonian cycle* is a cycle which meets every vertex exactly once (with the exception of the initial and final vertex, which coincide).

If we make G into a symmetric graph by giving each edge an orientation in each direction, then every Hamiltonian chain defines two Hamiltonian paths, and vice versa; and every Hamiltonian cycle defines two Hamiltonian circuits, and vice versa. Thus the problem of determining a Hamiltonian chain or cycle has already been investigated in an earlier chapter (Chapter 11) and we shall not return to it. On the other hand, we do encounter a new problem if we try to find a *semi-factor* of the graph $G = (X, U)$, that is to say a partial graph (X, V) in which every vertex is of degree 2 (and which therefore consists of one or more disjoint cycles). *A symmetric graph may very well have no semi-factors even though it possesses one or more factors*; to convince oneself of this it is sufficient to consider the graph in Fig. 19.1; the arcs which have been drawn with a heavy line form a factor, but there are no semi-factors. (Note that although the graph has 10 *arcs*, it has only 5 distinct *edges*.)

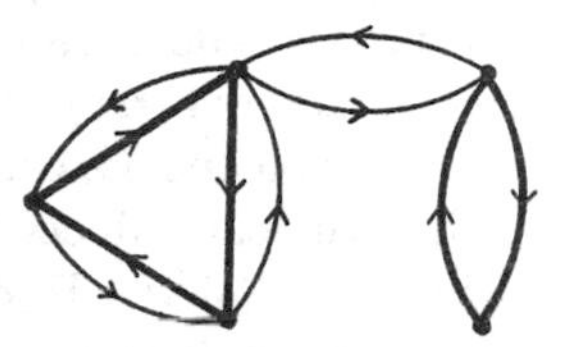

Fig. 19.1

Following the method developed in Chapter 18, the search for a semi-factor may be made by considering alternating chains of light and heavy edges between two unsaturated vertices; we shall merely set out to establish simple criteria whereby we can recognize directly whether a given graph has a semi-factor or not. We shall also be concerned with finding the maximum number of disjoint semi-factors, that is, semi-factors which do not have any edges in common, which can be extracted from a given graph. To begin with, we shall limit ourselves to homogeneous graphs of degree h.

EXAMPLE 1 (Kirkman). Eleven ministers meet every day at a round table: how should they be placed if each minister is to have different neighbours at every meeting? How many days can this be kept up?

The problem is one of finding the maximum number of disjoint Hamiltonian cycles in a complete graph with eleven vertices: $a, b, c, \ldots, i, j, k$. It can easily be seen that in this case there are only five disjoint cycles, as for example:

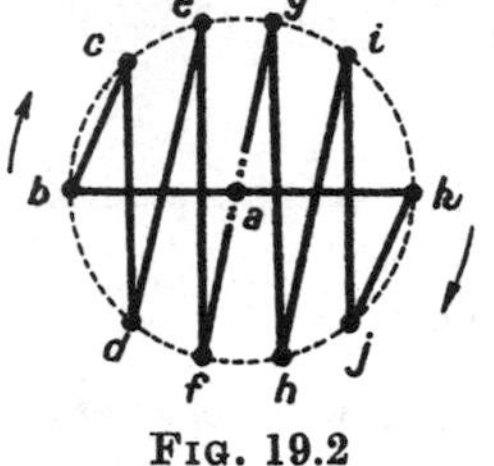

FIG. 19.2

$a\ b\ c\ d\ e\ f\ g\ h\ i\ j\ k\ a$

$a\ c\ e\ b\ g\ d\ i\ f\ k\ h\ j\ a$

$a\ e\ g\ c\ i\ b\ k\ d\ j\ f\ h\ a$

$a\ g\ i\ e\ k\ c\ j\ b\ h\ d\ f\ a$

$a\ i\ k\ g\ j\ e\ h\ c\ f\ b\ d\ a$

These cycles are found by rotating the pattern shown in Fig. 19.2 (in the direction of the arrow). More generally, in a complete graph with $2n+1$ vertices, there are n disjoint Hamiltonian cycles.

EXAMPLE 2 (LUCAS). Six boys, whom we shall denote by a, b, c, d, e, f, and six girls whom we shall denote by $\bar{a}, \bar{b}, \bar{c}, \bar{d}, \bar{e}, \bar{f}$ are dancing in a ring. How should we arrange the rings if we want each boy to have each girl as a neighbour on one occasion, and no two boys are ever to be neighbours?

The problem reduces to finding the maximum number of disjoint Hamiltonian cycles in a bichromatic graph with twelve vertices. We can find three disjoint cycles:

$a\ \bar{a}\ b\ \bar{b}\ c\ \bar{c}\ d\ \bar{d}\ e\ \bar{e}\ f\ \bar{f}\ a$

$a\ \bar{c}\ b\ \bar{d}\ c\ \bar{e}\ d\ \bar{f}\ e\ \bar{a}\ f\ \bar{b}\ a$

$a\ \bar{e}\ b\ \bar{f}\ c\ \bar{a}\ d\ \bar{b}\ e\ \bar{c}\ f\ \bar{d}\ a$

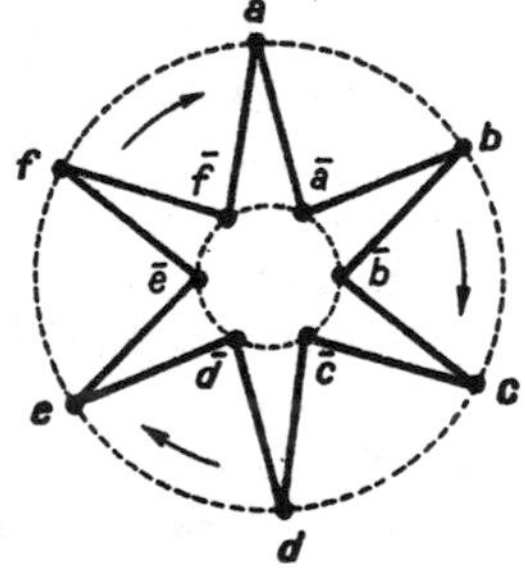

FIG. 19.3

These cycles are found by rotating the star drawn in Fig. 19.3 with respect to the letters on the outer circle: using the theory, we shall show that no further disjoint Hamiltonian cycles can be found, nor even any other disjoint semi-factors.

EXAMPLE 3 (HILBERT). Let us consider a set of positive integers p_{ij}, with $i \leqslant n, j \leqslant n, i<j$, and form the function of n complex variables:

$$f(z_1, z_2, \ldots z_n) = \prod_{i<j\leqslant n} (z_i - z_j)^{p_{ij}}$$

In his theory of invariants, HILBERT found that he needed to know if f could be expressed as a product of *primitive terms*, that is to say of terms which may be written in the form:

$$(z_1 - z_{i_1})(z_2 - z_{i_2}) \ldots (z_n - z_{i_n})$$

where $(i_1, i_2, \ldots, i_n)$ is some permutation of $(1, 2, \ldots, n)$. In order to see if such a decomposition is possible, construct an s-graph G by taking n vertices $x_1, x_2, \ldots, x_n$, and joining vertices x_i and x_j by p_{ij} distinct edges (if $i < j$); the problem then reduces to decomposing G into disjoint semi-factors.

THEOREM 1 (PETERSEN). *A homogeneous graph of even degree $h = 2k$ can be decomposed into k disjoint semi-factors.*

We may, in fact, assume the graph to be connected (if this is not so, we consider each component separately); therefore, from Theorem 1 (Chapter 17), the graph must possess an Eulerian cycle. By orienting each edge according to its sense in this cycle, we get a pseudo-symmetric graph G', in which both the internal and external demi-degree of every vertex is equal to k. From the corollary to Theorem 2 (Chapter 11), G' can be decomposed into k factors, each of which is equivalent to a semi-factor in the initial graph.

THEOREM 2 (BAEBLER). *A homogeneous graph of odd degree $h = 2k+1$ contains k disjoint semi-factors if we have:*

$$\min_{\substack{S \neq X \\ S \neq \emptyset}} |U_S| \geqslant 2k$$

This condition is sufficient but not necessary†.

In this case, the graph possesses a perfect matching (Theorem 6, Chapter 18). Those edges which are not a part of this matching form a homogeneous partial graph of degree $2k$, and from Theorem 1, this (partial) graph can be decomposed into k disjoint semi-factors.

Q.E.D.

EXAMPLE. How many different rings can we form containing $2k+2$ children, if no child is to have the same neighbour more than once? This problem reduces to finding the number of disjoint semi-factors in a complete homogeneous graph of degree $2k+1$. Since we have here $|U_S| \geqslant 2k$, k rings can be formed.

We are going to pay particular attention to the semi-factors of homogeneous graphs of degree 3 (because of the important role they play in the 'four colour problem'). To begin with, we observe that a homogeneous

† F. BAEBLER has further generalized this result, and proved that: *A homogeneous graph of odd degree $h = 2k+1$ contains q disjoint semi-factors (where $q \leqslant k$), if we have;*

$$\min_{\substack{S \neq X \\ S \neq \emptyset}} |U_S| \geqslant 2q$$

graph of degree 3 which contains no loops is 4-chromatic; for, like a symmetric graph, G has a Grundy function $g(x)$, and

$$\max g(x) \leqslant \max |\Gamma x| = 3$$

It follows from Theorem 4 (Chapter 4) that the chromatic number of G, $\gamma(G) = 3+1$ (at most).

What can now be said regarding the chromatic index of G? The reply here depends essentially on the existence of a semi-factor. Whenever G possesses a semi-factor V, it also must possess a perfect matching $W = U - V$ (and vice versa); in the following paragraphs, we shall speak of the edges of V as *light*, and of those of W as *heavy*. Any edge of a connected graph which, if deleted, divides the graph into two components (each of which contains at least one edge) is called an *isthmus*. Then we have:

Theorem 3. *Consider a homogeneous graph G of degree 3 which contains no isthmuses;*

1. *Its edges can be decomposed into a semi-factor V and a perfect matching W;*

2. *An alternating cycle (consisting of edges alternately in V and in W) of even length passes through any edge u_0.*

The first part of the theorem may be deduced directly from Theorem 2; and the second part may be deduced from the lemma on p. 183 (Chapter 18).

Corollary. *If G is a homogeneous graph of degree 3 with no isthmuses:*
1. *A perfect matching exists which contains any arbitrary given edge;*
2. *A semi-factor exists which contains any two arbitrary edges.*

1. If u_0 is a light edge for some perfect matching W, consider the alternating cycle μ which goes through u_0; $W' = (W - \mu) \cup (\mu - W)$ is also a perfect matching, and it certainly contains u_0.

2. Consider two arbitrarily chosen edges u_1 and u_2, and, starting from G, form a graph G' by adding a vertex a_1 in the middle of u_1, a vertex a_2 in the middle of u_2, and joining these vertices by an edge u_0. G' is a homogeneous graph of degree 3 which does not contain any isthmuses, and therefore it possesses a perfect matching which contains u_0. The light edges now define a semi-factor which contains both u_1 and u_2.

Theorem 4 (Errera). *If all the isthmuses of a homogeneous graph of degree 3 belong to one single elementary chain, the graph possesses a semi-factor.*

If every isthmus were deleted, we would get a series of disjoint components: C, D, etc...; the sets U_C, U_D, ..., consist of one or two isthmuses of G.

Let us suppose that U_C consists of the two isthmuses v and w, and consider the subgraph defined by C. We amalgamate the two edges adjacent to v by deleting the vertex at which these three edges meet, and we do the same for the two edges adjacent to w. Since the new graph is homogeneous of degree 3, the preceding corollary shows that a semi-factor can be constructed which contains the amalgamated edges. Such a semi-factor can be constructed for each of the components $C, D, \ldots$, thus giving a semi-factor of the graph G.

THEOREM 5. *The chromatic index of a homogeneous graph of degree* 3 *is less than or equal to* 5; *it is less than or equal to* 4 *if the graph has no isthmuses or if all the isthmuses are on one single elementary chain; and it is less than or equal to* 3 *if a Hamiltonian cycle exists.*

Let us consider a homogeneous graph of degree 3, and allot the first colour to a maximum matching of this graph. If there are any unsaturated vertices, use a second colour to colour an arbitrary edge incident to each unsaturated vertex. The only edges now uncoloured must belong to elementary disjoint chains and cycles, and three colours will obviously be sufficient to colour these.

If all the isthmuses are on one single elementary chain, the preceding result shows that there are no unsaturated points, and therefore that four colours will suffice.

If a Hamiltonian cycle exists, it must be even in length (since there are $3n/2$ edges, and n must therefore be even); therefore two colours will be sufficient to colour the Hamiltonian cycle, and three colours will be enough to colour all the edges of the graph.

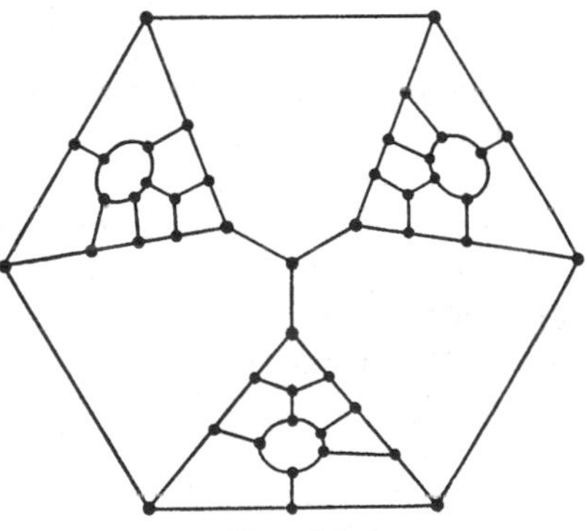

FIG. 19.4

It should be observed that a homogeneous graph of degree 3, even if it contains no isthmuses, does not necessarily possess a Hamiltonian cycle: for example, the graph of Fig. 19.4 (which has a chromatic index of 3).

COROLLARY (C. A. B. SMITH). *In a homogeneous graph of degree* 3, *the number of Hamiltonian cycles starting from any arbitrarily given edge is even.*

If no Hamiltonian cycles exist, the theorem is proved; if any do exist, the graph has a chromatic index of 3, and contains no isthmuses.

We shall consider on the one hand the set of possible colourings π (that is to say: the partitions of U into three sets of non-adjacent edges), and on the other the set of semi-factors $\boldsymbol{\mu}$ containing no cycles of uneven length. We can associate three bicoloured semi-factors $\boldsymbol{\mu}_1(\pi)$, $\boldsymbol{\mu}_2(\pi)$ and $\boldsymbol{\mu}_3(\pi)$ with each colouring π, with

$$(1) \qquad \boldsymbol{\mu}_1(\pi)+\boldsymbol{\mu}_2(\pi)+\boldsymbol{\mu}_3(\pi) \equiv \mathbf{0} \pmod{2}$$

Conversely, given a semi-factor $\boldsymbol{\mu}$ which contains k_μ disjoint elementary cycles of even length, we can associate with it exactly $2^{k_\mu-1}$ distinct colourings π. By summing the identities (1) with respect to all possible colourings π, we get

$$\sum_{\boldsymbol{\mu}} 2^{k_\mu-1}\boldsymbol{\mu} \equiv \mathbf{0} \pmod{2}$$

And hence:

$$\sum_{\boldsymbol{\mu}|k_\mu=1} \boldsymbol{\mu} \equiv \mathbf{0} \pmod{2}$$

from which we can deduce the required result.

A Necessary and Sufficient Condition for the Existence of a Semi-factor

In the preceding section, we were able to deduce very simply sufficient conditions for the existence of a semi-factor; we shall now look for somewhat stronger conditions.

Let G be a graph in which we are looking for a semi-factor, and let $\overline{G}$ be a graph built up from G in the following manner: each vertex x of G with for example $U_x = \{u_1, u_2, \ldots, u_k\}$ will be replaced by two sets containing k and $k-2$ elements, say:

$$\overline{X}(x) = \{\bar{x}_1, \bar{x}_2, \ldots, \bar{x}_k\}$$
$$\overline{Y}(x) = \{\bar{y}_1, \bar{y}_2, \ldots, \bar{y}_{k-2}\}$$

We shall join every $\bar{x}_i$ to every $\bar{y}_j$ by an edge, and in addition, for every edge $u_j = [x, x']$ we shall draw a corresponding edge $\bar{u}_j$ connecting $\bar{x}_j \in \overline{X}(x)$ and $\bar{x}' \in \overline{X}(x')$ (cf. Fig. 19.5).

Then every perfect matching of the graph $\overline{G}$ defines a semi-factor in G; and conversely, any semi-factor in G defines a perfect matching in $\overline{G}$. Thus we find ourselves with a known problem: to characterize the graphs $\overline{G}$ which possess a perfect matching.

We can eliminate the case in which G contains a vertex of degree 1, for obviously such a graph cannot contain a semi-factor; likewise, we can also eliminate the case in which G contains a vertex of degree 2, since we can delete this vertex (and unite the two incident edges). In other words, we may assume that for all $x \in X$, we have

$$\overline{Y}(x) \neq \emptyset$$

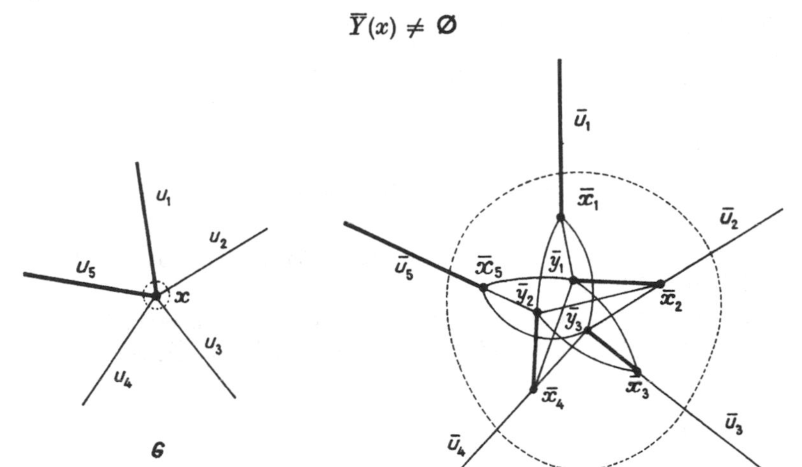

FIG. 19.5

LEMMA. *Let the set of vertices of $\overline{G}$ be $\overline{X}$, and let $p_i(\overline{S})$ be the number of components of odd order of the subgraph determined by $\overline{X}-\overline{S}$; then a set $\overline{S}$ exists which satisfies*

$$p_i(\overline{S})-|\overline{S}| = \max_{\overline{E}\subset\overline{X}} (p_i(\overline{E})-|\overline{E}|) \tag{0}$$

$$\overline{X}(a) \cap \overline{S} \neq \emptyset \Rightarrow \overline{X}(a) \subset \overline{S} \tag{1}$$

$$\overline{Y}(b) \cap \overline{S} \neq \emptyset \Rightarrow \overline{Y}(b) \subset \overline{S} \tag{2}$$

$$\{a|\overline{X}(a) \subset \overline{S}\} \quad \textit{and} \quad \{b|\overline{Y}(b) \subset \overline{S}\} \textit{ are disjoint} \tag{3}$$

We are going to show that of the sets $\overline{S}$ which satisfy (0), that for which $|\overline{S}|$ is minimum also satisfies (1), (2) and (3).

To prove (1): Let us assume that for some $a \in X$, we have $\overline{X}(a) \not\subset \overline{S}$, $\overline{X}(a) \cap \overline{S} \neq \emptyset$, and let us consider a vertex $\bar{x}_0 \in \overline{X}(a) \cap \overline{S}$; we shall put $\overline{S}_0 = \overline{S}-\{\bar{x}_0\}$.

In the subgraph determined by $\overline{X}-\overline{S}_0$, $\bar{x}_0$ can connect at most two disjoint components of the subgraph determined by $\overline{X}-\overline{S}$. If both these components are of even order (including the case in which one or both of them is of zero order) we have

$$p_i(\overline{S}_0) = p_i(\overline{S})+1$$

In all other cases

$$p_i(\bar{S}_0) = p_i(\bar{S})-1$$

Therefore we have ultimately:

$$p_i(\bar{S}_0) \geqslant p_i(\bar{S})-1$$
$$|\bar{S}_0| = |\bar{S}|-1$$

This gives

$$p_i(\bar{S}_0)-|\bar{S}_0| \geqslant p_i(\bar{S})-1-|\bar{S}|+1 = \max_{\bar{E}} (p_i(\bar{E})-|\bar{E}|)$$

Therefore $\bar{S}_0$ satisfies (0), which contradicts the assumption that $\bar{S}$ is the set satisfying (0) for which $|\bar{S}|$ is minimum.

To prove (2): Let us assume that for some $b \in X$, we have $\bar{Y}(b) \cap \bar{S} \neq \emptyset$ and $\bar{Y}(b) \not\subset \bar{S}$, and let us consider the set $\bar{T} = \bar{S}-(\bar{Y}(b) \cap \bar{S})$; we have

$$|\bar{T}| < |\bar{S}|$$

If $\bar{X}(b) \subset \bar{S}$, we have

$$p_i(\bar{T}) > p_i(\bar{S})$$

If $\bar{X}(b) \not\subset \bar{S}$, we have

$$p_i(\bar{T}) \geqslant p_i(\bar{S})-1$$

Therefore no matter what happens we have $p_i(\bar{T}) \geqslant p_i(\bar{S})-1$, and

$$p_i(\bar{T})-|\bar{T}| \geqslant p_i(\bar{S})-1-|\bar{S}|+1 \geqslant \max_{\bar{E}\subset\bar{X}} (p_i(\bar{E})-|\bar{E}|)$$

This again yields a contradiction.

To prove (3): Let us assume that, for some $a \in X$, we have $\bar{X}(a) \subset \bar{S}$, $\bar{Y}(a) \subset \bar{S}$; given $\bar{y}_0 \in \bar{Y}(a)$, let us put $\bar{T} = \bar{S}-\{\bar{y}_0\}$. Then we have

$$p_i(\bar{T}) = p_i(\bar{S})+1$$
$$|\bar{T}| = |\bar{S}|-1$$

Hence:

$$p_i(\bar{T})-|\bar{T}| = p_i(\bar{S})-|\bar{S}|+2 > \max_{\bar{E}\subset\bar{X}} (p_i(\bar{E})-|\bar{E}|)$$

Thus we are again led to a contradiction.

TUTTE'S THEOREM. *In a graph G, if A and B are two sets of disjoint vertices, let us denote the number of edges connecting A and B by $m(A, B)$; let $q(A, B)$ be the number of components C of the subgraph determined by $X-A$ for which $m(C, B)$ is odd. The necessary and sufficient condition for G to possess a semi-factor is that for every vertex x, the degree $d(x) \geqslant 2$, and that for every pair of disjoint sets A and B, we have:*

$$q(A, B)-\sum_{b\in B} d(b)-2|A|+2|B| \leqslant 0$$

We can use Theorem 5 (Chapter 18) to express the required condition in terms of the graph $\bar{G}$ defined above. It then becomes

$$\max_{\bar{S} \subset \bar{X}} (p_i(\bar{S}) - |\bar{S}|) = 0$$

From the lemma, we need only consider sets $\bar{S}$ satisfying (1), (2) and (3). Let us consider such a set $\bar{S}$, and put

$$A = \{a | \bar{X}(a) \subset \bar{S}\}$$
$$B = \{b | \bar{Y}(b) \subset \bar{S}\}$$

A and B are disjoint, and we have

$$|\bar{S}| = \sum_{a \in A} d(a) + \sum_{b \in B} (d(b) - 2) = \sum_{a \in A} d(a) + \sum_{b \in B} d(b) - 2|B|$$

Now let us determine $p_i(\bar{S})$; if a component $\bar{C}$ meets $\bar{Y}(A)$, it must be of the form $\bar{C} = \{\bar{y}\}$, with $\bar{y} \in \bar{Y}(a)$, $a \in A$. The total number of components of this sort is

$$\sum_{a \in A} (d(a) - 2) = \sum_{a \in A} d(a) - 2|A|$$

If $\bar{C}$ is a component which does not meet $\bar{Y}(A)$, let us put:

$$C = \{x | \bar{X}(x) \cap \bar{C} \neq \varnothing\}$$

We have $C \cap A = \varnothing$, and we can show that $m(C, B)$ and $|\bar{C}|$ are of the same parity. Therefore the number of components $\bar{C}$ of odd order which do not intersect with $\bar{Y}(A)$ is $q(A, B)$, which gives finally

$$p_i(\bar{S}) = q(A, B) + \sum_{a \in A} d(a) - 2|A|$$

The required condition now becomes

$$0 = \max_{S} (p_i(\bar{S}) - |\bar{S}|) = \max_{A,B} [q(A, B) - 2|A| + 2|B| - \sum_{b \in B} d(b)]$$

as stated in the theorem.

Clearly, by using a strictly analogous argument, we can obtain an existence condition for a partial graph with any given set of degrees for the vertices†.

† Unfortunately, this condition, derived by TUTTE, is too complicated to be useful: let $f(x)$ be the degree which we wish to assign to the vertex x, and let $d_A(x)$ be the degree of x in the subgraph determined by $X - A$; Tutte's condition is:

(1) $\quad d(x) \geqslant f(x) \qquad (x \in X)$

(2) $\quad \sum_{x \in A} f(x) \geqslant q(A, B) + \sum_{x \in B} [f(x) - d_A(x)] \qquad (A, B, A \cap B = \varnothing)$

Note that the same method may be used to find the deficiency of a maximum set W compatible with $f(x)$.

20. The Connectivity of a Graph

Articulation Points

A vertex x is said to be an *articulation point* of a connected graph (X, U) if the subgraph obtained by deleting x is not connected; further, if there is only one edge connecting the vertex x to each of the components of this subgraph, then we say that x is a *simple* articulation point.

EXAMPLE 1. Apart from pendant vertices, every vertex of a tree is a simple articulation point. On the other hand, a graph which contains a Hamiltonian cycle has no articulation points.

EXAMPLE 2 (communication network). Let X be a set of people belonging to some organization, and let us put $[x, y] \in U$, if x and y can communicate with one another. The points of articulation represent people whose connections are particularly important; they are especially vulnerable, since their loss can destroy the unity and cohesion of the organization.

THEOREM 1. *A vertex x_0 is an articulation point of a connected graph if and only if there exist two vertices a and b such that every chain joining a and b passes through x_0.*

1. If the subgraph determined by $X - \{x_0\}$ is not connected, it contains at least two components C and C'; let a be any vertex of C, and b any vertex of C': in the original connected graph, every chain joining a to b must necessarily have passed through x_0.

2. If every chain joining two vertices a and b passes through x_0, the subgraph determined by $X - \{x_0\}$ is no longer connected, and therefore x_0 must be a point of articulation.

The concept of an articulation point is closely tied to other concepts, analogous to those studied in Chapter 12. In a connected graph G, let us define the *distance between two vertices* x and y to be the length $\mathbf{d}(x, y)$ of the shortest chain going from x to y, and let us define the (*semi*) *associated number* of the vertex x to be

$$\mathbf{d}(x) = \max_{y \in X} \mathbf{d}(x, y)$$

A vertex of minimum associated number is called a *semi-centre*, while a vertex of maximum associated number is a *semi-peripheral* point. In a symmetric graph, these notions correspond exactly with those of centres and peripheral points. Let us put:

$$\rho = \min_{x \in X} \boldsymbol{d}(x) \qquad (\textit{semi-radius})$$

If $A \subset X$, we shall put:

$$\boldsymbol{d}_A(x_0) = \max_{y \in A} \boldsymbol{d}(x_0, y) \quad (\textit{associated number relative to } A)$$

THEOREM 2 (C. JORDAN). *If a connected graph G has a semi-centre which is a simple articulation point x_0, then the semi-centres of G consist either of this single vertex or else of two adjacent vertices.*

1. In the subgraph determined by $X - \{x_0\}$, there exists at least one component A such that

$$\boldsymbol{d}_A(x_0) = \rho$$

On the other hand, a component B exists such that

$$B \neq A; \qquad \boldsymbol{d}_B(x_0) = \rho \text{ or } \rho - 1$$

This must be so, since if for every component $B \neq A$ we had $\boldsymbol{d}_B(x_0) < \rho - 1$, the vertex a adjacent to x_0 which belongs to A would satisfy

$$\boldsymbol{d}(a) = \rho - 1 < \rho = \min_{x \in X} \boldsymbol{d}(x)$$

which contradicts the definition of ρ.

2. Let x_1 be a semi-centre $\neq x_0$ (if one exists), and let C_1 be the component of the subgraph determined by $X - \{x_0\}$ which contains x_1. Then there is no component $C \neq C_1$ for which $\boldsymbol{d}_C(x_0) = \rho$; for if there were, and if the vertex of C furthest removed from x_0 is c, then we would have

$$\boldsymbol{d}(x_1) \geqslant \boldsymbol{d}(x_1, c) = \boldsymbol{d}(x_1, x_0) + \boldsymbol{d}(x_0, c) \geqslant 1 + \rho$$

which is a contradiction, since x_1 is a semi-centre.

3. From this we deduce that $C_1 = A$, and that the component B defined in (1) satisfies $\boldsymbol{d}_B(x_0) = \rho - 1$; but then x_1 must be the vertex $a \in A$ which is adjacent to x_0, for otherwise, if we denote by b the vertex of B furthest removed from x_0, we would have:

$$\boldsymbol{d}(x_1) \geqslant \boldsymbol{d}(x_1, b) = \boldsymbol{d}(x_1, x_0) + \boldsymbol{d}(x_0, b) \geqslant 2 + (\rho - 1) = \rho + 1$$

which is a contradiction, since x_1 is a semi-centre.

COROLLARY. *A tree has either a single semi-centre or else two adjacent semi-centres.*

This follows since every semi-centre of a tree is also a simple point of articulation.

THEOREM 3. *A semi-peripheral point can never be an articulation point.*

Let z_0 be a semi-peripheral point; and let us consider a point z such that

$$\boldsymbol{d}(z_0, z) = \max_{x,y \in X} \boldsymbol{d}(x, y)$$

If z_0 is an articulation point, the subgraph determined by $X - \{z_0\}$ consists of several components; if we consider a point x in some component which does not contain z, we have

$$\boldsymbol{d}(x, z) = \boldsymbol{d}(x, z_0) + \boldsymbol{d}(z_0, z) > \max_{x,y} \boldsymbol{d}(x, y)$$

which contradicts the assumption that z_0 is a semi-peripheral point.

Biconnected Graphs

A graph G is said to be *biconnected* if it is connected and it contains no points of articulation. The principal properties of biconnected graphs are contained in several sensitive theorems, and the reader may, at first reading, prefer to pass over their proofs.

MENGER'S THEOREM (completed by DIRAC [1]). *Given any elementary chain $\mu = [a_0, a_1, \ldots, a_k]$ joining two distinct vertices a_0 and a_k of a biconnected graph G, we can associate with it two elementary chains μ' and μ'' such that:*

(1) *the terminals of μ' and μ'' are a_0 and a_k;*

(2) *the only vertices which μ' and μ'' have in common are a_0 and a_k;*

(3) *if μ' (or μ'') is followed from a_0 to a_k, the indices of the vertices of μ encountered en route are in increasing order.*

The theorem is true for a chain $\mu = [a_0, a_1]$ of length 1; for if a_0 is not an articulation point, a chain μ' exists going from a_0 to a_1 which does not make use of the edge $[a_0, a_1]$. Then by putting $\mu'' = [a_0, a_1]$, the conditions of the theorem are satisfied.

Let us assume the theorem to be true for elementary chains μ of length k, and deduce from this that it is also true for an elementary chain

$$\mu = [a_0, a_1, \ldots, a_k, a_{k+1}]$$

of length $k+1$.

Let us denote by μ' and μ'' the two disjoint chains joining a_0 and a_k which obey (1), (2) and (3); we now have to show that there exist two chains μ_0' and μ_0'' between a_0 and a_{k+1} with analogous properties.

There must be an elementary chain $\nu = \nu[a_0, a_{k+1}]$ joining a_0 to a_{k+1} which does not go through a_k (for otherwise, from Theorem 1, a_k is an articulation point). Let us denote by x the vertex of ν nearest to a_{k+1} which is also on $\mu[a_0,a_k]+\mu'[a_0,a_k]+\mu''[a_0,a_k]$. Several cases can now be distinguished.

First case: $x = a_0$. The required chains are:

$$\mu_0' = \mu$$
$$\mu_0'' = \nu$$

Second case: $x = a_{k+1}$. Since $x \notin \mu[a_0,a_k]$, we have either $x \in \mu'$, or $x \in \mu''$; if, for example, $x \in \mu'$, we take:

$$\mu_0' = \mu'[a_0,x]$$
$$\mu_0'' = \mu''[a_0,a_k]+[a_k,a_{k+1}]$$

Third case: $x \notin \mu$. In this case we either have $x \in \mu'$, or $x \in \mu''$; if, for example, $x \in \mu''$, we take:

$$\mu_0' = \mu'[a_0,a_k]+[a_k,a_{k+1}]$$
$$\mu_0'' = \mu''[a_0,x]+\nu[x,a_{k+1}]$$

Fourth case:

$x \in \mu$, $x \neq a_0$, $x \neq a_{k+1}$ Then we can write $x = a_q$, where $q < k$; let a_p be the vertex of μ, with $p \leqslant q$, as near as possible to a_q, which belongs also to $\mu'+\mu''$; if, for example, $a_p \in \mu''$ (cf. Fig. 20.1), we take:

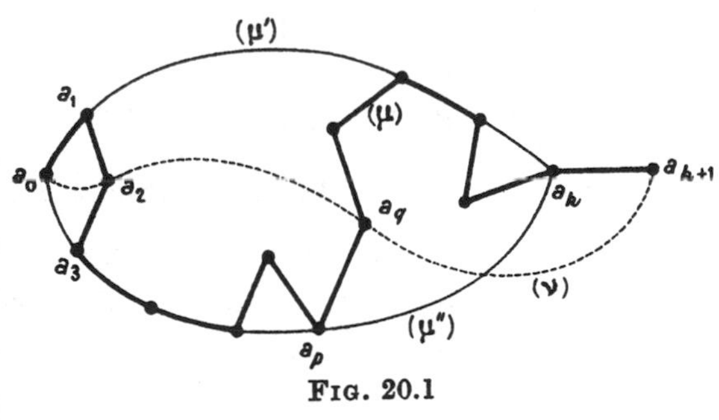

Fig. 20.1

$$\mu_0' = \mu'[a_0,a_k]+[a_k,a_{k+1}]$$
$$\mu_0'' = \mu''[a_0,a_p]+\mu[a_p,a_q]+\nu[a_q,a_{k+1}]$$

Since all possible situations are covered by these four cases, the theorem is proved.

LEMMA. *Given two arbitrary edges* **u** *and* **v** *of a connected graph without loops, we can construct an elementary chain which starts with* **u** *and finishes with* **v**.

If $\mathbf{u} = [a,x]$ and $\mathbf{v} = [b,y]$, then, since the graph is connected, the vertices a and b can be joined by an elementary chain $\mu = [a, a_1, a_2, \ldots, a_k = b]$.

If $x \notin \mu$, $y \notin \mu$, the required chain is:

$$\mu_0 = [x,a] + \mu + [b,y]$$

If $x \notin \mu$, $y \in \mu$, the required chain is:

$$\mu_0 = [x,a] + \mu[a,y] + [y,b]$$

If $x \in \mu$, $y \notin \mu$, the required chain is:

$$\mu_0 = [a,x] + \mu[x,b] + [b,y]$$

If $x \in \mu$, $y \in \mu$, the required chain is:

$$\mu_0 = [a,x] + \mu[x,y] + [y,b]$$

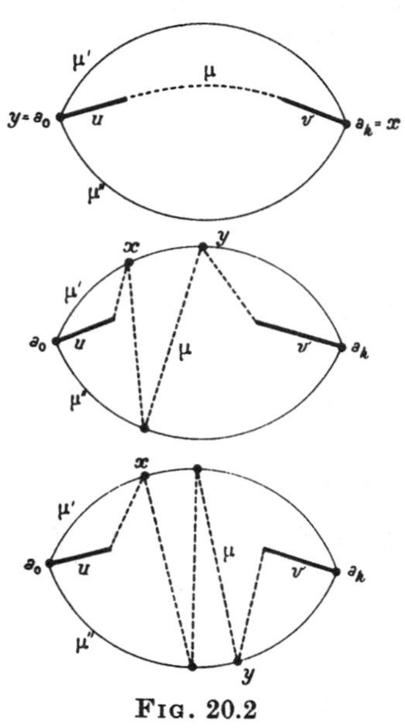

FIG. 20.2

THEOREM 4. *Given two arbitrary edges* **u** *and* **v** *of a biconnected graph without loops, an elementary cycle exists which passes through* **u** *and* **v**.

Let $\mu = [a_0, a_1, \ldots, a_k]$ be an elementary chain with $[a_0, a_1] = \mathbf{u}$, $[a_{k-1}, a_k] = \mathbf{v}$; from the preceding lemma, such a chain is known to exist. Now let us consider the two disjoint elementary chains μ' and μ'' which join a_0 and a_k and satisfy the conditions of MENGER's theorem. Let us denote by x the first vertex of μ (after a_0) which lies on $\mu' + \mu''$, and by y the last vertex of μ (before a_k) which lies on $\mu' + \mu''$.

First case: $x = a_k$, $y = a_0$. In this case (Fig. 20.2, top) the required cycle is:

$$\mu[a_0, a_k] + \mu'[a_k, a_0]$$

Second case: $x \neq a_k$, $y \neq a_0$, $x \in \mu'$, $y \in \mu'$. In this case (Fig. 20.2, centre), the required cycle is:

$$\mu[a_0, x] + \mu'[x, y] + \mu[y, a_k] + \mu''[a_k, a_0]$$

Third case: $x \neq a_k$, $y \neq a_0$, $x \in \mu''$, $y \in \mu''$. This case is analogous to the second case.

Fourth case: $x \neq a_k$, $y \neq a_0$, $x \in \mu'$, $y \in \mu''$. In this case (Fig. 20.2, bottom), the required cycle is:

$$\mu[a_0, x] + \mu'[x, a_k] + \mu[a_k, y] + \mu''[y, a_0]$$

Fifth case: $x \neq a_k$, $y \neq a_0$, $x \in \mu''$, $y \in \mu'$. This case is analogous to the fourth case.

Since all possible situations are covered by these five cases, the theorem is proved.

THEOREM 5. *For a graph G which contains neither loops nor isolated points, the following conditions are equivalent:*

(1) *given two arbitrary edges, an elementary cycle can always be found which contains these edges;*

(2) *given two arbitrary vertices, an elementary cycle can always be found which contains these vertices;*

(3) *the graph is biconnected.*

We shall show that (1) $\Rightarrow$ (2) $\Rightarrow$ (3) $\Rightarrow$ (1).

(1) $\Rightarrow$ (2), for if we take two arbitrary vertices x and y, they are the terminals of (at least) two edges (since G contains no isolated points), and an elementary cycle exists which contains these edges; therefore x and y belong to this elementary cycle.

(2) $\Rightarrow$ (3), for, from (2), the graph is connected; in addition, if an articulation point existed, two vertices x and y separated by this point of articulation could not be on the same elementary cycle, which contradicts (2).

(3) $\Rightarrow$ (1), from the preceding theorem.

Now let us consider any graph G whatever; if x is one of its articulation points, and C is one of the components isolated by deleting x, the subgraph of G determined by $C \cup \{x\}$ is defined to be a *piece* of G. A piece is attached to the rest of the graph by an articulation point. Any piece may contain other, smaller pieces; whenever a piece contains no other pieces, it is said to be a *minimal piece*.

These concepts allow us to establish a simple criterion for determining whether an elementary cycle exists which goes through any two arbitrarily given vertices; if, in fact, a and b are two distinct vertices on an elementary cycle, we can *strip* the graph by removing successively all pieces which do not contain both a and b, and we shall always be left with a subgraph which contains more than one edge. Conversely, if, after the graph has been stripped, the subgraph which remains contains more than one edge, it must be biconnected and will contain both a and b; therefore, from Theorem 5, there must be an elementary cycle which passes through a and b; to sum up, *the necessary and sufficient condition for an elementary cycle to exist which passes through two given vertices is that after stripping the graph, a subgraph is left which contains more than one edge.*

EXAMPLE. *The problem of 'Attila's huntsman'*. Let us put a white knight ('Attila's huntsman') and a black king on a chessboard as shown in Fig. 20.3 (on the left); the shaded squares will be called the 'burnt squares'. We wish to move the knight to the square occupied by the black king and then return the knight to his starting point, according to the standard rule for a knight's move; but he must never be placed on a burnt square, nor on a square which he has already visited (because everyone knows that Attila's huntsman destroys everything in his path).

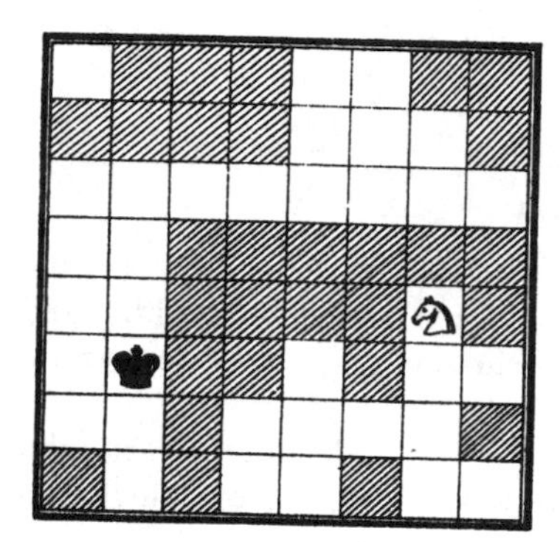

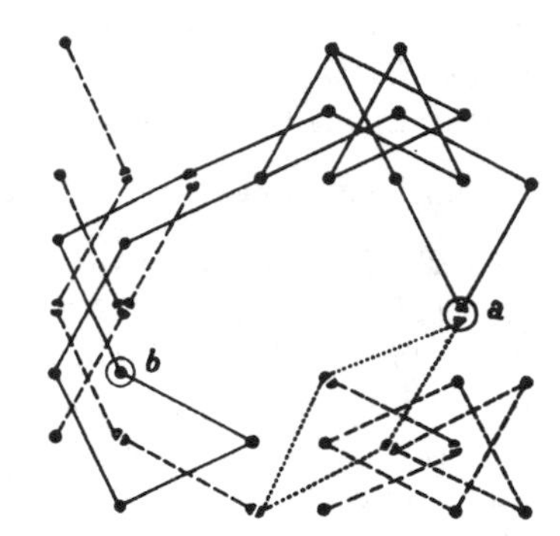

FIG. 20.3

This problem reduces to finding an elementary cycle passing through the two given vertices a and b of the graph shown on the right of Fig. 20.3; by stripping the graph, a number of pieces can be removed (shown in the figure with dotted lines) thus leaving a subgraph which contains more than one edge (shown in the figure by solid lines): therefore the problem is possible.

A connected graph G is said to be *bicoherent* if the deletion of any one edge does not disconnect it; it follows from MENGER's theorem that a biconnected graph is also bicoherent, but the converse is not necessarily true. We have:

THEOREM 6. *A necessary and sufficient condition for a connected graph to be bicoherent is that each articulation point should be joined to each adjacent piece by several edges.* Self-evident.

THEOREM 7 (H. E. ROBBINS). *If each edge of a graph G is given a unique orientation, the necessary and sufficient condition for the oriented graph to be strongly connected is that G be bicoherent.*

The condition is necessary, for if the oriented graph obtained from G is strongly connected, the deletion of any one edge cannot disconnect it.

The condition is sufficient: Let G be a bicoherent graph, x_0 be a vertex of G, $\mathscr{G}_0$ the family of subgraphs of G containing x_0 which, after suitable orientation of the edges, are strongly connected, and $G_0 = (X_0, U_0)$ a graph of $\mathscr{G}_0$ with a maximum number of vertices. We are going to show that if $X_0 \neq X$, we have a contradiction.

Consider a chain going from X_0 to $X - X_0$, and let $[x, y]$ be the first edge of this chain with $x \in X_0$ and $y \in X - X_0$; since the graph G is bicoherent, there exists a chain $\mu = [y, x_1, x_2, \ldots, x_p = x]$ going from y to x which does not use the edge $[x, y]$. Let x_k be the first vertex of μ belonging to X_0. Now let us form a subgraph G' by adding the vertices of μ to G_0; G' can be made strongly connected by giving the edges of G_0 suitable orientations, and by giving the new edges the orientations: $[x, y]$, $[y, x_1]$, $[x_1, x_2], \ldots$, $[x_{k-1}, x_k]$. Since G' has more vertices than G_0, we must have a contradiction.

These results have an obvious application in any attempt to schedule rush-hour traffic in a city, etc.

h-Connected Graphs

Let us consider a connected graph $G = (X, U)$ of order $n = |X|$; a non-empty set $A \subset X$, is defined to be an *articulation set* if the deletion of A leaves a subgraph which is not connected (or else consists of a single vertex); the *connection number* $\omega(G)$ of the graph G is the minimum number of elements contained in an articulation set.

A set V of edges is a *cut* if the deletion of these edges leaves a partial graph which is not connected; the *cohesion number* $\chi(G)$ of the graph G is the minimum number of elements in a cut.

Finally, if h is an integer $\geqslant 1$, we say that G is *h-connected* if $\omega(G) \geqslant h$, or that G is *h-coherent* if $\chi(G) \geqslant h$. If $h = 2$, we regain the notion of a biconnected graph (for $\omega(G) \geqslant 2$) or a bicoherent graph (for $\chi(G) \geqslant 2$). Our aim now is to generalize the principal results of the preceding paragraph.

EXAMPLE 1. *Arranging a tournament.* The concept of connectivity becomes important if a tournament in which n players are taking part is being organized, and the number of matches is to be limited to m. The tournament will be fair if a small clique of players cannot 'isolate' one or more of their opponents, that is to say, if the graph with n vertices and m edges which represents the competition has a sufficiently large connection number. With 5 players and 9 matches, for example, we can adopt the graph of Fig. 20.4, which is 3-connected; an articulation set of 3 players is indicated by the dotted circles. With 9 players and 20 matches, we can adopt the graph of Fig. 20.5, which is 4-connected, etc....

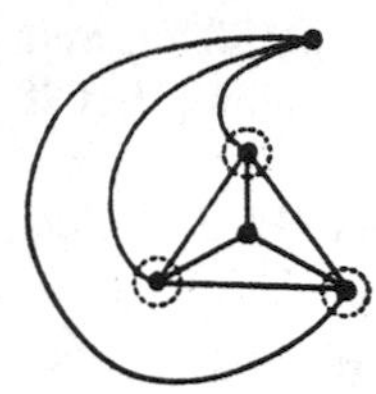

Fig. 20.4

EXAMPLE 2. *Bombardment of lines of communication.* It is generally considered that the economic and military power of an enemy country is effectively paralysed if, by bombarding their lines of communication, we can completely isolate one or more regions. The problem is: how many lines of communication (at the minimum) must be destroyed in order to attain this goal? Let us construct a graph (X, U) in which the vertices represent sectors which are well supplied with internal lines of communication, and the edges stand for vulnerable lines of communication connecting two distinct sectors (for example: bridges, tunnels, railways, etc.). Then the problem becomes one of finding the minimum number of edges which must be deleted in order to disconnect the graph.

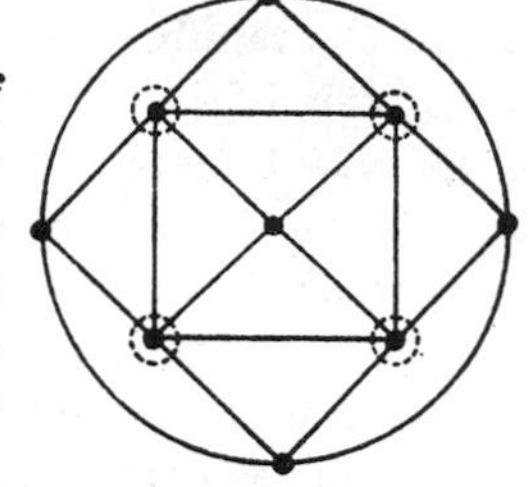

Fig. 20.5

This problem may be examined by means of the theory of connectivity if we construct the auxiliary graph (U, Δ) which has as its vertices the edges of U, and in which $u \in \Delta v$, if u and v are adjacent edges. If, for example, we

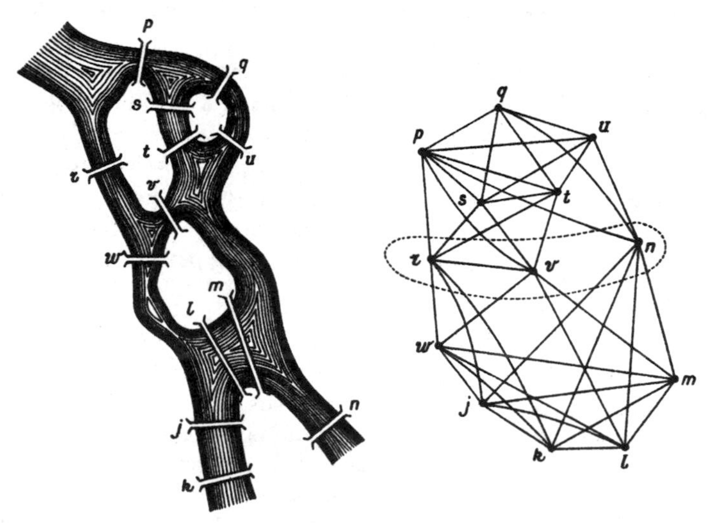

Fig. 20.6

propose to bombard the bridges shown in the map on the left of Fig. 20.6, we get the graph shown on the right, which is 3-connected; the destruction of the three bridges corresponding to the articulation set $\{r, v, n\}$ will be sufficient to enable us to achieve our goal.

The principal properties of h-coherent or h-connected graphs were formerly very difficult to establish, but they can now be derived with ease by making use of the theory of transport networks (Chapter 8).

THEOREM 8. *If a graph G is h-coherent, two arbitrary distinct vertices x and y can be joined by h simple chains with no edges in common.*

Let $q(x,y)$ be the minimum number of edges which must be deleted in order to disconnect the vertices x and y; we have

$$\min_{\substack{x,y \\ x \neq y}} q(x,y) = \chi(G) \geqslant h$$

More precisely, we are going to show that *there are in G at most $q(x,y)$ chains with no edges in common which join x to y.*

Let us construct a transport network $\bar{G}$ by taking x as source, y as sink, and representing each edge by two arcs of opposite orientations, each with a capacity of 1. From the theorem of FORD and FULKERSON (Chapter 8), the maximum value of the flow is $q(x,y)$; therefore $\bar{G}$ must contain $q(x,y)$ paths with no arcs in common going from x to y. These different paths are not necessarily chains with no edges in common, for the same edge may have been traversed in the two opposing directions.

Suppose, for example, we have two distinct paths $\mu = [x,\ldots,a,b,\ldots,y]$ and $\nu = [x,\ldots,b,a,\ldots,y]$, both of which use a certain edge $[a,b]$; by replacing these by $\mu[x,a]+\nu[a,y]$ and $\nu[x,b]+\mu[b,y]$ respectively, and doing the same for every edge which has been traversed in both directions, we shall at length be able to define $q(x,y)$ chains with no edges in common.

Q.E.D.

THEOREM 9. *If a graph G is h-connected, two arbitrary distinct vertices x and y can be joined by h elementary chains which have no vertices in common.*

Let $p(x,y)$ be the minimum number of vertices which must be eliminated in order to disconnect the vertices x and y; we have

$$\min_{\substack{x,y \\ x \neq y}} p(x,y) = \omega(G) \geqslant h$$

We shall again prove a result more precise than that stated in the theorem. We shall show that *in G there are at most $p(x,y)$ chains with no vertices in common joining x to y.*

Let us construct a transport network $\bar{G}$ by taking x as source, y as sink; we replace each vertex a by two distinct vertices a' and a'', and we join these vertices by an arc $[a', a'']$ of capacity 1; for any vertex z adjacent to the vertex a, we draw arcs $[z'', a']$ and $[a'', z']$ of infinite capacity. The maximum number of chains in G joining x to y which have no vertices in common is the maximum value of the flow in the transport network $\bar{G}$, and from the theorem of FORD and FULKERSON, this is equal to the capacity of a minimum cut, that is to say, to $p(x, y)$.

Q.E.D.

COROLLARY. *If a graph is h-connected, and if $k < h$, the graph obtained by deleting k edges is $(h-k)$-connected.*

In the new graph, two distinct arbitrary vertices can be joined by at least $(h-k)$ elementary chains with no vertices in common; therefore the graph is $(h-k)$-connected.

It will be observed that Theorem 9 is a direct generalization of MENGER's theorem (if an appropriate generalization is made of the conditions given in the statement of the theorem on p. 198).

21. Planar Graphs

General Properties

A graph (or an s-graph) G is said to be *planar* if it can be represented on a plane in such a fashion that the vertices are all distinct points, the edges are simple curves, and no two edges meet one another except at their terminals. A diagram of G on a plane which conforms with these conditions is called a *planar topological graph,* and will also be denoted by G; two planar topological graphs will not be regarded as distinct if they can be made to coincide with one another by elastic deformation of the plane.

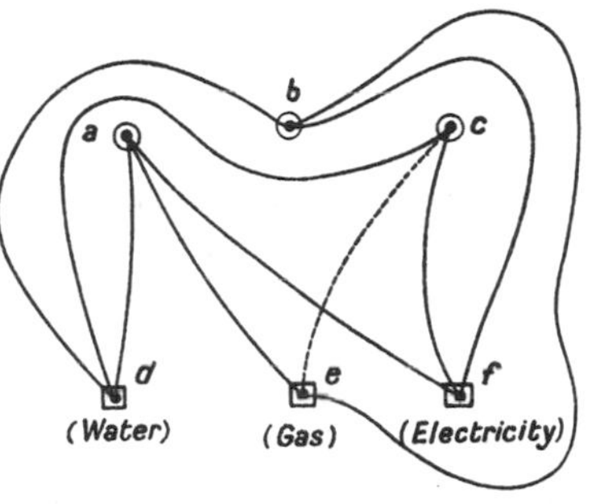

FIG. 21.1

EXAMPLE. *The problem of the three houses and the three public utilities.* We have three houses a, b and c which we wish to connect to the waterworks d, the gasworks e and the electricity supply substation f. Can we place (on a plane) the three houses, the three supply stations, and the pipelines connecting each house to each supply station, in such a way that no two pipelines cross one another except at their initial or terminal points? Experience shows that we can always place eight of the pipelines, but that the ninth always cuts one of the first eight. An explanation of this phenomenon will be given later.

Let G be a planar topological graph; a *face* of G is, by definition, an area of the plane bounded by edges of the graph, and which contains neither edges nor vertices in its interior; we shall denote faces by the letters r, s, t, ..., and the set of faces by R. The *frontier* of a face is the set of all arcs which 'touch' it; the *contour* of a finite face is the cycle formed by the edges which 'surround' it. The edges of the contour belong to the frontier, but the converse is not true (ex: a pendant edge, an isthmus, the boundary edges of an inner island, etc.). Two faces r and s are said to be *adjacent* if their frontier have at least one edge in common; two faces which meet only at a vertex are not adjacent.

EXAMPLE. In geography, a map is a planar topological graph (provided that it does not contain any islands); a special feature of this graph is that every vertex has a degree of 3 or more. A face may be adjacent to another face along more than one boundary; it should be noted that in Fig. 21.2, g is not adjacent to d, although these faces have a common vertex. In addition, the graph of Fig. 21.2 is an s-graph. Finally, it will be observed that every planar graph has exactly one unbounded face, which is the ***infinite face*** (this is face e in Fig. 21.2); the remaining faces c, d, f, g, h, i and j are ***finite faces***.

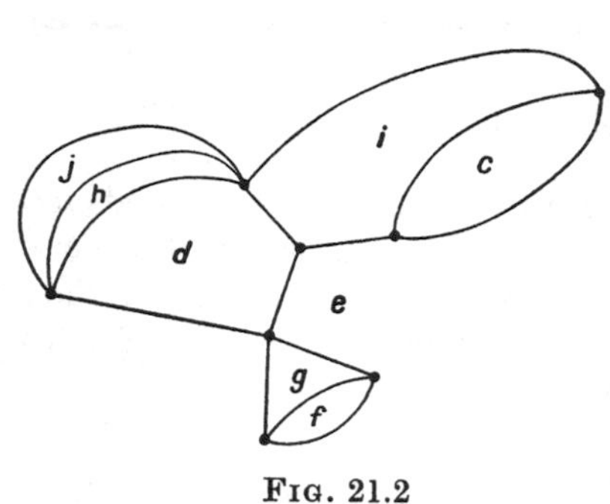

FIG. 21.2

THEOREM 1. *The contours of the different finite faces of a planar topological graph G make up a fundamental basis of independent cycles.*

The theorem is true if G contains only two finite faces; let us assume it to be true for every graph with $f-1$ finite faces, and show that it follows from this assumption that it must be true for a planar topological graph G with f finite faces.

If we delete an edge u of G, separating two different faces, we get a graph G' which has $f-1$ finite faces, and by hypothesis the contours of these faces make up a fundamental basis of independent cycles. If the edge u is now replaced, a new finite face is created, the contour of which is a cycle which must be independent of the preceding cycles (since it contains an edge which is not an element of any of the other cycles). Since by addition of one edge, the cyclomatic number can only be increased by one at most, the finite faces of G must determine a fundamental basis of independent cycles.

COROLLARY 1. *If a connected planar topological graph has n vertices, m edges and f faces, then*

$$n-m+f = 2 \qquad (\textit{Euler's formula})$$

Since the number of finite faces is equal to the cyclomatic number ν, we have:

$$f = \nu+1 = (m-n+1)+1 = m-n+2$$

which gives the required result.

COROLLARY 2. *In every planar graph (which is not an s-graph), there is a vertex x whose degree* $d(x) \leqslant 5$.

In the corresponding planar topological graph, every face is bounded by at least three distinct edges; if we form the simple graph of incident faces and edges†, the number of arcs $\leqslant 2m$ on the one hand, and $\geqslant 3f$ on the other: therefore we have $f \leqslant 2m/3$. If every vertex is the terminal of at least six edges, then in the same way we can deduce $n \leqslant 2m/6$, and therefore, from Euler's formula, we can write:

$$2 = n-m+f \leqslant \frac{m}{3}-m+\frac{2m}{3} = 0$$

which is absurd.

COROLLARY 3. *A map contains at least one face the contour of which consists of at most five edges.*

Every vertex of the map is the terminal of at least three edges; if we form the simple graph of incident vertices and edges, the number of arcs is, on the one hand, $\leqslant 2m$, and $\geqslant 3n$ on the other. Therefore $n \leqslant 2m/3$. If we suppose that the contour of every face contains at least six edges, and if we form the simple graph of incident faces and edges, the number of arcs $\leqslant 2m$ on the one hand, and $\geqslant 6f$ on the other. This gives

$$2 = n-m+f \leqslant \frac{2m}{3}-m+\frac{m}{3} = 0$$

which is a contradiction.

EULER's formula which was proved in Corollary 1 can be usefully employed on many occasions.

EXAMPLE 1 (EULER). Let us consider a convex polyhedron in 3-dimensional space with n vertices, m edges and f faces. Clearly, we can represent it on the surface of a sphere in such a way that no two edges cut one another except at their terminals; next, by making a stereographic projection of the surface from a centre in the middle of one of the faces, we can represent the polyhedron on a plane. The graph so obtained is planar, and thus we arrive at a fundamental relation concerning convex polyhedrons:

$$n-m+f = 2$$

EXAMPLE 2. We are now going to show, by means of EULER's formula, that the graph of the three houses and the three supply stations (Fig. 21.1) cannot be planar. If we assume it to be planar, we would have:

$$f = 2-n+m = 2-6+9 = 5$$

† That is to say the graph formed by a set X of points representing the faces, by a set Y of points representing the edges, and by arcs which go from $x \in X$ to $y \in Y$ whenever the face x is incident to the edge y.

The contour of each face must contain at least four edges (for if a face s was bordered by only three edges, it would also be bordered by three vertices, of which two must belong to the same category: houses or supply stations; however two vertices of the same category cannot be adjacent). If we form the simple incidence graph of faces and edges, then on the one hand, the number of arcs $\leqslant 2m$, and on the other $\geqslant 4f$, which gives

$$18 = 2m \geqslant 4f = 20$$

which is absurd.

EXAMPLE 3. We shall now use a similar method to show that the complete graph with five vertices is not planar. If we assume that this graph is planar, we would have:

$$f = 2-n+m = 2-5+10 = 7$$

The contour of each face contains at least three edges. If we form the simple incidence graph of faces and edges, the number of arcs $\leqslant 2m$ on the one hand, $\geqslant 3f$ on the other, and therefore

$$20 = 2m \geqslant 3f = 21$$

which is absurd.

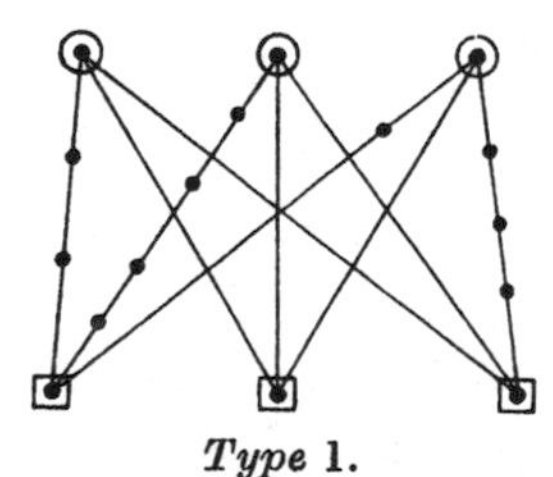

Type 1.

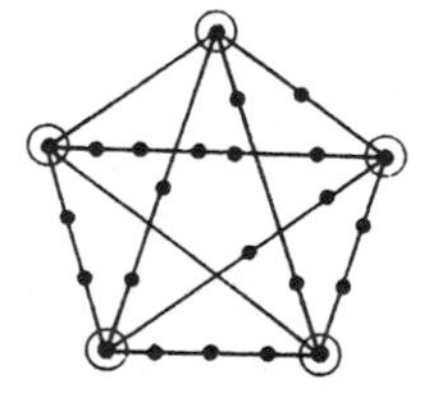

Type 2.

FIG. 21.3

The graph of the houses and the supply stations, and the complete graph of five vertices allow us to define a whole family of non-planar graphs: as can be seen in Fig. 21.3 we simply place as many vertices as we wish on each edge in order to define other non-planar graphs of type 1 or of type 2. This remark has a converse, which can be stated in the form of a difficult theorem due to KURATOWSKI, and which we now intend to prove. To begin with, we shall establish some conventions regarding terminology: if μ is an elementary cycle (to which we assign an arbitrary sense), and if $a, b \in \mu$, we shall denote by $\mu[a,b]$ the sequence of vertices of μ which are met in going from a to b in the positive sense (including a and b), while $\mu\,]a,b[$ denotes the same sequence but with a and b excluded. In addition, if G is a graph with an articulation set A, we call a connected component C of the subgraph determined by $X-A$, augmented by the edges incident to C and their terminals, a *piece* of the graph G (relative to A).

KURATOWSKI'S THEOREM. *The necessary and sufficient condition for a graph G to be planar is that it should possess no partial subgraphs of either type* 1 *or type* 2.

We have seen that a graph which possesses a partial subgraph of either type 1 or type 2 is not planar; conversely, let us now show that *a graph G which possesses no subgraphs of either type* 1 *or type* 2 *is planar.*

The theorem is true if G has one, two or three edges; therefore let us use the method of induction: let us assume the theorem to be true for every graph having fewer than m edges, and show that from this assumption we can prove the theorem to be true for every graph G with m edges.

Therefore let G be a graph with m edges which possesses no partial subgraphs of either type 1 or type 2, and let G be non-planar. This graph is connected, because, if it were not, it would be planar (since its components would be planar). Likewise, it must be biconnected; since otherwise, as the pieces of G relative to an articulation point a are planar, they could each be given a planar representation in such a way that the point a was on their infinite faces (using a stereographic projection if necessary), and G would be planar.

1. *We shall show that if we remove an edge $[a,b]$ from G, then an elementary cycle μ exists which passes through a and b.* If, in fact, this were not so, the graph G' which is obtained from G by deleting the edge $[a,b]$ would be articulated (from MENGER's theorem, Chapter 20); in addition, relative to some point of articulation c, the vertices a and b would be on two distinct pieces C_a and C_b.

The graph C'_a which is obtained from C_a by adding the edge $[a,c]$ is planar, because it contains no graphs of type 1 or type 2 (since it can be derived from G by appropriate shrinkages). We can therefore, by making a stereographic projection, represent C'_a on a plane in such a way that the edge $[a,c]$ lies on the contour of the infinite face. In the same way, the planar graph C'_b which is obtained by adding the edge $[c,b]$ to C_b, can also be given a planar representation in which the edge $[c,b]$ lies on the contour of the infinite face. By joining a and b, we get a planar topological graph containing G, which is a contradiction.

2. Consider now the planar topological graph G' obtained by deleting the edge $[a,b]$. In this graph we find the elementary cycle μ which passes through a and b, and which includes the greatest possible number of faces in its interior. Let us give μ an arbitrary orientation. μ divides the plane into two regions, and the pieces of G' with vertices in the interior (exterior) will

be called the *internal pieces* (*external pieces*). *An external piece cannot contain more than one vertex of* $\mu[a, b]$ (for otherwise we would be able to construct a new cycle μ which included a larger number of faces in its interior); since the same is true for $\mu[b, a]$, *an external piece can meet* μ *at only one or two points*. Further, there is at least one external piece and one internal piece which meet both $\mu\,]a, b[$ and $\mu\,]b, a[$ (since the edge $[a, b]$ cannot be represented on the plane).

3. *Now we shall show that an internal piece C and an external piece D exist which meet both* $\mu\,]a, b[$ *and* $\mu\,]b, a[$, *and are such that the points of contact c and d of D with* μ, *and the points of contact e and f of C with* μ, *are on* μ *in alternating order: c, e, d, f.*

Let us assume that this is not the case; let C_1 be an internal piece which meets $\mu\,]a, b[$ and $\mu\,]b, a[$ and let e_1 and f_1 be two consecutive points of

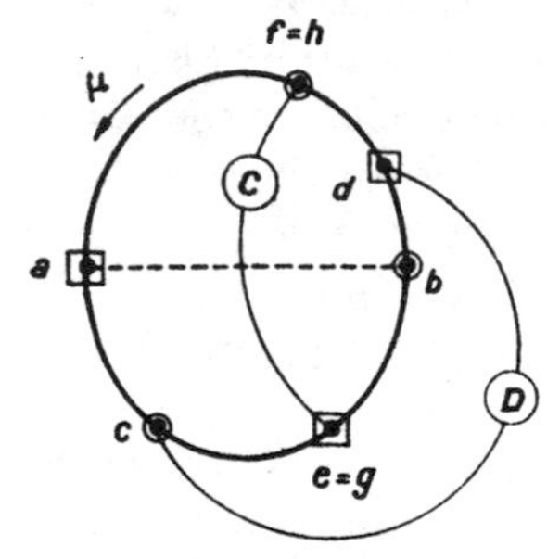

FIG. 21.4

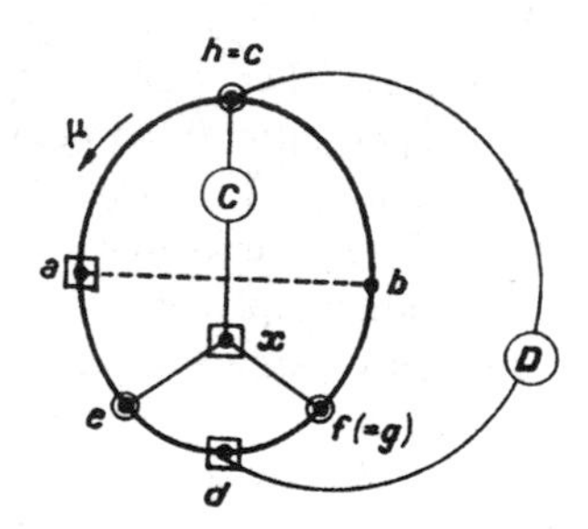

FIG. 21.5

contact on μ. We can draw a continuous line ν joining f_1 and e_1 which lies outside μ, and which does not meet any existing edges (since, by hypothesis, there are no external pieces which meet $\mu\,]e_1, f_1[$ and $\mu\,]f_1, e_1[$).

Every internal piece which only meets $\mu[f_1, e_1]$ can be transferred to the exterior, in the face which is limited by ν and μ; in particular, this may be done for the piece C_1. There must remain at least one internal piece C_2 which cannot be so transferred (for otherwise it would be possible to make a planar connection between a and b), and this piece has two consecutive points of contact e_2 and f_2 on $\mu[f_1, e_1]$, at least one of which is on $\mu\,]f_1, e_1[$.

The same transfer procedure can now be continued with C_2 replacing C_1, etc....; but, since this procedure does not terminate, and since G is finite, we must have a contradiction.

In the next steps, we shall denote by e, f, g, h the points of contact of C and of μ which satisfy:

$$e \in \mu\,]c, d[, \qquad f \in \mu\,]d, c[, \qquad g \in \mu\,]a, b[, \qquad h \in \mu\,]b, a[$$

Clearly $e \neq f$, and $g \neq h$, but we could have $e = g$, $e = h$, etc....

4. If one of the vertices e, f is on $\mu\,]a,b[$, and the other is on $\mu\,]b,a[$, we shall write, for example, $e = g$, $f = h$; we can see immediately (Fig. 21.4) that G contains a graph of type 1, which is contrary to our hypotheses.

5. If both the vertices e and f are on $\mu\,]a,b[$, we may assume $h = c$ (because if $h \neq c$, $h \in \mu\,]b,a[$, we regain one of the figures which was eliminated at step 4). Then we get Fig. 21.5, which also contains a graph of type 1, and hence we have again reached a contradiction. For the same reason, we can eliminate the case in which e and f are both on $\mu\,]b,a[$.

6. If $e = a$, $f \neq b$ (say, for example, $f \in \mu\,]a,b[$) we get Fig. 21.6 which contains a graph of type 1: which leads to a contradiction.

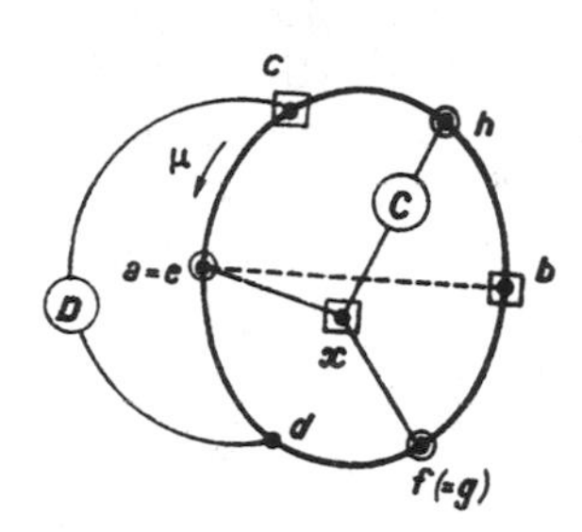

Fig. 21.6

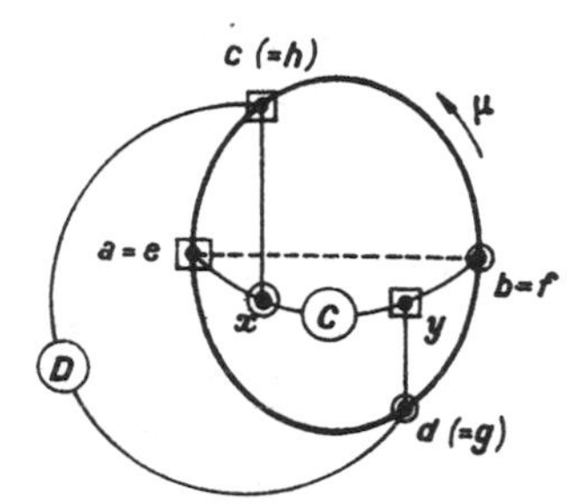

Fig. 21.7

7. If $e = a$, $f = b$ we shall assume that $g = d$, and $h = c$ (for otherwise we are led to a figure which has been eliminated in (4) and in (6)). There are two cases to be considered: if the chains of C joining cd and ef have more than one vertex in common, we get the graph shown in Fig. 21.7 which contains a graph of type 1, and hence leads to a contradiction. If the chains of C joining cd and ef have only one vertex in common, we get the graph of Fig. 21.8, which contains a graph of type 2, and hence also leads to a contradiction.

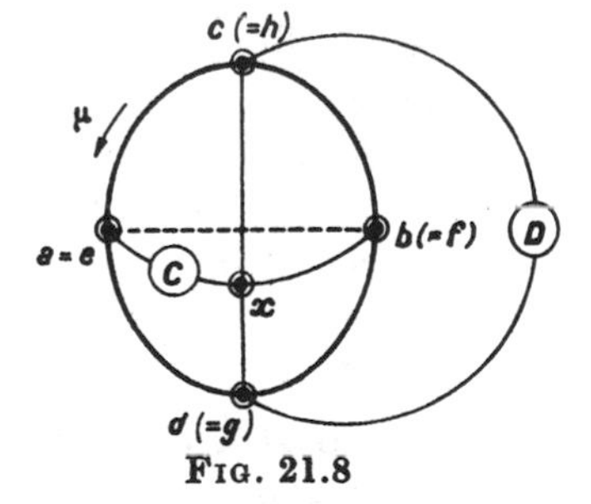

Fig. 21.8

All positions for e and f have now been considered, and therefore the graph G cannot exist in the form in which it was defined above. This proves the theorem.

Theorem 3. *A planar graph is 5-chromatic.*

The theorem is true for planar graphs with 1, 2, 3, 4 or 5 vertices; let us assume that it is true for planar graphs with $n-1$ vertices, and prove from this that it must also be true for any arbitrarily chosen planar graph G with n vertices.

From Corollary 2, G contains at least one vertex x which is adjacent to 0, 1, 2, 3, 4 or 5 vertices. Since the subgraph obtained by deleting x is 5-chromatic, let us colour its vertices with five colours: α, β, γ, δ, ϵ. Then we can easily colour x, except in the case where x is adjacent to five distinct vertices, each coloured with a different colour.

In this case, let a, b, c, d, e be the five vertices adjacent to x (taken in, say, anticlockwise order about x), and let their colours be α, β, γ, δ and ϵ, respectively. Let us denote by $G_{\alpha,\beta}$ the subgraph determined by the vertices coloured α or β, etc....

1. If a and c are not connected in the graph $G_{\alpha,\gamma}$, let us consider the component containing a, and interchange the colours of the vertices in it (that is, those which were coloured α are now coloured γ, and vice versa). The new colouring is compatible with the graph G, and since the vertex x is now surrounded by the colours γ, β, γ, δ and ϵ, it may itself be coloured α.

2. If a and c are connected in the graph $G_{\alpha,\gamma}$, the vertices b and d cannot be connected in the graph $G_{\beta,\delta}$. By permuting the colours β and δ in the component of $G_{\beta,\delta}$ which contains the vertex b, we shall be able to colour x with the colour β.

It seems that we should be able to improve this result: the following theorem which has never yet been proved is known as the *four-colour problem: every planar graph is 4-chromatic.*

By employing the notion of *duality*, we may restate the theorem in another form. Let us consider a connected planar topological graph G, with no isolated vertices; we now construct a planar topological graph G^* to correspond to G in the following manner:

We place a vertex x^* of G^* inside every face s of G; for every edge u of G, we construct a corresponding edge u^* of G^* which connects the vertices x^* and y^* corresponding to the faces s and t lying on either side of the edge u. The topological graph G^* thus defined is planar (cf. Fig. 21.9), connected and possesses no isolated vertices: it is called the *dual graph* of G. It will be noted that:

1. A loop of G corresponds to a pendant edge of G^*, and vice versa.
2. If G possesses an articulation point, then so also does G^*†.
3. If G has any antinodes, G^* is an s-graph, and vice versa.

EXAMPLE. We shall consider a (non-oriented) planar transport network G, in which each edge u has an associated capacity $c(u) \geqslant 0$, and the source

† The links between the connectivity of G and that of G^* have been studied by WHITNEY [9].

x_0 and sink z are placed on the same horizontal line. We form the dual graph G^*, in which we place two distinct vertices x_0^* and z^* to correspond to the infinite face of G: x_0^* corresponds to the lower half-plane and z^* to the

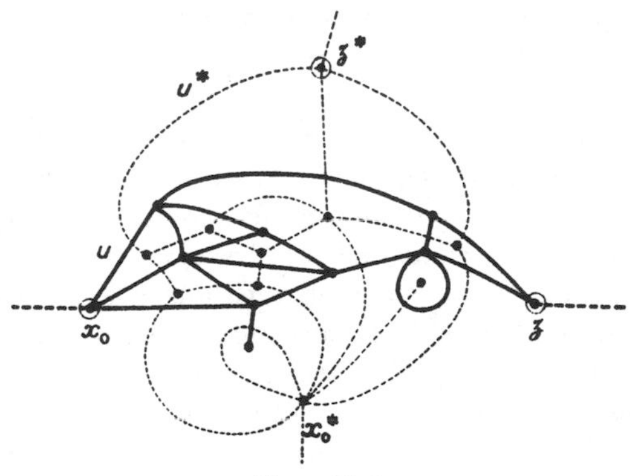

Fig. 21.9

upper half-plane. The edge u^* which corresponds to the edge u, is given a length $l(u^*) = c(u)$. Now let us consider the following problems:

Problem 1. *To find a cut V of minimum capacity in G; that is to say, to minimize*

$$c(V) = \sum_{v \in V} c(v)$$

Problem 2. *To find the shortest path V^* in G^* going from x_0^* to z^*; that is to say, to minimize*

$$l(V^*) = \sum_{v^* \in V^*} l(v^*)$$

Problem 1 is equivalent to the maximum flow problem, and Problem 2 to the problem of the shortest path. It is clear that, for a planar graph, these are simply different aspects of the same problem.

If we state the four-colour problem in its dual form, it becomes: *in a planar graph G, it is possible to colour the faces with four colours in such a way that no two adjacent faces are of the same colour.*

If we wish to colour the graph G, we can limit ourselves to the case of a connected graph (since each component may be coloured separately), and even to a biconnected graph. We may assume that there are no vertices of degree 0, 1 or 2. We can also limit ourselves to homogeneous graphs of degree 3, since every vertex x with a degree greater than 3 can be covered during the colouring process by a new face which is small in area, and which has an edge in common with every face which touches x. Thus the graph G in finally reduced to a homogeneous graph of degree 3, with no isthmuses (because it is biconnected), and with a chromatic index of 4 or less (from Theorem 5, Chapter 19).

THEOREM 4. *Let G be a planar, homogeneous graph of degree 3 with no isthmuses; the necessary and sufficient condition for it to be possible to colour its faces with four colours in such a way that no two adjacent faces are of the same colour is that the chromatic index of G be equal to 3.*

1. If the faces of G can be coloured with four colours α, β, γ and δ, we shall give the symbol 1 to edges separating a face α and a face β, or a face γ and a face δ; we give the symbol 2 to edges separating a face α and a face γ, or a face β and a face δ; and we give the symbol 3 to edges separating a face α and a face δ, or a face β and a face γ. Two adjacent edges cannot have the same symbol (because if this were so, we would have two adjacent faces of the same colour); therefore the graph G has chromatic index 3.

2. Now let us assume that G has a chromatic index of 3, and that we can therefore divide the edges of G into three types: 1, 2 or 3. The edges of types 1 and 2 form a homogeneous graph of degree 2, and therefore we need only two colours, p and q, to colour the regions of this graph; similarly, the edges of types 1 and 3 form a graph whose regions may be coloured with two colours, r and s. Now let us superpose these colours on our graph G: each face will receive one of the four combinations pr, ps, qr or qs. Two faces which are separated by an edge of type 1 will certainly have received different combinations, and the same is true for faces separated by edges of type 2 or type 3; thus we have achieved the required colouring of the graph.

HEAWOOD'S THEOREM. *For a planar, homogeneous graph of degree 3 with no isthmuses, the following two conditions are equivalent:*

(1) *the chromatic index of G is 3;*

(2) *each vertex x may be assigned a coefficient $p(x)$ which equals $+1$ or -1, in such a way that the contour of each face μ satisfies*

$$\sum_{x \in \mu} p(x) \equiv 0 \pmod{3}$$

(1) $\Rightarrow$ (2): If the chromatic index of G is 3, we can divide the edges into three types: 1, 2 and 3. Let us put $p(x) = +1$ if the three edges incident to the vertex x are in the order 1, 2, 3 (considering an anticlockwise movement), and $p(x) = -1$ in the opposite case. Now let us follow the contour of a face

μ in the anticlockwise direction, starting from any arbitrary edge: each time that we encounter a coefficient of $+1$, the index of the edge being followed undergoes a negative cyclic permutation ($3 \to 2$, $2 \to 1$, $1 \to 3$); and each time that we encounter a coefficient of -1, the permutation is positive ($1 \to 2$, $2 \to 3$, $3 \to 1$). Therefore, by the time we have returned to the initial edge, the algebraic sum of the coefficients of the vertices of the contour must be a multiple of 3.

(2) $\Rightarrow$ (1): Given coefficients $p(x)$ which satisfy (2), we shall show first of all that the algebraic sum of the $p(x)$ along any arbitrary cycle ν is also a multiple of 3; clearly, we can limit ourselves to the case in which ν is an elementary cycle, and contains, say, the faces $\mu_1, \mu_2, \ldots$. We have

$$\sum_{x \in \nu} p(x) = \sum_{x \in \mu_1} p(x) + \sum_{x \in \mu_2} p(x) + \ldots - 3 \sum_{x \text{ is inside } \nu} p(x)$$

Since the sum of the terms on the right-hand side is a multiple of 3, so also must be the algebraic sum of the $p(x)$ along any arbitrary cycle.

Starting from some arbitrary edge to which we assign the index 1, we shall now give each edge an index of either 1 or 2 or 3; this is done step by step with the help of the coefficients $p(x)$ in such a way that the three edges incident to a vertex x are in anticlockwise order 1, 2, 3 if $p(x) = +1$, and in the reverse order if $p(x) = -1$. It is impossible for a contradiction to arise, because if an edge could be given two different indices, the algebraic sum of the $p(x)$ along a cycle ν would not be a multiple of 3. The chromatic index of G must therefore be 3.

COROLLARY 1. *If the number of edges bordering each face is a multiple of* 3, *four colours are sufficient to colour the faces of G.*

If we take $p(x) = +1$ for every vertex x, condition (2) of Heawood's theorem is satisfied.

COROLLARY 2. *If the number of edges bordering each face is a multiple of* 2, *four colours are sufficient to colour the faces of G.*

Let ν be an elementary cycle which contains the faces μ_1, μ_2, etc....; its length $l(\nu)$ satisfies:

$$l(\nu) = l(\mu_1) + l(\mu_2) + \ldots - 2 \text{ (the number of edges interior to } \nu)$$

It follows from this formula that $l(\nu)$ is even, and this is true for any arbitrary cycle (since it is the sum of elementary cycles). Therefore, from KÖNIG's theorem (p. 32), each vertex x may be assigned a coefficient $p(x) = +1$ or

-1 in such a way that no two adjacent vertices have the same coefficient. With this definition of the $p(x)$, it is clear that condition (2) of Heawood's theorem will be satisfied.

Generalization

The preceding concepts may be rederived, and given a more general form, if use is made of more advanced topological theory which would go beyond the scope of this work†. We shall content ourselves simply with some indications of what can be done.

Every graph G may be represented in three-dimensional space in such a way that no two edges cut one another; such a representation is called a *topological graph*, and we shall denote it also by G. If S denotes a surface of the space $\boldsymbol{R}^3$, and if there exists a single-valued, doubly-continuous function σ which maps G on to S, then it is known that no two edges of σG will cut one another; σG is, by definition, an *S-topological graph*. If S is a plane, we regain the concept of a *planar topological graph*.

We shall now limit ourselves to considering closed, finite surfaces which are orientable‡ and which can easily be visualized; the *genus* g of such a surface S is defined to be the maximum number of closed curves which do not intersect one another, which may be traced on the surface without disconnecting it. Thus, a sphere is of genus 0; a torus (a motor-car tyre) is of genus 1, etc....

It follows from certain general theorems (cf. [1]) that every finite graph can be represented on a surface S of sufficiently large genus; further, given an S-topological graph G, we can construct an S-topological graph G^* in exactly the same way as we construct the *dual* of a planar graph. If $\gamma(G)$ denotes the chromatic number of the graph G, then we shall write:

$$\gamma(S) = \max\{\gamma(G) | G \text{ is } S\text{-topological}\}$$

$$\gamma'(S) = \max\{n | \text{the complete graph of order } n \text{ is } S\text{-topological}\}$$

Clearly,
$$\gamma'(S) \leqslant \gamma(S)$$

For a surface S of genus g, we shall write:

$$n_g = \text{the integral part of } [\tfrac{7}{2} + \tfrac{1}{2}\sqrt{(48g+1)}]$$

† Cf. for example S. Lefschetz, *Introduction to Topology*, Princeton University Press, Princeton, 1949.

‡ It will be recalled that a topological space S is a *surface* if every point $x \in S$ possesses a neighbourhood which is homeomorphic to the unit circle; if the orientations of the different neighbourhoods are compatible one with another, the surface is said to be *orientable*. Finally, if S is a compact space, the surface is said to be *closed*; thus a sphere is a closed surface, but a plane is not. Certain results, analogous to those given here, have been established for non-orientable surfaces (cf. [6]).

Then we have:

$$g = 0 \quad \text{(sphere)} \quad \rightarrow n_g = 4$$
$$g = 1 \quad \text{(torus)} \quad \rightarrow n_g = 7$$
$$g = 2 \quad \text{(a pretzel with two holes)} \rightarrow n_g = 8$$
$$g = 3 \quad \rightarrow n_g = 9, \text{etc.} \ldots$$

HEAWOOD has shown† that *for a surface S of genus $g > 0$* (i.e. other than the sphere),

$$\gamma(S) \leqslant n_g$$

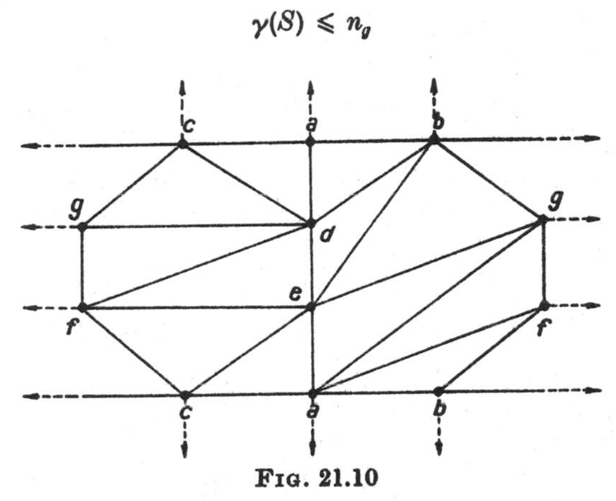

FIG. 21.10

Consider now the number $\gamma'(S)$; RINGEL has shown‡ for a surface S of genus $g > 0$,

$$n_g - 2 \leqslant \gamma'(S) \leqslant n_g$$

HEAWOOD himself thought that it was always true that $\gamma'(S) = n_g$; this conjecture has recently been proved by HEFFTER for $g = 1, 2, \ldots, 7$, and by RINGEL for $g = 8, 9$, as well as in a large number of special cases. Therefore for the torus T where $g = 1$, we have $\gamma'(T) = 7$, and a complete graph with seven vertices can be drawn on a torus as indicated in Fig. 21.10; on the other hand, we have $\gamma(T) \leqslant 7$ (from HEAWOOD's formula) and $\gamma(T) \geqslant \gamma'(T) = 7$, which gives finally $\gamma(T) = 7$.

† For a proof, see for example: D. HILBERT and S. COHN VOSSEN, *Geometry and the Imagination* (trans.), p. 337 (New York, 1952).

‡ Cf. G. RINGEL, 'Farbensatz für orientare Flächen vom Geschlechte $p > 0$', *J. reine angew. Math.*, 193, 1954, p. 11.

Appendix I

Note on the General Theory of Games

In Chapter 6 we gave the general definition of a game *with perfect information*; if such restrictions are not imposed on the information players may have about their positions, the definition of a game becomes more complicated, but the graph defined by the rules of the game can still be studied according to the same general principles.

For two players, called for simplicity (A) and (B), a *game* is defined by:

(1) An abstract set X, the elements of which are the *positions* of the game;

(2) A function Γ mapping X into itself, called the *rules* of the game;

(3) A partition (X_0, X_A, X_B) of X; we assume that

$$X_0 = \{x \mid \Gamma x = \emptyset\}$$

In a position of X_A we say that it is player (A)'s *move* (or that A has the *right to play*), and in a position of X_B we shall say that it is player (B)'s *move*;

(4) A partition $\quad \mathscr{U} = (U_1, U_2, \ldots)$ of X_A

and a partition $\quad \mathscr{V} = (V_1, V_2, \ldots)$ of X_B

We say that $\mathscr{U}$ is the *scheme of information* of player (A) and that $U_1 \in \mathscr{U}$ is an *information set* of player (A). For every position x of the same information set, we shall assume that Γx has the same cardinality.

(5) For any information set U, a family $(\nu_h \mid h \in H_U)$ of single-valued functions of U in X such that

$$h \neq k \Rightarrow \nu_h x \neq \nu_k x \tag{1}$$

$$x \in U \Rightarrow \{\nu_h x \mid h \in H_U\} = \Gamma x \tag{2}$$

If $y = \nu_h x$, we shall say that h is the *index* of position y relative to x.

(6) The *fundamental probability law* $\pi_0 = (\pi_0(z) \mid z \in Z)$ which is defined on the set $Z = \{x \mid \Gamma^{-1} x = \emptyset\}$.

(7) Two numerical functions $f(x)$ and $g(x)$ defined on X, called the *preference functions* of players (A) and (B).

Play takes place in the following way: an initial position x_0 is chosen at random from Z according to the known probability law π_0; if the information set U containing x_0 belongs to $\mathscr{U}$, player (A) will not know the position x_0, but only the set U to which it belongs. We have

$$\Gamma x_0 = (\nu_h x_0 | h \in H_U)$$

Player (A) then chooses an index h from H_U. This determines a position of the game $x_1 = \nu_h x_0$. If $x_1 \in X_B$, player (B), who does not know anything of what has happened so far, will only be informed that the position of the game belongs to the set $V \in \mathscr{V}$. He chooses an index k from H_V; from this arises a new position of the game $x_2 = \nu_k x_1$, etc. The game stops if the position x of the game is such that $\Gamma x = \varnothing$.

For player (A) a *strategy* σ is, by definition, a single-valued function which determines for each set U of $\mathscr{U}$ an index $h = \sigma U$ such that $\sigma U \in H_U$; the set of strategies will be denoted by Σ_A; we shall say that (A) *adopts* σ if he undertakes to choose the index $h = \sigma U$, whenever the position of the game belongs to $U \in \mathscr{U}$.

If the two players adopt the strategies σ and τ respectively, the set of positions arising from an initial position x_0 is well defined and will be denoted by $\langle x_0; \sigma, \tau\rangle$; then, by definition, the *gain* of player (A) is

$$f(x_0; \sigma, \tau) = \sup \{f(x) | x \in \langle x_0; \sigma, \tau\rangle\}$$

The *expected gain* of player (A) is

$$F(\sigma, \tau) = \sum_{z \in Z} \pi_0(z) f(z; \sigma, \tau)$$

Similarly, the *expected gain* of player (B) is

$$G(\sigma, \tau) = \sum_{z \in Z} \pi_0(z) g(z; \sigma, \tau)$$

A player aims to choose a strategy σ so as to get the largest possible expected gain (or payoff), taking into account the possible choices of his opponent.

As in Chapter 6, we shall say that a pair (σ_0, τ_0) is a *point of equilibrium* if

$$F(\sigma, \tau_0) \leqslant F(\sigma_0, \tau_0) \qquad (\sigma \in \Sigma_A)$$
$$G(\sigma_0, \tau) \leqslant G(\sigma_0, \tau_0) \qquad (\tau \in \Sigma_B)$$

If the pair (σ_0, τ_0) represents the strategies adopted by the two players, it is evident that neither player can benefit by modifying his strategy in isolation.

The situation in which players (A) and (B) make a simultaneous choice of strategies σ and τ, with a gain to (A) of $F(\sigma, \tau)$, and a gain to (B) of $G(\sigma, \tau)$ is also a game which is known as the *normalized form* of the initial game.

EXAMPLE 1. Every game with *complete* information is a game having as its scheme of information the discrete partition:

$$\mathscr{U} = (\{x\} | x \in X_A)$$

For example, chequers, chess, etc.

EXAMPLE 2. The game of bridge. Consider a game after the bidding has been completed: South is dummy. If (A) denotes North–South and (B) denotes East–West, we have a two-person game; although they are considered to be a single player, (B) consists of two *agents* who are unable to transmit their information but who have identical aims. Z is the set of different possible deals. Let us denote by N, S, E, W, the set of cards held respectively by different hands at a given moment, by $\bar{R}$ the set of cards already played with an indication of their origin, and by $\bar{T}$ the cards exposed on the table at the current trick with an indication of their origin (if $\bar{T} = \emptyset$, then $\bar{T}$ denotes the position of the lead).

The position x of the game will be defined by the sets N, S, E, W, and by the 'oriented' sets $\bar{R}$, $\bar{T}$.

We have $f(x) = h_1(\bar{R})$, $g(x) = h_2(\bar{R})$, h_1 and h_2 being two increasing functions dependent on the system of notation adopted. If, for example, it is West's turn to play, the information set U_0 which contains the position x_0 of the game is $(N_0, S_0, E_0, W_0, \bar{R}_0, \bar{T}_0)$ and this will be the set of old positions $(N, S, E, W, \bar{R}, \bar{T})$, with

$$W = W_0$$

$$S = S_0$$

$$E \cup N = E_0 \cup N_0$$

$$\bar{T} = \bar{T}_0$$

$$\bar{R} = \bar{R}_0$$

EXAMPLE 3. *A game with perfect information about the player* (B). We say that a player has *perfect* information about a player (B) if he knows, at the time of play, the information sets which player (B) has encountered previously as well as the indices chosen by (B).

We say that a game has *recall* for (A) if player (A) has perfect information about himself. The above game of bridge has recall for North–South but not for East–West.

Mixed Strategies

The concept of a *mixed strategy* was introduced by E. Borel in order to eliminate the effect of deliberate tactics at the time of making a decision. To be more definite, let us suppose that the strategies of player (A) form a finite set $\Sigma_A = \sigma^1, \sigma^2, \ldots, \sigma^m$, and consider a probability distribution $s = \{s^k, k = 1, 2, \ldots, m\}$ on Σ_A; then we have

$$s_i^k \geqslant 0, \qquad (k = 1, 2, \ldots, m)$$

$$\sum_{k=1}^{m} s_i^k = 1$$

The vector $\boldsymbol{s} = (s^1, s^2, \ldots, s^m)$ is, by definition, *a mixed strategy of the player* (A). A pure strategy, such as σ^k, is thus also a mixed strategy of the form $(0, 0, \ldots, 0, 1, 0, \ldots, 0)$.

We say player (A) *adopts the mixed strategy* $\boldsymbol{s} = (s^1, s^2, \ldots, s^m)$ if he adopts a simple strategy from Σ_A at random by assigning to σ^k the probability s^k.

It can be seen that this kind of action makes any clever guesswork on the part of one's opponent completely ineffective; the best way to prevent him from learning too much is not to know oneself what one is going to do up to the moment of decision. We shall see that there are other, even more important, advantages.

If (A) and (B) adopt the mixed strategies $\boldsymbol{s}$ and $\boldsymbol{t}$ respectively, the expected gain to (A) will be

$$F(\boldsymbol{s}, \boldsymbol{t}) = \sum_{i,j} s^i t^j F(\sigma^i, \tau^j)$$

If we denote the sets of mixed strategies of the two players by S and T, the following fundamental results can be proved:

The von Neumann–Nash theorem. *An equilibrium point* $(\boldsymbol{s}_0, \boldsymbol{t}_0)$ *in mixed strategies always exists such that:*

$$F(\boldsymbol{s}, \boldsymbol{t}_0) \leqslant F(\boldsymbol{s}_0, \boldsymbol{t}_0) \qquad (\boldsymbol{s} \in S)$$

$$G(\boldsymbol{s}_0, \boldsymbol{t}) \leqslant G(\boldsymbol{s}_0, \boldsymbol{t}_0) \qquad (\boldsymbol{t} \in T)$$

The Kuhn–Birch theorem. *If player* (A) *has perfect information about his opponent, there exists an equilibrium point of the form* $(\sigma_0, \boldsymbol{t}_0)$, *where* $\sigma_0 \in \Sigma_A$, $\boldsymbol{t}_0 \in T$.

The greatest expected payoff which player (A) can guarantee with a mixed strategy is

$$\alpha_0 = \max_{s} \min_{\tau} F(s, \tau)$$

With a pure strategy he can guarantee

$$\alpha_1 = \max_{\sigma} \min_{\tau} F(\sigma, \tau)$$

Since $\Sigma_A \subset S$, we have $\alpha_0 \geqslant \alpha_1$, and in many cases $\alpha_0 > \alpha_1$; *the choice of a mixed strategy can therefore guarantee more than a pure strategy.* Furthermore, *no method of play can guarantee a higher payoff than a mixed strategy* (as can be shown by the von Neumann–Nash theorem with $G = -F$).

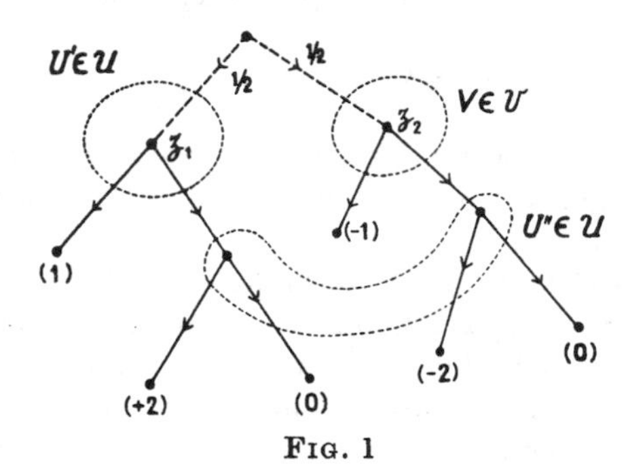

FIG. 1

EXAMPLE (KUHN). Consider the game defined by the above figure: in this case, the graph (X, Γ) is an arborescence; $f(x) = -g(x)$, and the value of $f(x)$ at the various terminal vertices is 0, 1, -1, 2 or -2, as shown in Fig. 1 [the value of $f(x)$ at the other vertices can be assumed to be equal to $-\infty$ and, since the game is finite, will be of no importance]. We choose $h = 1$ or 2 according to whether we move down to the left or to the right.

The fundamental probability law is $\pi_0 = (\frac{1}{2}, \frac{1}{2})$, i.e. vertices z_1 and z_2 are chosen at random with equal probability.

In concrete terms, this game can be described in the following manner: player (A) consists of two agents (A') and (A'') who cannot communicate directly, player (B) has only a single agent. Two cards marked 'high' and 'low' are dealt at random to agent (A') and to (B); the agent holding the 'high' card receives a penny from the agent holding the 'low' card and can either continue or stop the game. If the game continues, (A''), without knowing the deal, can instruct (A') either to keep his card or to exchange cards with player (B). The holder of the 'high' card again receives a penny from the holder of the 'low' card.

(A)'s expected payoff is

$$F(\sigma, \tau) = \tfrac{1}{2}f(z_1; \sigma, \tau) + \tfrac{1}{2}f(z_2; \sigma, \tau)$$

The value of the payoff for the different possible strategies is given by the following table:

$(A) \rightarrow$; $(B) \downarrow$	$\tau^1\ V = 1$	$\tau^2\ V = 2$
$(A) \rightarrow \sigma^1\ U' = 1;\quad \sigma^1\ U'' = 1$	0	$-\frac{1}{2}$
$\sigma^2\ U' = 1;\quad \sigma^2\ U'' = 2$	0	$+\frac{1}{2}$
$\sigma^3\ U' = 2;\quad \sigma^3\ U'' = 1$	$+\frac{1}{2}$	0
$\sigma^4\ U' = 2;\quad \sigma^4\ U'' = 2$	$-\frac{1}{2}$	0

If player (A) adopts a pure strategy, he should choose σ^2 or σ^3, and thus guarantee a payoff of zero (since σ^1 and σ^4 guarantee only $-\frac{1}{2}$). Suppose now that he adopts the mixed strategy $\boldsymbol{s}_0 = (0, \frac{1}{2}, \frac{1}{2}, 0)$, that is he lets a machine choose σ^2 or σ^3 with equal probabilities; instead of guaranteeing a payoff of zero, he will now guarantee $\frac{1}{4}$, since

$$F(\boldsymbol{s}_0, \tau^1) = 0+0+\tfrac{1}{2}\cdot\tfrac{1}{2}+0 = \tfrac{1}{4}$$
$$F(\boldsymbol{s}_0, \tau^2) = 0+\tfrac{1}{2}\cdot\tfrac{1}{2}+0+0 = \tfrac{1}{4}$$

He cannot guarantee more than this since if (B) adopts $\boldsymbol{t}_0 = (\frac{1}{2}, \frac{1}{2})$, the expected payoff will be

$$F(\boldsymbol{s}_0, \boldsymbol{t}_0) = \tfrac{1}{4}$$

Remark. The determination of a good (mixed) strategy is simply a problem in linear programming. In fact, player (A) can guarantee at least α if and only if there exists a vector $\boldsymbol{s} = (s^1, s^2, \ldots, s^m)$ such that

$$\begin{cases} s^i \geqslant 0 \quad (i = 1, 2, \ldots, m) \\ \sum_{i=1}^{m} s^i = 1 \\ F(\boldsymbol{s}, \tau^j) \geqslant \alpha \quad (j = 1, 2, \ldots, n) \end{cases}$$

If we put $F(\sigma^i, \tau^j) = a_j^i$, $s^i/\alpha = x_i$, and if we assume $\alpha > 0$, we get

$$\begin{cases} x_i \geqslant 0 \quad (i = 1, 2, \dots, m) \\ \sum_{i=1}^{m} x_i = \dfrac{1}{\alpha} \\ \sum_{i=1}^{m} a_j^i x_i \geqslant 1 \quad (j = 1, 2, \dots, n) \end{cases}$$

We want to find the largest number α which is compatible with these equations; the problem is thus reduced to finding x_i such that

$$\begin{cases} x_i \geqslant 0 \quad (i = 1, 2, \dots, m) \\ \sum_{i=1}^{m} a_j^i x_i \geqslant 1 \quad (j = 1, 2, \dots, n) \\ \sum x_i \text{ is minimized.} \end{cases}$$

Appendix II

Note on Transport Problems

We call a 'transport problem' any problem which can be expressed analytically in the following form: given the integers a_i, b_j, c_j^i and d_j^i (with $i \leqslant m, j \leqslant n$), we wish to determine integers ξ_j^i such that

$$(1) \qquad \begin{cases} \sum_{i=1}^{m} \xi_j^i = b_j & (j \leqslant n) \\ \sum_{j=1}^{n} \xi_j^i = a_i & (i \leqslant m) \\ 0 \leqslant \xi_j^i \leqslant c_j^i & (i \leqslant m; j \leqslant n) \end{cases}$$

$$(2) \qquad \sum_{i=1}^{m} \sum_{j=1}^{n} d_j^i \xi_j^i \text{ is a maximum}$$

This type of problem is found very frequently in the distribution of commodities, deliveries, supply to consumers, etc...., as well as in questions of a more theoretical nature.

The transport problem was attributed successively to HITCHCOCK [7] who formulated it in 1941, to T. C. KOOPMANS [9], who treated it independently several years later, and to KANTOROVICH [8], who studied the continuous case. In fact, MONGE [11] had already formulated it in 1781, and his very geometric enquiries were continued by APPELL [1] in 1928.

Remark 1. If the system given by (1) and (2) has a solution ξ_j^i, we must have

$$(3) \qquad a_i \geqslant 0, \qquad b_j \geqslant 0, \qquad c_j^i \geqslant 0$$

Further, by summing all the ξ_j^i, we get

$$(4) \qquad \sum_{j=1}^{n} b_j = \sum_{i=1}^{m} a_i$$

Finally, we shall associate with the problem an auxiliary transport network with vertices $x_1, x_2, \ldots, x_m, y_1, y_2, \ldots, y_n, x_0, z$, where the source x_0 is linked to a point x_i by an arc of capacity a_i, x_i is linked to y_j by an arc of capacity c_j^i, and y_j is linked to the sink z by an arc of capacity b_j. If the problem is possible, then according to GALE's theorem (Chapter 8) we have:

$$(5) \qquad \sum_{i=1}^{m} \max \left\{a_i, \sum_{j \in J} c_j^i\right\} \geqslant \sum_{j \in J} b_j \qquad (J \subset \{1, 2, \ldots, n\})$$

Conditions (3), (4) and (5) are necessary conditions for the problem to have a solution; it can be shown that they are also sufficient.

Remark 2. The sign of the numbers d_j^i is unimportant and we can always assume that they are non-negative; but if this is not so, we put

$$h = -\min_{i,j} d_j^i > 0$$

$$e_j^i = d_j^i + h > 0$$

We now have

$$\sum_i \sum_j e_j^i \xi_j^i = \sum_i \sum_j d_j^i \xi_j^i + h \sum_i a_i$$

The new problem then consists of maximizing $\sum \sum e_j^i \xi_j^i$.

It follows from this remark that the problem of finding a minimum is the same as that of finding a maximum.

EXAMPLE 1 (HITCHCOCK). We wish to establish an itinerary for petrol tankers, given that there are m ports of supply $x_1, x_2, \ldots, x_m$; n ports $y_1, y_2, \ldots, y_n$ where there is a demand; we assume that the cost of a trip from x_i to y_j is d_j^i, that there are a_i ships at a supply port x_i, and that the demand at y_j is for b_j ships.

If we denote the number of ships sent from the port of supply x_i to the port of demand y_j by ξ_j^i, and if we wish to minimize the total cost of the ship movements $\sum_{i,j} d_j^i \xi_j^i$, we again obtain the analytic form given above (with $c_j^i = +\infty$).

EXAMPLE 2. *The soil removal problem* (MONGE). There are two surfaces S and T; certain heaps of earth which are distributed over the first surface S are to be transported to the surface T in such a way as to minimize the total cost of transport.

If we put $\xi_j^i = 1$ for each unit of soil which is transported from $x_i \in S$ to $y_j \in T$, and if d_j^i denotes the distance from x_i to y_j, we again want to minimize an expression of the form

$$\sum_i \sum_j d_j^i \xi_j^i$$

(However, since the problem in this case is continuous, we should replace the summations by integrations.)

From purely geometric reasoning, we see that:

1. The optimum path of movement of a heap from $x_i \in S$ to $y_j \in T$ is the segment of a straight line which joins x_i and y_j, and no two such segments should intersect (except at their extreme points).

2. The segments $x_i y_j$ form an optimum system of routes if there exists a convex surface R such that all segments $x_i y_j$ are normal to R.

The problem is analogous to one in which a certain container holds material which we wish to transport piecemeal into another equivalent container.

EXAMPLE 3. *Optimal assignment of personnel* (VOTAW and ORDEN). Consider n workers $x_1, x_2, \ldots, x_n$ who are to be assigned to n machines $y_1, y_2, \ldots, y_n$. The output of the worker x_i on machine y_j is a number d_j^i which can be easily determined by tests. Taking $\xi_j^i = 1$ if worker x_i is assigned to machine y_j, and $\xi_j^i = 0$ otherwise, the problem consists of maximizing total output $\sum_{i,j} d_j^i \xi_j^i$ subject to the constraints:

$$\sum_{j=1}^{n} \xi_j^i = 1, \qquad \sum_{i=1}^{n} \xi_j^i = 1$$

Clearly, we again have to solve a transport problem.

EXAMPLE 4. *Distribution of a product over a transport network* (A. ORDEN). Let us consider an oriented graph with n vertices $x_1, x_2, \ldots, x_n$, with a capacity $c_j^i = c(x_i, x_j)$ on each of its arcs, and some given integer $p \leqslant n$; we assume that at every vertex x_i with $i \leqslant p$ there is a surplus amount a_i of a certain product, and that at every vertex x_i with $i > p$ there is a demand for a quantity b_i of the same product.

We assume further that

$$\sum_{i \leqslant p} a_i = \sum_{j > p} b_j = h$$

If (x_i, x_j) is an arc of the graph, we can transport a certain quantity $\xi_j^i \geqslant 0$ of the product from x_i to x_j, the cost of transporting unit amount being d_j^i. Then we wish to satisfy all the demands with a minimum total cost of transport, that is, we wish to minimize

$$\sum_{i=1}^{n} \sum_{j=1}^{n} d_j^i \xi_j^i$$

To translate this problem to the canonical form, we shall want to find numbers η_j^i, where

$$\begin{aligned} \eta_j^i &= \xi_j^i && \text{if } i \neq j, \quad (x_i, x_j) \in U \\ \eta_j^i &= 0 && \text{if } i \neq j, \quad (x_i, x_j) \notin U \\ \eta_j^i &= h - \sum_{k \neq j} \xi_j^k && \text{if } i = j. \end{aligned}$$

These η_j^i obey the following conditions:

$$(1) \quad \begin{cases} \sum_{i=1}^{n} \eta_j^i = h & (j = 1, 2, \ldots, n \\ \sum_{j=1}^{n} \eta_j^i = h + a_i & (i \leqslant p) \\ \sum_{j=1}^{n} \eta_j^i = h - b_i & (i > p) \\ 0 \leqslant \eta_j^i \leqslant c_j^i & (i \neq j) \\ 0 \leqslant \eta_j^i \leqslant h & (i = j) \end{cases}$$

We wish to minimize $\sum_{i=1}^{n} \sum_{j=1}^{n} e_j^i \eta_j^i$, where

$$e_j^i \begin{cases} = 0 & \text{if } i = j \\ = d_j^i & \text{if } i \neq j, \quad (x_i, x_j) \in U \\ = +\infty & \text{if } i \neq j, \quad (x_i, x_j) \notin U \end{cases}$$

In this way, we have succeeded in stating the problem in the desired form.

EXAMPLE 5. *The warehouse problem* (A. S. CAHN). We consider the level of stock of some product stored in a warehouse at different instants of time $1, 2, \ldots, n$, and we assume that we know

k = the total capacity of the warehouse;
t_0 = the initial stock deposited in the warehouse;
p_i = the selling price (per unit) at time i;
q_i = the purchasing price (per unit) at time i;
r_i = the storage price (per unit) at time i.

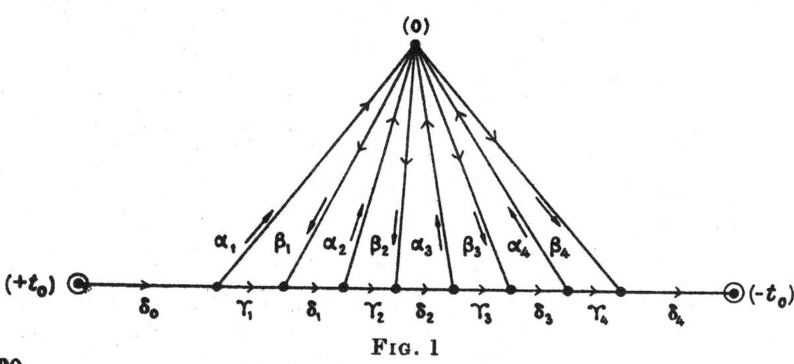

FIG. 1

We intend to speculate over a given period n, and then to restore the stock to its original level t_0. Therefore we require (for all i):

α_i = the amount sold at time i;
β_i = the amount bought at time i;
γ_i = the amount in the warehouse after the sale of the old stock;
δ_i = the amount in the warehouse after the purchase of the new stock.

The problem consists of maximizing the total gain

$$\sum_{i=1}^{n} (p_i \alpha_i - q_i \beta_i - r_i \gamma_i)$$

subject to the constraints:

$$(1)\quad \begin{cases} \delta_i - \alpha_{i+1} - \beta_{i+1} = 0 & (i = 1, 2, \ldots, n-1) \\ -\alpha_1 - \gamma_1 = -t_0 & \\ \gamma_i + \beta_i - \delta_i = 0 & (i = 1, 2, \ldots, n-1) \\ \gamma_n + \beta_n = t_n & \\ 0 \leqslant \alpha_i \leqslant \infty & \\ 0 \leqslant \beta_i \leqslant \infty & \\ 0 \leqslant \gamma_i \leqslant \infty & \\ 0 \leqslant \delta_i \leqslant k & \end{cases}$$

This problem can be immediately restated as a problem relating to the distribution of a product over a transport network (Example 4); for $n = 4$ the corresponding network and the flows in the several arcs are shown in Fig. 1; if there is some sort of rationing such that the total amount of the product which can be bought during a time n is a given amount r_0, we get the graph shown in Fig. 2.

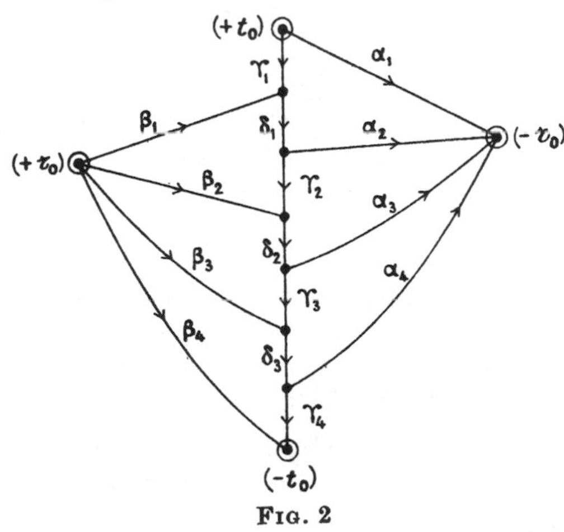

Fig. 2

The Hungarian Method

As the reader will have been able to deduce from conditions (1) and (2), the transport problem is simply a special case of 'linear programming'; as such, it can be solved by the simplex method (G. DANTZIG [3]), and we can use the theory of graphs in order to simplify the usual method considerably (cf. [6]). We intend here to give a brief exposition of an entirely different method, which makes direct use of the theory of transport networks (Chapter 8) and which has shown itself in practice to be much more effective than the simplex method. This method, first developed by H. KUHN [10], has been extended by FORD and FULKERSON [5], M. FLOOD [4], etc....

§ 1. It is unnecessary here to turn to analytic methods, and throughout we shall keep to the intuitive approach of the theory of graphs.

Consider a transport network G, with two disjoint sets of vertices $X = \{x_1, x_2, \ldots, x_m\}$ and $Y = \{y_1, y_2, \ldots, y_n\}$. We connect a source x_0 to every vertex x_i by an arc of capacity $c(x_0, x_i) = a_i$; we also connect every vertex y_j to a sink z by an arc of capacity $c(y_j, z) = b_j$; and, finally, we connect every x_i to every y_j by an arc of capacity $c(x_i, y_j) = c^i_j$. We require a flow which saturates the terminal arcs, that is a flow such that:

$$\begin{cases} 0 \leqslant \phi(u) \leqslant c(u) & (u \in U) \\ \phi(x_0, x_i) = a_i & (i \leqslant m) \\ \phi(y_j, z) = b_j & (j \leqslant n) \end{cases}$$

The *work done* by such a flow ϕ is, by definition:

$$d[\phi] = \sum_{i=1}^{m} \sum_{j=1}^{n} d^i_j \, \phi(x_i, y_j)$$

The general transport problem which was defined above consists of *finding a flow ϕ which saturates the terminal arcs and for which the work done is a maximum.*

In addition to this *primal problem*, we also consider a *dual problem*, which can be defined as follows: we define a function $\gamma(u) \geqslant 0$, called the *budget*, over the arcs of the transport network G, such that

(2) $$\gamma(x_0, x_i) + \gamma(x_i, y_j) + \gamma(y_j, z) \geqslant d^i_j \qquad (i \leqslant m; j \leqslant n)$$

$\gamma(u)$ may be likened to a sum of money which we allot to every vehicle before it travels over the arc u; any vehicle which follows the path $[x_0, x_i, y_j, z]$ will receive as its budget an amount which is greater than or equal to the amount of work d_j^i which it will have done in traversing that path.

For simplicity, let us put

$$\gamma(x_0, x_i) = \alpha_i$$
$$\gamma(y_j, z) = \beta_j$$
$$\gamma(x_i, y_j) = \gamma_j^i$$

We define the *capacity* of the budget γ to be:

$$c[\gamma] = \sum_{i=1}^{m} a_i \alpha_i + \sum_{i=1}^{m} \sum_{j=1}^{n} c_j^i \gamma_j^i + \sum_{j=1}^{n} b_j \beta_j$$

Our aim is *to find a budget γ with a minimum capacity.*

LEMMA. *For every budget γ and for every flow ϕ, we have*

$$d[\phi] \leqslant c[\gamma]$$

In fact, we can write

$$d_j^i \phi_j^i \leqslant \alpha_i \phi_j^i + \gamma_j^i \phi_j^i + \beta_j \phi_j^i$$

This gives

$$d[\phi] = \sum_{i=1}^{m} \sum_{j=1}^{n} d_j^i \phi_j^i \leqslant \sum_{i=1}^{m} \alpha_i \sum_{j=1}^{n} \phi_j^i + \sum_{i=1}^{m} \sum_{j=1}^{n} \gamma_j^i \phi_j^i + \sum_{j=1}^{n} \beta_j \sum_{i=1}^{m} \phi_j^i$$
$$\leqslant \sum_{i=1}^{m} a_i \alpha_i + \sum_{i=1}^{m} \sum_{j=1}^{n} c_j^i \gamma_j^i + \sum_{j=1}^{n} b_j \beta_j = c[\gamma]$$

which proves the lemma.

It follows from this lemma that if a flow ϕ and a budget γ exist such that $d[\phi] = c[\gamma]$, then we can be sure that the flow ϕ is the one we require.

§ 2. For simplicity, we shall make the assumption that all the c_j^i are infinite; this implies that the γ_j^i must all be zero. The dual problem then becomes one of finding numbers $\alpha_i \geqslant 0$ and $\beta_j \geqslant 0$ such that

$$\alpha_i + \beta_j \geqslant d_j^i \qquad (i = 1, 2, \ldots, m; j = 1, 2, \ldots, n)$$

In addition to the primal problem and the dual problem, we also consider an *auxiliary problem* which may be stated in the following way: given any budget γ, construct an auxiliary transport network $\overline{G}_\gamma$ which is derived from G by eliminating all arcs (x_i, y_j) for which

$$\alpha_i + \beta_j > d_j^i$$

Now determine a maximum flow for the transport network $\overline{G}_\gamma$.

THEOREM 1. *If, for some budget γ, a flow $\bar{\phi}$ can be found which saturates the terminal arcs of the graph $\bar{G}_\gamma$, then γ is a budget of minimum capacity, and $\bar{\phi}$ is a flow of G for which the work done is a maximum.*

Clearly $\bar{\phi}$ is a flow for the network G, and it saturates the terminal arcs of G. The work done is

$$d[\bar{\phi}] = \sum_{\alpha_i+\beta_j=d_j^i} d_j^i \bar{\phi}_j^i = \sum_{i=1}^{m} \sum_{j=1}^{n} (\alpha_i+\beta_j)\bar{\phi}_j^i = \sum_{i=1}^{m} a_i \alpha_i + \sum_{j=1}^{n} b_j \beta_j = c[\gamma]$$

Therefore, from the lemma, $\bar{\phi}$ must be a flow of G which does a maximum amount of work, and γ must be a budget of minimum capacity.

THEOREM 2. *If, for some budget γ, a flow cannot be found which saturates the terminal arcs of the network $\bar{G}_\gamma$, then a new budget γ' can be found with a capacity $c[\gamma'] < c[\gamma]$.*

Consider a maximum flow $\bar{\phi}$ which does not saturate the terminal arcs of $\bar{G}_\gamma$; from the theorem of Ford and Fulkerson, the value of this flow must equal the capacity of a cut U_A^- which it saturates. Let us assume that:

$$A = \{x_1, x_2, \ldots, x_r, y_1, y_2, \ldots, y_s, z\}$$

Since U_A^- is a cut of minimum capacity, it cannot contain any arcs of infinite capacity; therefore $i > r$, $j \leqslant s$ implies that (x_i, y_j) is not an arc of the graph $\bar{G}_\gamma$, that is:

$$\alpha_i + \beta_j > d_j^i$$

On the other hand, we can also assume that all the $\alpha_i > 0$, or that all the $\beta_j > 0$, for if this were not so, then for some pair (i, j) the equation

$$d_j^i = \alpha_i + \beta_j = 0$$

would hold.

To be definite, let us assume:

$$\alpha_i > 0 \qquad (i = 1, 2, \ldots, m)$$

Now put

$$\alpha_i' = \begin{cases} \alpha_i & \text{if } i \leqslant r \\ \alpha_i - 1 & \text{if } i > r \end{cases}$$

$$\beta_j' = \begin{cases} \beta_j & \text{if } j \leqslant s \\ \beta_j + 1 & \text{if } j > s \end{cases}$$

We must now show that these α_i' and β_j' constitute a budget γ'; if $i > r$, $j \leqslant s$, we know that $\alpha_i + \beta_j > d_j^i$, and therefore

$$\alpha_i' + \beta_j' \geqslant d_j^i$$

On the other hand, if $i > r$, we have $\alpha_i > 0$, and therefore

$$\alpha'_i \geqslant 0$$

The capacity of the new budget is

$$c[\gamma'] = \sum_{i=1}^{m} a_i \alpha'_i + \sum_{j=1}^{n} b_j \beta'_j = c[\gamma] - \sum_{i=r+1}^{m} a_i + \sum_{j=s+1}^{n} b_j$$

This gives

$$c[\gamma] - c[\gamma'] = \sum_{i=1}^{m} a_i - \sum_{i=1}^{r} a_i + \sum_{j=s+1}^{n} b_j = \sum_{i=1}^{m} a_i - \bar{\phi}_z > 0$$

Therefore the capacity of the new budget is strictly smaller than the capacity of the original budget.

Q.E.D.

We can therefore summarize the algorithm for finding a flow which does the maximum amount of work as follows: we start with a budget γ defined by

$$\alpha_i = \max_j d^i_j$$

$$\beta_j = \max_i (d^i_j - \alpha_i)$$

$$\gamma^i_j = 0$$

and find the maximum flow for the network $\bar{G}_\gamma$. If this flow does not saturate the terminal arcs of the network, we use it to determine a new budget γ', and begin again; if the flow does saturate the terminal arcs, it is the solution required.

§ 3. If we do not make the assumption that the c^i_j are all infinite, the numbers γ^i_j must be taken into account. We can always assume:

$$\gamma^i_j = \max \{0, d^i_j - \alpha_i - \beta_j\}$$

Then we shall have three possibilities to consider:

1. $\gamma^i_j = 0,\quad d^i_j = \alpha_i + \beta_j \quad$ (*or:* $(i,j) \in P$)
2. $\gamma^i_j = 0,\quad d^i_j < \alpha_i + \beta_j \quad$ (*or:* $(i,j) \in Q$)
3. $\gamma^i_j > 0,\quad d^i_j = \alpha_i + \beta_j + \gamma^i_j, \quad$ (*or:* $(i,j) \in R$)

As before, define a transport network $\bar{G}_\gamma$, with the same arcs as G, and in which the capacities are:

$$\bar{a}_i = a_i - \sum_{j|(i,j) \in R} c^i_j \quad (i = 1, 2, \ldots, m)$$

$$\bar{b}_j = b_j - \sum_{i|(i,j) \in R} c^i_j \quad (j = 1, 2, \ldots, n)$$

$$\bar{c}^i_j = \begin{cases} c^i_j & \text{if } (i,j) \in P \\ 0 & \text{if } (i,j) \in Q \cup R \end{cases}$$

THEOREM 1′. *If, for some budget γ, a flow $\bar{\phi}$ can be found which saturates the terminal arcs of the network $\bar{G}_\gamma$, then γ is a budget of minimum capacity; in addition, a flow ϕ which does a maximum amount of work can be defined over the network G in the following way:*

$$\phi = \begin{cases} \bar{\phi}_j^i & \text{if } (i,j) \in P \cup Q \\ c_j^i & \text{if } (i,j) \in R \end{cases}$$

We have, in fact,

$$\begin{aligned} d[\phi] &= \sum_{(i,j)\in P\cup Q} d_j^i \bar{\phi}_j^i + \sum_{(i,j)\in R} d_j^i c_j^i = \sum_{i=1}^m \sum_{j=1}^n (\alpha_i + \beta_j)\bar{\phi}_j^i + \sum_{(i,j)\in R} d_j^i c_j^i \\ &= \sum_{i=1}^m \alpha_i \bar{a}_i + \sum_{j=1}^n \beta_j \bar{b}_j + \sum_{(i,j)\in R} d_j^i c_j^i \\ &= c[\gamma] - \sum_{i=1}^m \alpha_i \sum_{j|(i,j)\in R} c_j^i - \sum_{j=1}^n \beta_j \sum_{i|(i,j)\in R} c_j^i - \sum_{(i,j)\in R} \gamma_j^i c_j^i \\ &\qquad + \sum_{(i,j)\in R} (\alpha_i + \beta_j + \gamma_j^i) c_j^i \\ &= c[\gamma] \end{aligned}$$

Therefore $d[\phi] = c[\gamma]$, and, from the lemma, ϕ is the flow required.

THEOREM 2′. *If, for some budget γ, there is no flow $\bar{\phi}$ which saturates the terminal arcs of the network $\bar{G}_\gamma$, a new budget γ' can be found for which* $c[\gamma'] < c[\gamma]$.

Let $\bar{\phi}$ be a maximum flow of $\bar{G}_\gamma$, which, by hypothesis, does not saturate the terminal arcs of $\bar{G}_\gamma$. We shall consider a set

$$A = \{x_1, x_2, \ldots, x_r, y_1, y_2, \ldots, y_s, z\}$$

such that

$$c(U_A^-) = \bar{\phi}_z < \sum_{i=1}^m \bar{a}_i$$

As in Theorem 2, we can again assume that all the α_i are greater than zero. Now let us put:

$$\alpha_i' = \begin{cases} \alpha_i & \text{if } i \leqslant r \\ \alpha_i - 1 & \text{if } i > r \end{cases}$$

$$\beta_j' = \begin{cases} \beta_j & \text{if } j \leqslant s \\ \beta_j + 1 & \text{if } j > s \end{cases}$$

$$\gamma'^i_j = \begin{cases} \gamma_j^i - 1 & \text{if } i \leqslant r, j > s, \quad (i,j) \in R \\ \gamma_j^i + 1 & \text{if } i > r, j \leqslant s, \quad (i,j) \in R \cup P \\ \gamma_j^i \text{ in all other cases} \end{cases}$$

First we shall show that the new values define a budget; we have $\alpha'_i \geqslant 0$, $\beta'_j \geqslant 0$, $\gamma'^i_j \geqslant 0$. In order to show that $\alpha_i+\beta_j+\gamma^i_j-d^i_j$ remains >0, we construct Table 1:

TABLE 1

	P $i\leqslant r$ $j\leqslant s$	P $i\leqslant r$ $j>s$	P $i>r$ $j\leqslant s$	P $i>r$ $j>s$	Q $i\leqslant r$ $j\leqslant s$	Q $i\leqslant r$ $j>s$	Q $i>r$ $j\leqslant s$	Q $i>r$ $j>s$	R $i\leqslant r$ $j\leqslant s$	R $i\leqslant r$ $j>s$	R $i>r$ $j\leqslant s$	R $i>r$ $j>s$
$\Delta\alpha_i$	0	0	-1	-1	0	0	-1	-1	0	0	-1	-1
$\Delta\beta_j$	0	$+1$	0	$+1$	0	$+1$	0	$+1$	0	$+1$	0	$+1$
$\Delta\gamma^i_j$	0	0	$+1$	0	0	0	0	0	0	-1	$+1$	0
$\Delta=$	0	$+1$	0	0	0	$+1$	-1	0	0	0	0	0

The only case in which $\Delta = -1$ makes no difference, since for $(i,j) \in Q$, we have

$$\alpha_i+\beta_j+\gamma^i_j-d^i_j > 0$$

Now we must prove that the capacity of the budget has been diminished; we have

$$\begin{aligned} c[\gamma'] &= \sum_i a_i\alpha'_i + \sum_j b_j\beta'_j + \sum_{i,j} c^i_j\gamma'^i_j \\ &= c[\gamma] - \sum_{i>r} a_i + \sum_{j>s} b_j - \sum_{\substack{i\leqslant r\\ j>s\\ R}} c^i_j + \sum_{\substack{i>r\\ j\leqslant s\\ P}} c^i_j + \sum_{\substack{i>r\\ j\leqslant s\\ R}} c^i_j; \end{aligned}$$

This gives, finally:

$$c[\gamma']-c[\gamma] = -\sum_{i>r}\bar{a}_i + \sum_{j>s}\bar{b}_j + \sum_{\substack{i>r\\ j\leqslant s}}\bar{c}^i_j = -\sum_i \bar{a}_i + \bar{c}(U_A^-) < 0$$

This completes the proof of the theorem.

Remark. Given a set of numbers $c''^i_j \geqslant c'^i_j \geqslant 0$, we can generalize the problem by imposing the condition:

$$c'^i_j \leqslant \phi^i_j \leqslant c''^i_j \quad (i \leqslant m, j \leqslant n)$$

In this case, we have to deal with a classical transport problem where the auxiliary network $\bar{G}$ is of the type defined on p. 81; the procedure to be followed is again that given on p. 235.

Bibliography

CHAPTER 1. *General definitions*

[1] C. BERGE, Sur l'isovalence et la régularité des transformateurs, *C. R. Acad. Sciences*, **231**, 1950, p. 1404; Sur une théorie ensembliste des jeux alternatifs, *C. R. Acad. Sciences*, **232**, 1951, p. 294.
[2] D. KÖNIG, *Theorie der Endlichen und Unendlichen Graphen*, Leipzig, 1936 (Akad. Verl. M. B. H.), and New York, 1950 (Chelsea).
[3] J. RIGUET, Relations binaires, fermetures, correspondances de Galois, *Bull. Soc. Math. France*, **76**, 1948, p. 114.
[4] A. SAINTE-LAGÜE, Les réseaux (ou graphes), *Mém. des Sciences Math.*, **18**, Paris, 1926.

CHAPTER 2. *Descendance relations*

[1] R. CROISOT, M. L. DUBREIL-JACOTIN and L. LESIEUR, *Leçons sur la Théorie des Treillis*, Paris, 1953 (Gauthier-Villars).
[2] M. RICHARDSON, Relativization and extension of solutions of irreflexive relations, *Pacific J. Math.*, **5**, 1955, p. 551.

CHAPTER 3. *The ordinal function and the Grundy function on an infinite graph*

[1] C. BERGE, La fonction de Grundy d'un graphe infini, *C. R. Acad. Sciences*, **242**, 1956, p. 1404.
[2] C. BERGE, A Problem on Infinite Graphs; Lecture at the joint Harvard-M.I.T. Colloquium, May 1957.
[3] N. G. DE BRUIJN and P. ERDÖS, A colour problem for infinite graphs and a problem in the theory of relations, *Proc. Kon. Ned. Akad. Wetensch.*, **54**, 1951, p. 371.
[4] W. H. GOTTSCHALK, Choice functions and Tychonoff's theorem, *Proc. Amer. Math. Soc.*, **2**, 1951, p. 172.
[5] D. KÖNIG, Über eine Schlussweise aus dem Endlichen ins Unendliche, *Acta Szeged.*, **3**, 1927, p. 121.
[6] R. RADO, Axiomatic treatment of rank in infinite sets, *Canad. J. Math.*, **1**, 1949, p. 337.

CHAPTER 4. *The fundamental numbers of the theory of graphs*

[1] C. BERGE, Two theorems in graph theory, *Proc. Nat. Acad. Sci. U.S.A.*, **43**, 1957, p. 842.
[2] R. E. GOMORY, Outline of an algorithm for the solution of integer programs, *Bull. Amer. Math. Soc.*, **64**, 1958, p. 275.
[3] M. KRAITCHIK, *Le Problème des Reines*, Brussels, 1926.
[4] G. KREWERAS, Peut-on former un réseau donné avec des parties finies d'un ensemble dénombrable ?, *C. R. Acad. Sciences*, **222**, 1946, p. 1025.

[5] A. H. LAND and A. G. DOIG, An automatic method for the solution of discrete programming problems, *Econometrica*, **28**, 1960, p. 497.
[6] G. POLYA, Sur le nombre des isomètres de certains composés chimiques, *C. R. Acad. Sciences*, **202**, 1936, p. 1554.
[7] C. E. SHANNON, The zero-error capacity of a noisy channel, Composium on Information Theory, *Trans. Inst. Elect. Engr.*, **3**, 1956, p. 3.

CHAPTER 5. *Kernels of a graph*

[1] C. BERGE, Théorie générale des jeux à n personnes, Chapter 5: Coalitions, p. 82, *Mém. des Sciences Math.*, **138**, Paris, 1957 (Gauthier-Villars).
[2] G. T. GUILBAUD, La théorie des jeux, *Écon. Appl.*, 1949, p. 18.
[3] J. VON NEUMANN and O. MORGENSTERN, *Theory of Games and Economic Behavior*, Princeton, 1944, 1947 and 1953 (Princeton University Press).
[4] M. RICHARDSON, Solutions of irreflexive relations, *Ann. Math.*, (2) **58**, 1953, p. 573; On weakly ordered systems, *Bull. Amer. Math. Soc.*, **52**, 1946, p. 113; Extension theorems for solutions of irreflexive relations, *Proc. Nat. Acad. Sci. U.S.A.*, **39**, 1953, p. 649.

CHAPTER 6. *Games on a graph*

[1] E. W. ADAMS and D. C. BENSON, Nim type games, Carnegie Inst. Technology, *Tech. Report*, **13**, 1956.
[2] C. BERGE, Topological games with perfect information. (Contributions to the theory of games, 3), *Ann. Math. Studies, Princeton*, **39**, 1957, p. 165; Théorie générale des jeux à n personnes, *Mém. des Sciences Math.*, **138**, Paris, 1957.
[3] C. BERGE and M. P. SCHÜTZENBERGER, Jeux de Nim et solutions, *C. R. Acad. Sciences*, **242**, 1956, p. 1672.
[4] P. M. GRUNDY, Mathematics and games, *Eureka*, **2**, 1939, p. 6.
[5] P. M. GRUNDY and C. A. B. SMITH, Disjunctive games with the last player losing, *Proc. Camb. Phil. Soc.*, **52**, 1956, p. 527.
[6] J. C. HOLLADAY, Cartesian product of termination games. (Contributions to the theory of games, 3), *Ann. Math. Studies, Princeton*, **39**, 1957, p. 189.
[7] E. H. MOORE, A generalization of the game called Nim, *Ann. Math.*, **11**, 1909, p. 93.
[8] J. NASH, Equilibrium points in n-person games, *Proc. Nat. Acad. Sci. U.S.A.*, **36**, 1950, p. 48.
[9] R. SPRAGUE, Bemerkungen über eine spezielle Abelsche Gruppe, *Math. Zeit.*, **51**, 1947, p. 82; Über mathematische Kampfspiele, *Tohoku J. Math.*, **41**, 1936, p. 438.
[10] C. P. WELTER, The theory of a class of games on a sequence of squares, in terms of the advancing operation in a special group, *Proc. Kon. Ned. Akad. Wetensch. A*, **57**, 1954, p. 194.

CHAPTER 7. *The problem of the shortest route*

[1] M. BECKMANN, C. B. MCGUIRE and C. B. WINSTEN, *Studies in the economics of transportation*, New Haven, 1956 (Yale University Press).
[2] R. BELLMAN, The theory of dynamic programming, *Bull. Amer. Math. Soc.*, **60**, 1954, p. 503.
[3] L. R. FORD, Network flow theory, *RAND Corp. Paper* P-923, 1956.
[4] E. LUCAS, *Récreations mathématiques*, **1**, Paris, 1921, p. 47.
[5] G. TARRY, Le problème des labyrinthes, *Nouvelles Annales de Math.*, **14**, 1895, p. 187.

CHAPTER 8. *Transport networks*

[1] C. BERGE, Sur la déficience d'un réseau infini, *C. R. Acad. Sciences*, **245**, 1957, p. 1206.
[2] L. R. FORD and D. R. FULKERSON, Maximal flow through a network, *Canad. J. Math.*, **8**, 1956, p. 399.
[3] L. R. FORD and D. R. FULKERSON, Dynamic network flow, *RAND Corp. Paper* P-967, 1956.
[4] D. GALE, A theorem on flows in networks, *Pacific J. Math.*, **7**, 1957, p. 1073.
[5] A. J. HOFFMAN and H. W. KUHN, On systems of distinct representatives. (Linear inequalities and related systems), *Ann. Math. Studies, Princeton*, **38**, 1956, p. 199.
[6] G. KREWERAS, Extension d'un théorème sur les répartitions en classes, *C. R. Acad. Sciences*, **222**, 1946, p. 431.
[7] J. MARCZEWSKI and G. T. GUILBAUD, Essai d'analyse graphique d'une comptabilité nationale, *Écon. Appl.*, **2**, 1949, p. 138.

CHAPTER 9. *The theorem of the demi-degrees*

[1] D. GALE, A theorem on flows in networks, *Pacific J. Math.*, **7**, 1957, p. 1073.
[2] G. H. HARDY, J. E. LITTLEWOOD and G. POLYA, *Inequalities*, Cambridge, 1952.
[3] R. F. MUIRHEAD, Some methods applicable to identities and inequalities of symmetric algebraic functions of n letters, *Proc. Edin. Math. Soc.*, **21**, 1903, p. 144.

CHAPTER 10. *Matching of a simple graph*

[1] C. BERGE, Sur la déficience d'un réseau infini, *C. R. Acad. Sciences*, **245**, 1957, p. 1206.
[2] G. BIRKHOFF, Tres observaciones sobre el algebra lineal, *Rev. Univ. nac. Tucuman*, Ser. A, **6**, 1941, p. 147.
[3] E. EGERVÁRY, Matrixok kombinatórius tulajdonságairól, *Mat. Fiz. Lapok*, **38**, 1931, p. 16. (On combinatorial properties of matrices, trans. by H. W. KUHN, Office of Naval Research Logistics Project Report, Dept. Math., Princeton University, 1953).
[4] P. HALL, On representations of subsets, *J. London Math. Soc.*, **10**, 1934, p. 26.
[5] P. R. HALMOS and H. E. VAUGHAN, The marriage problem, *Amer. J. Math.*, **72**, 1950, p. 26.
[6] D. KÖNIG and S. VALKO, Über mehrdeutige Abbildungen von Mengen, *Math. Annalen*, **95**, 1926, p. 135.
[7] H. W. KUHN, The Hungarian method for the assignment problem, *Naval Res. Logist. Quart.*, **2**, 1955, p. 83.
[8] H. B. MANN and H. J. RYSER, Systems of distinct representatives, *Amer. Math. Monthly*, **60**, 1953, p. 397.
[9] O. ORE, Graphs and matching theorems, *Duke Math. J.*, **22**, 1955, p. 625.
[10] R. RADO, Factorization of even graphs, *Quart. J. Math.*, **20**, 1949, p. 95.

CHAPTER 11. *Factors*

[1] W. AHRENS, Mathematische Unterhaltungen und Spiele, Berlin, 1910, p. 319.
[2] G. A. DIRAC, The structure of k-chromatic graphs, *Fund. Math.*, **40**, 1953, p. 50.
[3] L. EULER, Commentationes Arithmeticae Collectae, St. Petersburg, 1766, p. 337.
[4] FLYE SAINTE-MARIE, Note sur un problème relatif à la marche du cavalier sur l'échiquier, *Bull. Soc. Math., France*, **5**, 1876, p. 144.

[5] S. JOHNSON, Optimal two- and three-stage production schedules with set up times included, *Naval Res. Logist. Quart.*, **1**, 1954, p. 61.
[6] M. KRAITCHIK, *La mathématique des jeux*, Brussels, 1930, p. 357.
[7] G. RINGEL, Über drei kombinatorische Probleme am n-dim. Würfel und Würfelgitter, *Abh. Math. Sem. Univ., Hamburg*, **20**, 1955, p. 10.
[8] A. SAINTE-LAGÜE, *Avec des nombres et des lignes*, Paris, 1943 (Vuibert), p. 291.

CHAPTER 13. *The diameter of a strongly connected graph*

[1] D. BRATTON, Efficient communication networks, *Cowles Comm. Disc. Paper*, 2119, 1955.
[2] L. S. CHRISTIE, R. D. LUCE and J. MACY, Communication and learning in a task-oriented group, *Mass. Inst. Tech. Res. Lab. Electronics, Tech. Report*, **231**, 1952, p. 238.
[3] R. RADNER and A. TRITTER, Communication in networks, *Cowles Comm. Paper*, 2098, 1954.

CHAPTER 14. *The matrix associated with a graph*

[1] L. KATZ, An application of matrix algebra to the study of human relations, *Univ. North Carolina*, mim. report, 1950.
[2] M. G. KENDALL, *Rank Correlation Methods*, London, 1948 (Griffin); Further contributions to the theory of paired comparisons, *Biometrics*, **11**, 1955, p. 43.
[3] C. LÉVI-STRAUSS, *Les structures élémentaires de la parenté*, Paris, 1949 (Presses universitaires).
[4] R. D. LUCE and A. D. PERRY, A method of matrix analysis of group structure, *Psychometrika*, **14**, 1949, p. 95.
[5] R. D. LUCE, A note on Boolean matrix theory, *Proc. Amer. Math. Soc.*, **3**, 1952, p. 382.
[6] J. L. MORENO, *Fondements de la sociométrie*, (trans. LESAGE-MAUCORPS), Paris, 1954 (Presses universitaires).
[7] I. C. ROSS and F. HARARY, On the determination of redundancies in sociometric chains, *Psychometrika*, **17**, 1952, p. 195.
[8] T. H. WEI, *The algebraic foundations of ranking theory*, thesis, Cambridge, 1952.

CHAPTER 15. *Incidence matrices*

[1] J. CHUARD, Questions d'analysis situs, *R. C. Circ. Mat.*, **46**, 1922, p. 185.
[2] I. HELLER and C. B. TOMPKINS, An extension of a theorem of Dantzig's. (Linear inequalities and related systems), *Ann. Math. Studies, Princeton*, **38**, 1956, p. 247.
[3] A. J. HOFFMAN and J. B. KRUSKAL, Integral boundary points of convex polyhedra. (Linear inequalities and related systems), *Ann. Math. Studies, Princeton*, **38**, 1956, p. 223.
[4] G. KIRCHHOFF, *Poggendorf Annalen*, **72**, 1847, p. 497.
[5] S. OKADA, Algebraic and topological foundations of network synthesis, *Proc. Symp. on Modern Network Synthesis*, New York, 1955, p. 283.
[6] H. POINCARÉ, Complément à l'analysis situs, *R. C. Circ. Mat.*, **13**, 1899, p. 285.
[7] O. VEBLEN and J. W. ALEXANDER, Manifolds of n dimensions, *Ann. Math.*, **14**, 1913, p. 163.

CHAPTER 16. *Trees and arborescences*

[1] T. VAN AARDENNE-EHRENFEST and N. G. DE BRUIJN, Circuits and trees in oriented linear graphs, *Simon Stevin*, **28**, 1951, p. 203.
[2] R. BOTT and J. P. MAYBERRY, Matrices and trees, *Economic activity analysis*, New York, 1954 (J. Wiley and Sons), p. 391.

[3] A. CAYLEY, Collected mathematical papers, Cambridge, **3**, p. 242; **13**, p. 26, etc.
[4] C. JORDAN, Sur les assemblages des lignes, *J. reine angew. Math.*, **70**, 1869, p. 185.
[5] P. J. KELLY, A congruence theorem for trees, *Pacific J. Math.*, **7**, 1957, p. 961.
[6] J. B. KRUSKAL, On the shortest spanning subtree of a graph, *Proc. Amer. Math. Soc.*, **7**, 1956, p. 48.
[7] C. DE POLIGNAC, Théorie des ramifications, *Bull. Soc. Math., France*, **8**, 1880, p. 120.
[8] H. M. TRENT, A note on the enumeration and listing of all possible trees in a connected linear graph, *Proc. Nat. Acad. Sci. U.S.A.*, **40**, 1954, p. 1004.

CHAPTER 17. *Euler's problem*

[1] T. VAN AARDENNE-EHRENFEST and N. G. DE BRUIJN, Circuits and trees in oriented linear graphs, *Simon Stevin*, **28**, 1951, p. 203.
[2] N. G. DE BRUIJN, A combinatorial problem, *Proc. Kon. Ned. Akad. Wetensch.*, **49**, 1946, p. 758.
[3] R. DAWSON and I. J. GOOD, Exact Markov probabilities from oriented linear graphs, *Ann. Math. Statist.*, **28**, 1957, p. 838.
[4] M. H. MARTIN, A problem in arrangements, *Bull. Amer. Math. Soc.*, **40**, 1934, p. 859.

CHAPTER 18. *Matching in the general case*

[1] C. BERGE, Two theorems in graph theory, *Proc. Nat. Acad. Sci. U.S.A.*, **43**, 1957, p. 842.
[2] C. BERGE, Sur le couplage maximum d'un graphe, *C. R. Acad. Sciences*, **247**, 1958, p. 258.
[3] T. GALLAI, On factorization of graphs, *Acta Math. Acad. Sci., Hung.*, **1**, 1950, p. 133.
[4] F. G. MAUNSELL, A note on Tutte's paper, *J. London Math. Soc.*, **27**, 1952, p. 127.
[5] R. Z. NORMAN and M. O. RABIN, An algorithm for a minimum cover of a graph (abstract), *Notices of the Amer. Math. Soc.*, **5**, February 1958, p. 36.
[6] J. PÉTERSEN, Die Theorie der regulären Graphs, *Acta Math.*, **15**, 1891, p. 193.
[7] J. P. ROTH, Combinatorial topological methods in the synthesis of switching circuits, *Int. Bus. Mach. Res. Report*, RC-11, Poughkeepsie, 1957.
[8] W. T. TUTTE, The factorization of linear graphs, *J. London Math. Soc.*, **22**, 1947, p. 107.

CHAPTER 19. *Semifactors*

[1] F. BAEBLER, Über die Zerlegung regularen Streckenkomplexe ungerader Ordnung, *Comment Math. Helvetia*, **10**, 1938, p. 275.
[2] H. B. BELCK, Reguläre Faktoren von Graphen, *J. reine angew. Math.*, **188**, 1950, p. 228.
[3] A. SAINTE-LAGÜE, Les réseaux unicursaux et bicursaux, *C. R. Acad. Sciences*, **182**, 1926, p. 747.
[4] W. T. TUTTE, The factors of graphs, *Canadian J. Math.*, **4**, 1952, p. 314; A short proof of the factor theorem for finite graphs, *Canadian J. Math.*, **6**, 1954, p. 347.
[5] W. T. TUTTE, On Hamiltonian circuits, *J. London Math. Soc.*, **21**, 1946, p. 99.

CHAPTER 20. *The connectivity of a graph*

[1] G. A. DIRAC, The structure of k-chromatic graphs, *Fund. Math.*, **40**, 1953, p. 42; Some theorems on abstract graphs, *Proc. London Math. Soc.*, **2**, 1952, p. 69.
[2] G. HAJÓS, Zum Mengerschen Graphensatz, *Acta Lith. Acad. Sci. (Math.)*, **7**, 1934, p. 44.

[3] J. B. KELLY and L. M. KELLY, Paths and circuits in critical graphs, *Amer. J. Math.*, **76**, 1954, p. 790.
[4] K. MENGER, Zur allgemeinen Kurventheorie, *Fund. Math.*, **10**, 1926, p. 96.
[5] H. WHITNEY, Non-separable and planar graphs, *Trans. Amer. Math. Soc.*, **34**, 1932, p. 339; Congruent graphs and the connectivity of graphs, *Amer. J. Math.*, **54**, 1932, p. 150.

CHAPTER 21. *Planar graphs*

[1] G. A. DIRAC, Map colour theorems related to the Heawood colour formula, *J. London Math. Soc.*, **31**, 1956, p. 460.
[2] G. A. DIRAC and S. SCHUSTER, A theorem of Kuratowski, *Proc. Kon. Ned. Akad. Wetensch. A*, **57**, 1954, p. 343.
[3] A. ERRERA, Du coloriage des cartes, thesis, Brussels, 1921, *Mathesis*, **36**, 1922, p. 56.
[4] A. ERRERA, Exposé historique sur le problème des 4 couleurs, *Periodico di Mat.*, **17**, 1927, p. 20; Sur le problème des 4 couleurs, *Acad. R. Belg.*, **33**, 1947, p. 807: **34**, 1948, p. 65; **36**, 1950, p. 594; Une vue d'ensemble sur le problème des 4 couleurs, *R. C. Sem. Mat. Turin*, **11**, 1951–52, p. 5.
[5] G. KURATOWSKI, Sur le problème des courbes gauches en topologie, *Fund. Math.*, **15–16**, 1930, p. 271.
[6] G. RINGEL, Bestimmung der Maximalzahl der Nachbargebiete auf nichtorientierbaren Flächen, *Math. Ann.*, **127**, 1954, p. 181.
[7] P. G. TAIT, Note on a theorem in geometry of position, *Trans. R. Soc. Edinburgh*, **29**, 1880, p. 657.
[8] H. WHITNEY, Planar graphs, *Fund. Math.*, **21**, 1933, p. 73.
[9] H. WHITNEY, Congruent graphs and the connectivity of graphs, *Amer. J. Math.*, **54**, 1932, p. 150.

APPENDIX I. *Note on the general theory of games*

[1] C. BERGE, Théorie générale des jeux à *n* personnes, *Mém. des Sciences Math.*, **138**, Paris, 1957.
[2] H. W. KUHN, Extensive games and the problem of information. (Contributions to the theory of games, 2), *Ann. Math. Studies, Princeton*, **28**, 1953, p. 193.
[3] J. VILLE, in *Traité du calcul des probabilités et ses applications* (E. Borel), **4**, Paris, 1938, p. 105.

APPENDIX II. *Note on transport problems*

[1] P. APPELL, Le problème géométrique des déblais et remblais, *Mém. des Sciences Math.*, **27**, Paris, 1928 (Gauthier-Villars).
[2] A. S. CAHN, The warehouse problem, *Bull. Amer. Math. Soc.*, **54**, 1948, p. 1073.
[3] G. B. DANTZIG, Application of the simplex method to a transportation problem. (Activity analysis of production and allocation), *Cowles Comm. Monograph*, **13**, New York, 1951 (J. Wiley and Sons), p. 359.
[4] M. M. FLOOD, On the Hitchcock distribution problem, *Pacific J. Math.*, **3**, 1953, p. 369.
[5] L. R. FORD and D. R. FULKERSON, Solving the transportation problem, *Management Sci.*, **3**, 1956, p. 24; A primal-dual algorithm for the capacitated Hitchcock problem, *RAND Res. Memo.*, P-827, Santa Monica, 1956.
[6] D. R. FULKERSON, On the Hitchcock transportation problem, *RAND Paper* P-890, Santa Monica, 1956.

[7] F. L. Hitchcock, The distribution of a product from several sources to numerous localities, *J. Math. Phys.*, **20**, 1941, p. 224.
[8] L. Kantorovich, On the translocation of masses, *Dokl. Akad. Nauk SSSR*, **37**, 1942, p. 199.
[9] T. C. Koopmans and S. Reiter, A model of transportation. (Activity analysis of production and allocation), *Cowles Comm. Monograph* **13**, New York, 1951 (J. Wiley and Sons), p. 222.
[10] H. W. Kuhn, The Hungarian method for the assignment problem, *Naval Res. Logist. Quart.*, **2**, 1955, p. 83.
[11] G. Monge, Déblai et remblai, *Mém. de l'Acad. des Sciences*, 1781.
[12] A. Orden, The transshipment problem, *Management Sci.*, **2**, 1956, p. 276.
[13] D. F. Votaw and A. Orden, The personnel assignment problem, *SCOOP Symp. on Linear Inequalities and Programming*, Washington, 1952, p. 155.

NOTE. For general results in analysis, the reference [E.T.F.M.] is used throughout the book. This stands for C. Berge, *Espaces topologiques et Fonctions multivoques*, No. III in the Collection Universitaire de Mathématiques.

List of Symbols

(The page reference indicates where the symbol is first used)

(X, Y denote sets; A, B, subsets of X; x, y, z, ..., elements of X; Γ, Δ multivalued functions (defined over X); k an integer > 0)

Index of Terms Used

(*The page reference indicates where the term is first defined*)

A CATALOG OF SELECTED

DOVER BOOKS

IN SCIENCE AND MATHEMATICS

A CATALOG OF SELECTED

DOVER BOOKS

IN SCIENCE AND MATHEMATICS

Astronomy

BURNHAM'S CELESTIAL HANDBOOK, Robert Burnham, Jr. Thorough guide to the stars beyond our solar system. Exhaustive treatment. Alphabetical by constellation: Andromeda to Cetus in Vol. 1; Chamaeleon to Orion in Vol. 2; and Pavo to Vulpecula in Vol. 3. Hundreds of illustrations. Index in Vol. 3. 2,000pp. 6⅛ x 9¼.
23567-X, 23568-8, 23673-0 Pa., Three-vol. set $46.85

THE EXTRATERRESTRIAL LIFE DEBATE, 1750–1900, Michael J. Crowe. First detailed, scholarly study in English of the many ideas that developed between 1750 and 1900 regarding the existence of intelligent extraterrestrial life. Examines ideas of Kant, Herschel, Voltaire, Percival Lowell, many other scientists and thinkers. 16 illustrations. 704pp. 5⅜ x 8½. 40675-X Pa. $19.95

A HISTORY OF ASTRONOMY, A. Pannekoek. Well-balanced, carefully reasoned study covers such topics as Ptolemaic theory, work of Copernicus, Kepler, Newton, Eddington's work on stars, much more. Illustrated. References. 521pp. 5⅜ x 8½.
65994-1 Pa. $15.95

AMATEUR ASTRONOMER'S HANDBOOK, J. B. Sidgwick. Timeless, comprehensive coverage of telescopes, mirrors, lenses, mountings, telescope drives, micrometers, spectroscopes, more. 189 illustrations. 576pp. 5⅝ x 8¼. (Available in U.S. only.)
24034-7 Pa. $13.95

STARS AND RELATIVITY, Ya. B. Zel'dovich and I. D. Novikov. Vol. 1 of *Relativistic Astrophysics* by famed Russian scientists. General relativity, properties of matter under astrophysical conditions, stars, and stellar systems. Deep physical insights, clear presentation. 1971 edition. References. 544pp. 5⅝ x 8¼.
69424-0 Pa. $14.95

Chemistry

CHEMICAL MAGIC, Leonard A. Ford. Second Edition, Revised by E. Winston Grundmeier. Over 100 unusual stunts demonstrating cold fire, dust explosions, much more. Text explains scientific principles and stresses safety precautions. 128pp. 5⅜ x 8½. 67628-5 Pa. $5.95

THE DEVELOPMENT OF MODERN CHEMISTRY, Aaron J. Ihde. Authoritative history of chemistry from ancient Greek theory to 20th-century innovation. Covers major chemists and their discoveries. 209 illustrations. 14 tables. Bibliographies. Indices. Appendices. 851pp. 5⅜ x 8½. 64235-6 Pa. $24.95

CATALYSIS IN CHEMISTRY AND ENZYMOLOGY, William P. Jencks. Exceptionally clear coverage of mechanisms for catalysis, forces in aqueous solution, carbonyl- and acyl-group reactions, practical kinetics, more. 864pp. 5⅜ x 8½.
65460-5 Pa. $19.95

THE HISTORICAL BACKGROUND OF CHEMISTRY, Henry M. Leicester. Evolution of ideas, not individual biography. Concentrates on formulation of a coherent set of chemical laws. 260pp. 5⅜ x 8½. 61053-5 Pa. $8.95

A SHORT HISTORY OF CHEMISTRY, J. R. Partington. Classic exposition explores origins of chemistry, alchemy, early medical chemistry, nature of atmosphere, theory of valency, laws and structure of atomic theory, much more. 428pp. 5⅜ x 8½. (Available in U.S. only.) 65977-1 Pa. $12.95

GENERAL CHEMISTRY, Linus Pauling. Revised 3rd edition of classic first-year text by Nobel laureate. Atomic and molecular structure, quantum mechanics, statistical mechanics, thermodynamics correlated with descriptive chemistry. Problems. 992pp. 5⅜ x 8½. 65622-5 Pa. $19.95

Engineering

DE RE METALLICA, Georgius Agricola. The famous Hoover translation of greatest treatise on technological chemistry, engineering, geology, mining of early modern times (1556). All 289 original woodcuts. 638pp. 6¾ x 11. 60006-8 Pa. $21.95

FUNDAMENTALS OF ASTRODYNAMICS, Roger Bate et al. Modern approach developed by U.S. Air Force Academy. Designed as a first course. Problems, exercises. Numerous illustrations. 455pp. 5⅜ x 8½. 60061-0 Pa. $12.95

DYNAMICS OF FLUIDS IN POROUS MEDIA, Jacob Bear. For advanced students of ground water hydrology, soil mechanics and physics, drainage and irrigation engineering and more. 335 illustrations. Exercises, with answers. 784pp. 6⅛ x 9¼. 65675-6 Pa. $24.95

ANALYTICAL MECHANICS OF GEARS, Earle Buckingham. Indispensable reference for modern gear manufacture covers conjugate gear-tooth action, gear-tooth profiles of various gears, many other topics. 263 figures. 102 tables. 546pp. 5⅜ x 8½. 65712-4 Pa. $16.95

MECHANICS, J. P. Den Hartog. A classic introductory text or refresher. Hundreds of applications and design problems illuminate fundamentals of trusses, loaded beams and cables, etc. 334 answered problems. 462pp. 5⅜ x 8½. 60754-2 Pa. $13.95

MECHANICAL VIBRATIONS, J. P. Den Hartog. Classic textbook offers lucid explanations and illustrative models, applying theories of vibrations to a variety of practical industrial engineering problems. Numerous figures. 233 problems, solutions. Appendix. Index. Preface. 436pp. 5⅜ x 8½. 64785-4 Pa. $15.95

STRENGTH OF MATERIALS, J. P. Den Hartog. Full, clear treatment of basic material (tension, torsion, bending, etc.) plus advanced material on engineering methods, applications. 350 answered problems. 323pp. 5⅜ x 8½. 60755-0 Pa. $11.95

ANALYTICAL FRACTURE MECHANICS, David J. Unger. Self-contained text supplements standard fracture mechanics texts by focusing on analytical methods for determining crack-tip stress and strain fields. 336pp. 6⅛ x 9¼. 41737-9 Pa. $19.95

A HISTORY OF MECHANICS, René Dugas. Monumental study of mechanical principles from antiquity to quantum mechanics. Contributions of ancient Greeks, Galileo, Leonardo, Kepler, Lagrange, many others. 671pp. 5⅜ x 8½.
65632-2 Pa. $18.95

STATISTICAL MECHANICS: Principles and Applications, Terrell L. Hill. Standard text covers fundamentals of statistical mechanics, applications to fluctuation theory, imperfect gases, distribution functions, more. 448pp. 5⅜ x 8½.
65390-0 Pa. $14.95

THE VARIATIONAL PRINCIPLES OF MECHANICS, Cornelius Lanczos. Graduate level coverage of calculus of variations, equations of motion, relativistic mechanics, more. First inexpensive paperbound edition of classic treatise. Index. Bibliography. 418pp. 5⅜ x 8½. 65067-7 Pa. $14.95

THE VARIOUS AND INGENIOUS MACHINES OF AGOSTINO RAMELLI: A Classic Sixteenth-Century Illustrated Treatise on Technology, Agostino Ramelli. One of the most widely known and copied works on machinery in the 16th century. 194 detailed plates of water pumps, grain mills, cranes, more. 608pp. 9 x 12.
28180-9 Pa. $24.95

ORDINARY DIFFERENTIAL EQUATIONS AND STABILITY THEORY: An Introduction, David A. Sánchez. Brief, modern treatment. Linear equation, stability theory for autonomous and nonautonomous systems, etc. 164pp. 5⅜ x 8¼.
63828-6 Pa. $6.95

ROTARY WING AERODYNAMICS, W. Z. Stepniewski. Clear, concise text covers aerodynamic phenomena of the rotor and offers guidelines for helicopter performance evaluation. Orignially prepared for NASA. 537 figures. 640pp. 6⅛ x 9¼.
64647-5 Pa. $16.95

INTRODUCTION TO SPACE DYNAMICS, William Tyrrell Thomson. Comprehensive, classic introduction to space-flight engineering for advanced undergraduate and graduate students. Includes vector algebra, kinematics, transformation of coordinates. Bibliography. Index. 352pp. 5⅜ x 8½. 65113-4 Pa. $10.95

HISTORY OF STRENGTH OF MATERIALS, Stephen P. Timoshenko. Excellent historical survey of the strength of materials with many references to the theories of elasticity and structure. 245 figures. 452pp. 5⅜ x 8½. 61187-6 Pa. $14.95

CONSTRUCTIONS AND COMBINATORIAL PROBLEMS IN DESIGN OF EXPERIMENTS, Damaraju Raghavarao. In-depth reference work examines orthogonal Latin squares, incomplete block designs, tactical configuration, partial geometry, much more. Abundant explanations, examples. 416pp. 5⅜ x 8¼.
65685-3 Pa. $10.95

Mathematics

HANDBOOK OF MATHEMATICAL FUNCTIONS WITH FORMULAS, GRAPHS, AND MATHEMATICAL TABLES, edited by Milton Abramowitz and Irene A. Stegun. Vast compendium: 29 sets of tables, some to as high as 20 places. 1,046pp. 8 x 10½. 61272-4 Pa. $32.95

FUNCTIONAL ANALYSIS (Second Corrected Edition), George Bachman and Lawrence Narici. Excellent treatment of subject geared toward students with background in linear algebra, advanced calculus, physics and engineering. Text covers introduction to inner-product spaces, normed, metric spaces, and topological spaces; complete orthonormal sets, the Hahn-Banach Theorem and its consequences, and many other related subjects. 1966 ed. 544pp. 6⅛ x 9¼. 40251-7 Pa. $18.95

ASYMPTOTIC EXPANSIONS OF INTEGRALS, Norman Bleistein & Richard A. Handelsman. Best introduction to important field with applications in a variety of scientific disciplines. New preface. Problems. Diagrams. Tables. Bibliography. Index. 448pp. 5⅜ x 8½. 65082-0 Pa. $13.95

FAMOUS PROBLEMS OF GEOMETRY AND HOW TO SOLVE THEM, Benjamin Bold. Squaring the circle, trisecting the angle, duplicating the cube: learn their history, why they are impossible to solve, then solve them yourself. 128pp. 5⅜ x 8½. 24297-8 Pa. $6.95

VECTOR AND TENSOR ANALYSIS WITH APPLICATIONS, A. I. Borisenko and I. E. Tarapov. Concise introduction. Worked-out problems, solutions, exercises. 257pp. 5⅝ x 8¼. 63833-2 Pa. $10.95

THE ABSOLUTE DIFFERENTIAL CALCULUS (CALCULUS OF TENSORS), Tullio Levi-Civita. Great 20th-century mathematician's classic work on material necessary for mathematical grasp of theory of relativity. 452pp. 5⅝ x 8¼. 63401-9 Pa. $14.95

AN INTRODUCTION TO ORDINARY DIFFERENTIAL EQUATIONS, Earl A. Coddington. A thorough and systematic first course in elementary differential equations for undergraduates in mathematics and science, with many exercises and problems (with answers). Index. 304pp. 5⅜ x 8½. 65942-9 Pa. $9.95

FOURIER SERIES AND ORTHOGONAL FUNCTIONS, Harry F. Davis. An incisive text combining theory and practical example to introduce Fourier series, orthogonal functions and applications of the Fourier method to boundary-value problems. 570 exercises. Answers and notes. 416pp. 5⅜ x 8½. 65973-9 Pa. $13.95

COMPUTABILITY AND UNSOLVABILITY, Martin Davis. Classic graduate-level introduction to theory of computability, usually referred to as theory of recurrent functions. New preface and appendix. 288pp. 5⅜ x 8½. 61471-9 Pa. $12.95

ASYMPTOTIC METHODS IN ANALYSIS, N. G. de Bruijn. An inexpensive, comprehensive guide to asymptotic methods–the pioneering work that teaches by explaining worked examples in detail. Index. 224pp. 5⅜ x 8½ 64221-6 Pa. $9.95

THEORY OF MATRICES, Sam Perlis. Outstanding text covering rank, nonsingularity and inverses in connection with the development of canonical matrices under the relation of equivalence, and without the intervention of determinants. Includes exercises. 237pp. 5⅜ x 8½. 66810-X Pa. $8.95

INTRODUCTION TO ANALYSIS, Maxwell Rosenlicht. Unusually clear, accessible coverage of set theory, real number system, metric spaces, continuous functions, Riemann integration, multiple integrals, more. Wide range of problems. Undergraduate level. Bibliography. 254pp. 5⅜ x 8½. 65038-3 Pa. $11.95

MODERN NONLINEAR EQUATIONS, Thomas L. Saaty. Emphasizes practical solution of problems; covers seven types of equations. ". . . a welcome contribution to the existing literature...."–*Math Reviews*. 490pp. 5⅜ x 8½. 64232-1 Pa. $13.95

MATRICES AND LINEAR ALGEBRA, Hans Schneider and George Phillip Barker. Basic textbook covers theory of matrices and its applications to systems of linear equations and related topics such as determinants, eigenvalues and differential equations. Numerous exercises. 432pp. 5⅜ x 8½. 66014-1 Pa. $12.95

MATHEMATICS APPLIED TO CONTINUUM MECHANICS, Lee A. Segel. Analyzes models of fluid flow and solid deformation. For upper-level math, science and engineering students. 608pp. 5⅜ x 8½. 65369-2 Pa. $18.95

ELEMENTS OF REAL ANALYSIS, David A. Sprecher. Classic text covers fundamental concepts, real number system, point sets, functions of a real variable, Fourier series, much more. Over 500 exercises. 352pp. 5⅜ x 8½. 65385-4 Pa. $11.95

AN INTRODUCTION TO MATRICES, SETS AND GROUPS FOR SCIENCE STUDENTS, G. Stephenson. Concise, readable text introduces sets, groups, and most importantly, matrices to undergraduate students of physics, chemistry, and engineering. Problems. 164pp. 5⅜ x 8½. 65077-4 Pa. $7.95

SET THEORY AND LOGIC, Robert R. Stoll. Lucid introduction to unified theory of mathematical concepts. Set theory and logic seen as tools for conceptual understanding of real number system. 496pp. 5⅜ x 8¼. 63829-4 Pa. $14.95

TENSOR CALCULUS, J.L. Synge and A. Schild. Widely used introductory text covers spaces and tensors, basic operations in Riemannian space, non-Riemannian spaces, etc. 324pp. 5⅜ x 8¼. 63612-7 Pa. $13.95

ORDINARY DIFFERENTIAL EQUATIONS, Morris Tenenbaum and Harry Pollard. Exhaustive survey of ordinary differential equations for undergraduates in mathematics, engineering, science. Thorough analysis of theorems. Diagrams. Bibliography. Index. 818pp. 5⅜ x 8½. 64940-7 Pa. $19.95

INTEGRAL EQUATIONS, F. G. Tricomi. Authoritative, well-written treatment of extremely useful mathematical tool with wide applications. Volterra Equations, Fredholm Equations, much more. Advanced undergraduate to graduate level. Exercises. Bibliography. 238pp. 5⅜ x 8½. 64828-1 Pa. $8.95

FOURIER SERIES, Georgi P. Tolstov. Translated by Richard A. Silverman. A valuable addition to the literature on the subject, moving clearly from subject to subject and theorem to theorem. 107 problems, answers. 336pp. 5⅜ x 8½. 63317-9 Pa. $11.95

POPULAR LECTURES ON MATHEMATICAL LOGIC, Hao Wang. Noted logician's lucid treatment of historical developments, set theory, model theory, recursion theory and constructivism, proof theory, more. 3 appendixes. Bibliography. 1981 edition. ix + 283pp. 5⅜ x 8½. 67632-3 Pa. $10.95

CALCULUS OF VARIATIONS, Robert Weinstock. Basic introduction covering isoperimetric problems, theory of elasticity, quantum mechanics, electrostatics, etc. Exercises throughout. 326pp. 5⅜ x 8½. 63069-2 Pa. $12.95

THE CONTINUUM: A Critical Examination of the Foundation of Analysis, Hermann Weyl. Classic of 20th-century foundational research deals with the conceptual problem posed by the continuum. 156pp. 5⅜ x 8½. 67982-9 Pa. $8.95

CHALLENGING MATHEMATICAL PROBLEMS WITH ELEMENTARY SOLUTIONS, A. M. Yaglom and I. M. Yaglom. Over 170 challenging problems on probability theory, combinatorial analysis, points and lines, topology, convex polygons, many other topics. Solutions. Total of 445pp. 5⅜ x 8½. Two-vol. set.
Vol. I: 65536-9 Pa. $9.95
Vol. II: 65537-7 Pa. $8.95

A SURVEY OF NUMERICAL MATHEMATICS, David M. Young and Robert Todd Gregory. Broad self-contained coverage of computer-oriented numerical algorithms for solving various types of mathematical problems in linear algebra, ordinary and partial, differential equations, much more. Exercises. Total of 1,248pp. 5⅜ x 8½. Two volumes. Vol. I: 65691-8 Pa. $16.95
Vol. II: 65692-6 Pa. $16.95

INTRODUCTION TO PARTIAL DIFFERENTIAL EQUATIONS WITH APPLICATIONS, E. C. Zachmanoglou and Dale W. Thoe. Essentials of partial differential equations applied to common problems in engineering and the physical sciences. Problems and answers. 416pp. 5⅜ x 8½. 65251-3 Pa. $13.95

THE THEORY OF GROUPS, Hans J. Zassenhaus. Well-written graduate-level text acquaints reader with group-theoretic methods and demonstrates their usefulness in mathematics. Axioms, the calculus of complexes, homomorphic mapping, *p*-group theory, more. Many proofs shorter and more transparent than older ones. 276pp. 5⅜ x 8½. 40922-8 Pa. $12.95

DISTRIBUTION THEORY AND TRANSFORM ANALYSIS: An Introduction to Generalized Functions, with Applications, A. H. Zemanian. Provides basics of distribution theory, describes generalized Fourier and Laplace transformations. Numerous problems. 384pp. 5⅜ x 8½. 65479-6 Pa. $13.95

Math–Decision Theory, Statistics, Probability

ELEMENTARY DECISION THEORY, Herman Chernoff and Lincoln E. Moses. Clear introduction to statistics and statistical theory covers data processing, probability and random variables, testing hypotheses, much more. Exercises. 364pp. 5⅜ x 8½. 65218-1 Pa. $12.95

STATISTICS MANUAL, Edwin L. Crow et al. Comprehensive, practical collection of classical and modern methods prepared by U.S. Naval Ordnance Test Station. Stress on use. Basics of statistics assumed. 288pp. 5⅜ x 8½. 60599-X Pa. $8.95

SOME THEORY OF SAMPLING, William Edwards Deming. Analysis of the problems, theory and design of sampling techniques for social scientists, industrial managers and others who find statistics important at work. 61 tables. 90 figures. xvii +602pp. 5⅜ x 8½. 64684-X Pa. $16.95

STATISTICAL ADJUSTMENT OF DATA, W. Edwards Deming. Introduction to basic concepts of statistics, curve fitting, least squares solution, conditions without parameter, conditions containing parameters. 26 exercises worked out. 271pp. 5⅜ x 8½. 64685-8 Pa. $9.95

LINEAR PROGRAMMING AND ECONOMIC ANALYSIS, Robert Dorfman, Paul A. Samuelson and Robert M. Solow. First comprehensive treatment of linear programming in standard economic analysis. Game theory, modern welfare economics, Leontief input-output, more. 525pp. 5⅜ x 8½. 65491-5 Pa. $17.95

DICTIONARY/OUTLINE OF BASIC STATISTICS, John E. Freund and Frank J. Williams. A clear concise dictionary of over 1,000 statistical terms and an outline of statistical formulas covering probability, nonparametric tests, much more. 208pp. 5⅜ x 8½. 66796-0 Pa. $8.95

PROBABILITY: An Introduction, Samuel Goldberg. Excellent basic text covers set theory, probability theory for finite sample spaces, binomial theorem, much more. 360 problems. Bibliographies. 322pp. 5⅜ x 8½. 65252-1 Pa. $11.95

GAMES AND DECISIONS: Introduction and Critical Survey, R. Duncan Luce and Howard Raiffa. Superb nontechnical introduction to game theory, primarily applied to social sciences. Utility theory, zero-sum games, n-person games, decision-making, much more. Bibliography. 509pp. 5⅜ x 8½. 65943-7 Pa. $14.95

FIFTY CHALLENGING PROBLEMS IN PROBABILITY WITH SOLUTIONS, Frederick Mosteller. Remarkable puzzlers, graded in difficulty, illustrate elementary and advanced aspects of probability. Detailed solutions. 88pp. 5⅜ x 8½. 65355-2 Pa. $5.95

PROBABILITY THEORY: A Concise Course, Y. A. Rozanov. Highly readable, self-contained introduction covers combination of events, dependent events, Bernoulli trials, etc. 148pp. 5⅜ x 8¼. 63544-9 Pa. $8.95

STATISTICAL METHOD FROM THE VIEWPOINT OF QUALITY CONTROL, Walter A. Shewhart. Important text explains regulation of variables, uses of statistical control to achieve quality control in industry, agriculture, other areas. 192pp. 5⅜ x 8½. 65232-7 Pa. $8.95

THE COMPLEAT STRATEGYST: Being a Primer on the Theory of Games of Strategy, J. D. Williams. Highly entertaining classic describes, with many illustrated examples, how to select best strategies in conflict situations. Prefaces. Appendices. 268pp. 5⅜ x 8½. 25101-2 Pa. $9.95

Math–Geometry and Topology

ELEMENTARY CONCEPTS OF TOPOLOGY, Paul Alexandroff. Elegant, intuitive approach to topology from set-theoretic topology to Betti groups; how concepts of topology are useful in math and physics. 25 figures. 57pp. 5⅜ x 8½. 60747-X Pa. $4.95

COMBINATORIAL TOPOLOGY, P. S. Alexandrov. Clearly written, well-organized, three-part text begins by dealing with certain classic problems without using the formal techniques of homology theory and advances to the central concept, the Betti groups. Numerous detailed examples. 654pp. 5⅜ x 8½. 40179-0 Pa. $18.95

EXPERIMENTS IN TOPOLOGY, Stephen Barr. Classic, lively explanation of one of the byways of mathematics. Klein bottles, Moebius strips, projective planes, map coloring, problem of the Koenigsberg bridges, much more, described with clarity and wit. 43 figures. 210pp. 5⅜ x 8½. 25933-1 Pa. $8.95

CONFORMAL MAPPING ON RIEMANN SURFACES, Harvey Cohn. Lucid, insightful book presents ideal coverage of subject. 334 exercises make book perfect for self-study. 55 figures. 352pp. 5⅝ x 8¼. 64025-6 Pa. $11.95

THE GEOMETRY OF RENÉ DESCARTES, René Descartes. The great work founded analytical geometry. Original French text, Descartes's own diagrams, together with definitive Smith-Latham translation. 244pp. 5⅜ x 8½. 60068-8 Pa. $9.95

THE THIRTEEN BOOKS OF EUCLID'S ELEMENTS, translated with introduction and commentary by Sir Thomas L. Heath. Definitive edition. Textual and linguistic notes, mathematical analysis. 2,500 years of critical commentary. Unabridged. 1,414pp. 5⅜ x 8½. Three-vol. set.
Vol. I: 60088-2 Pa. $11.95
Vol. II: 60089-0 Pa. $11.95
Vol. III: 60090-4 Pa. $12.95

GEOMETRY OF COMPLEX NUMBERS, Hans Schwerdtfeger. Illuminating, widely praised book on analytic geometry of circles, the Moebius transformation, and two-dimensional non-Euclidean geometries. 200pp. 5⅝ x 8¼. 63830-8 Pa. $8.95

DIFFERENTIAL GEOMETRY, Heinrich W. Guggenheimer. Local differential geometry as an application of advanced calculus and linear algebra. Curvature, transformation groups, surfaces, more. Exercises. 62 figures. 378pp. 5⅜ x 8½. 63433-7 Pa. $11.95

Physics

OPTICAL RESONANCE AND TWO-LEVEL ATOMS, L. Allen and J. H. Eberly. Clear, comprehensive introduction to basic principles behind all quantum optical resonance phenomena. 53 illustrations. Preface. Index. 256pp. 5⅜ x 8½.
65533-4 Pa. $10.95

ULTRASONIC ABSORPTION: An Introduction to the Theory of Sound Absorption and Dispersion in Gases, Liquids and Solids, A. B. Bhatia. Standard reference in the field provides a clear, systematically organized introductory review of fundamental concepts for advanced graduate students, research workers. Numerous diagrams. Bibliography. 440pp. 5⅜ x 8½. 64917-2 Pa. $11.95

QUANTUM THEORY, David Bohm. This advanced undergraduate-level text presents the quantum theory in terms of qualitative and imaginative concepts, followed by specific applications worked out in mathematical detail. Preface. Index. 655pp. 5⅜ x 8½. 65969-0 Pa. $16.95

ATOMIC PHYSICS (8th edition), Max Born. Nobel laureate's lucid treatment of kinetic theory of gases, elementary particles, nuclear atom, wave-corpuscles, atomic structure and spectral lines, much more. Over 40 appendices, bibliography. 495pp. 5⅜ x 8½. 65984-4 Pa. $14.95

AN INTRODUCTION TO HAMILTONIAN OPTICS, H. A. Buchdahl. Detailed account of the Hamiltonian treatment of aberration theory in geometrical optics. Many classes of optical systems defined in terms of the symmetries they possess. Problems with detailed solutions. 1970 edition. xv + 360pp. 5⅜ x 8½.
67597-1 Pa. $10.95

THIRTY YEARS THAT SHOOK PHYSICS: The Story of Quantum Theory, George Gamow. Lucid, accessible introduction to influential theory of energy and matter. Careful explanations of Dirac's anti-particles, Bohr's model of the atom, much more. 12 plates. Numerous drawings. 240pp. 5⅜ x 8½. 24895-X Pa. $8.95

ELECTRONIC STRUCTURE AND THE PROPERTIES OF SOLIDS: The Physics of the Chemical Bond, Walter A. Harrison. Innovative text offers basic understanding of the electronic structure of covalent and ionic solids, simple metals, transition metals and their compounds. Problems. 1980 edition. 582pp. 6⅛ x 9¼.
66021-4 Pa. $19.95

HYDRODYNAMIC AND HYDROMAGNETIC STABILITY, S. Chandrasekhar. Lucid examination of the Rayleigh-Benard problem; clear coverage of the theory of instabilities causing convection. 704pp. 5⅜ x 8¼. 64071-X Pa. $17.95

INVESTIGATIONS ON THE THEORY OF THE BROWNIAN MOVEMENT, Albert Einstein. Five papers (1905–8) investigating dynamics of Brownian motion and evolving elementary theory. Notes by R. Fürth. 122pp. 5⅜ x 8½.
60304-0 Pa. $7.95

INTRODUCTION TO QUANTUM MECHANICS With Applications to Chemistry, Linus Pauling & E. Bright Wilson, Jr. Classic undergraduate text by Nobel Prize winner applies quantum mechanics to chemical and physical problems. Numerous tables and figures enhance the text. Chapter bibliographies. Appendices. Index. 468pp. 5⅜ x 8½. 64871-0 Pa. $13.95

METHODS OF THERMODYNAMICS, Howard Reiss. Outstanding text focuses on physical technique of thermodynamics, typical problem areas of understanding, and significance and use of thermodynamic potential. 1965 edition. 238pp. 5⅜ x 8½. 69445-3 Pa. $8.95

TENSOR ANALYSIS FOR PHYSICISTS, J. A. Schouten. Concise exposition of the mathematical basis of tensor analysis, integrated with well-chosen physical examples of the theory. Exercises. Index. Bibliography. 289pp. 5⅜ x 8½. 65582-2 Pa. $13.95

RELATIVITY IN ILLUSTRATIONS, Jacob T. Schwartz. Clear nontechnical treatment makes relativity more accessible than ever before. Over 60 drawings illustrate concepts more clearly than text alone. Only high school geometry needed. Bibliography. 128pp. 6⅛ x 9¼. 25965-X Pa. $7.95

THE ELECTROMAGNETIC FIELD, Albert Shadowitz. Comprehensive undergraduate text covers basics of electric and magnetic fields, builds up to electromagnetic theory. Also related topics, including relativity. Over 900 problems. 768pp. 5⅝ x 8¼. 65660-8 Pa. $19.95

GREAT EXPERIMENTS IN PHYSICS: Firsthand Accounts from Galileo to Einstein, edited by Morris H. Shamos. 25 crucial discoveries: Newton's laws of motion, Chadwick's study of the neutron, Hertz on electromagnetic waves, more. Original accounts clearly annotated. 370pp. 5⅜ x 8½. 25346-5 Pa. $12.95

RELATIVITY, THERMODYNAMICS AND COSMOLOGY, Richard C. Tolman. Landmark study extends thermodynamics to special, general relativity; also applications of relativistic mechanics, thermodynamics to cosmological models. 501pp. 5⅜ x 8½. 65383-8 Pa. $15.95

LIGHT SCATTERING BY SMALL PARTICLES, H. C. van de Hulst. Comprehensive treatment including full range of useful approximation methods for researchers in chemistry, meteorology and astronomy. 44 illustrations. 470pp. 5⅜ x 8½. 64228-3 Pa. $14.95

STATISTICAL PHYSICS, Gregory H. Wannier. Classic text combines thermodynamics, statistical mechanics and kinetic theory in one unified presentation of thermal physics. Problems with solutions. Bibliography. 532pp. 5⅜ x 8½. 65401-X Pa. $14.95

Prices subject to change without notice.

Available at your book dealer or online at **www.doverpublications.com**. Write for free Dover Mathematics and Science Catalog (59065-8) to Dept. Gl, Dover Publications, Inc., 31 East 2nd St., Mineola, NY 11501. Dover publishes more than 400 books each year on science, elementary and advanced mathematics, biology, music, art, literature, history, social sciences, and other subjects.